MODERN HUMANITIES RESEARCH ASSOCIATION

TUDOR & STUART TRANSLATIONS

VOLUME 17

General Editors
ANDREW HADFIELD
NEIL RHODES

RICHARD CAREW

THE EXAMINATION OF MEN'S WITS

RICHARD CAREW
THE EXAMINATION OF MEN'S WITS

Edited by

Rocío G. Sumillera

MODERN HUMANITIES RESEARCH ASSOCIATION
2014

Published by
The Modern Humanities Research Association,
1 Carlton House Terrace
London SW1Y 5AF

© The Modern Humanities Research Association, 2014

Rocío G. Sumillera has asserted her right under the Copyright, Designs and Patents Act 1988 to be identified as the author of this work.

Parts of this work may be reproduced as permitted under legal provisions for fair dealing (or fair use) for the purposes of research, private study, criticism, or review, or when a relevant collective licensing agreement is in place. All other production requires the written permission of the copyright holder who may be contacted at rights@mhra.org.uk

First published 2014

ISBN 978-1-907322-81-5 (hbk)
ISBN 978-1-78188-161-3 (pbk)

Copies may be ordered from www.tudor.mhra.org.uk

MHRA TUDOR AND STUART TRANSLATIONS

GENERAL EDITORS

Andrew Hadfield (University of Sussex)
Neil Rhodes (University of St Andrews)

ASSOCIATE EDITORS

Guyda Armstrong (University of Manchester)
Fred Schurink (University of Northumbria)
Louise Wilson (University of St Andrews)

ADVISORY BOARD

Warren Boutcher (Queen Mary, University of London); Colin Burrow (All Souls College, Oxford); A. E. B. Coldiron (Florida State University); José María Pérez Fernández (University of Granada); Robert S. Miola (Loyola College, Maryland); Alessandra Petrina (University of Padua); Anne Lake Prescott (Barnard College, Columbia University); Quentin Skinner (Queen Mary, London); Alan Stewart (Columbia University)

For details of published and forthcoming volumes please visit our website:

www.tudor.mhra.org.uk

TABLE OF CONTENTS

General Editors' Foreword viii

Acknowledgements ix

Introduction 1

Further Reading 68

THE EXAMINATION OF MEN'S WITS 71

Textual Notes 329

Glossary 346

Neologisms 350

Bibliography 351

Index 368

GENERAL EDITORS' FOREWORD

The aim of the *MHRA Tudor & Stuart Translations* is to create a representative library of works translated into English during the early modern period for the use of scholars, students and the wider public. The series will include both substantial single works and selections of texts from major authors, with the emphasis being on the works that were most familiar to early modern readers. The texts themselves will be newly edited with substantial introductions, notes, and glossaries, and will be published both in print and online.

The series aims to restore to view a major part of English Renaissance literature which has become relatively inaccessible and to present these texts as literary works in their own right. For that reason it will follow the same principle of modernisation adopted by other scholarly editions of canonical literature from the period. The series will have a similar scope to that of the original *Tudor Translations* published early in the last century, and while the great majority of the works presented will be from the sixteenth century, like the original series it will not be rigidly bound by the end-date of 1603. There will, however, be a very different range of texts with new and substantial scholarly apparatus.

The *MHRA Tudor & Stuart Translations* will extend our understanding of the English Renaissance through its representation of the process of cultural transmission from the classical to the early modern world and the process of cultural exchange within the early modern world.

Andrew Hadfield
Neil Rhodes

ACKNOWLEDGEMENTS

I would like to express my deep gratitude to Andrew Hadfield and Neil Rhodes, General Editors of the series, for having accepted my proposal for this edition, and to José María Pérez Fernández, of the University of Granada, for providing the context for such proposal, as well as for sharing useful information throughout the writing process.

I am grateful to the Residencia de Estudiantes in Madrid, which awarded me a generous seven-month Young Researcher in the Humanities Scholarship; to the Vicerectorat d'Investigació i Política Científica of the Universitat de València, which funded a three-month research stay at the University of Cambridge, and to the JoinEU-SEE III Erasmus Mundus scholarship scheme, which allowed me to work for a month at the welcoming English Department of Ss. Cyril and Methodius University in Skopje.

For their insightful readings of this edition, I am indebted to Neil Rhodes, Warren Boutcher, of Queen Mary, University of London, and José Luis Peset Reig, of the Centro de Ciencias Humanas y Sociales, CSIC, in Madrid. To Simon Davies, of the University of Sussex, and to Louise Wilson, of the University of St Andrews, I have to thank for their thorough copy-editing. Josefina Rodríguez Arribas, of the Warburg Institute, Felice Gambin, of the University of Verona, and Héctor Felipe Pastor Andrés, of the University of Granada, were so kind as to help me with diverse philological queries. I would also like to acknowledge the assistance provided by the Reference Specialists of the Rare Books Section of the British Library.

My greatest debt remains to my parents; to them I dedicate this edition.

INTRODUCTION

'To distinguish and discern these natural differences of man's wit, and to apply to each by art that science wherein he may profit, is the intention of this my work' (P3. 106–108).[1] This sentence concisely summarizes the ultimate purpose of one of the most successful and influential Spanish scientific books published in the early modern period, one with a long-lasting influence upon the European intellectual world: the *Examen de ingenios para las ciencias* (Baeza, 1575), by the Spanish physician and natural philosopher Juan Huarte de San Juan (1529–1588).[2] Juan Huarte did not belong to princely circles, and was not one of those international physicians of the sixteenth century that study and practise abroad, travel throughout Europe, publish their treatises in Latin, and eventually become physicians of kings or popes. Instead, the *Examen* was Huarte's only work and was written in Spanish, and Huarte never travelled abroad nor worked for the king or any other member of the royal family. His life modestly revolved around three geographical areas within Spain: San Juan de Pie de Puerto (Navarre), Baeza-Linares (Andalusia) and Alcalá (Castile). And yet, this was no impediment to the meteoric success that his treatise would enjoy from the time of its publication, and the international recognition as a natural philosopher that he would gain.

Huarte is now hailed as the precursor of several branches of pedagogy and psychology, including differential pedagogy and differential psychology, and their fundamental practical applications, professional orientation and selection.[3] Recently

---

[1] References to the text of this edition are to chapters and line numbers. 'P' stands for 'Proems', i.e. the dedicatory epistle (P1) and the two proems (P2 and P3) that precede Chapter I.

[2] Luis S. Granjel, *La medicina española renacentista* (Salamanca: Universidad de Salamanca, 1980), p. 35; José María López Piñero, *Diccionario histórico de la ciencia moderna en España*, II vols (Barcelona: Península, 1983), I, p. 459.

[3] See Luis Calatayud Buades, *La pedagogía y los clásicos españoles. Simón de Abril, Huarte de San Juan, Saavedra Fajardo, Sabuco de Nantes* (Madrid: Imprenta de Juan Pueyo, 1925); César A. Figuerido, *La orientación profesional y el médico navarro Juan de Huarte* (Bilbao: Casa Dochao, 1930); Vicente Peset Llorca, 'Tres figuras de la psicología médica del Renacimiento español (Vives, Huarte, Sabuco de Nantes)', in *La historia de la psicología y de la psiquiatría en*

too, Noam Chomsky recognized in Huarte a forerunner of the rationalist innatism and the linguistic theory of seventeenth-century French scholars, notably Descartes. In the eyes of Chomsky, the *Examen* is the first scientific treatise to define human wit as a generative power that reveals the creative capacities of human imagination.[4]

The understanding of 'wit' in early modernity often varied from author to author. Frequently wit refers to the capacity to fully engage in rhetorical amplification in the composition of essays, which is closely related to the culture of the commonplace book

---

España: desde los más remotos tiempos hasta la actualidad, ed. by J. B. Ullersperger (Madrid: Alhambra, 1954), pp. 166–78; A. Vázquez Fernández, 'Un tratado de psicología diferencial para una selección y orientación profesionales en la España del siglo XVI', *Cuadernos Salmantinos de Filosofía*, 2.1 (1975), 185–216; Klemens Dieckhöfer, 'Juan Huarte de San Juan, precursor de la psicología diferencial', *XXVII Congreso Internacional de Historia de la Medicina*, II vols (Barcelona: Acadèmia de Ciències Mèdiques de Catalunya i Balears, 1981), I, 64–68; J. Moya and L. García Vega, 'Juan Huarte de San Juan: Padre de la Psicología Diferencial', *Revista de Historia de la Psicología*, 2 (1990), 123–58; Luis García Vega and José Moya Santoyo, *Juan Huarte de San Juan: patrón de la psicología española* (Madrid: Ediciones Académicas, 1991); H. Carpintero, 'Huarte de San Juan: Reflexiones en torno a su influencia en la psicología española', *Investigaciones Psicológicas*, 11 (1992), 9–20; José María Gondra Rezola, *Huarte de San Juan: precursor de la moderna psicología de la inteligencia. Lección inaugural del curso academico 1993–1994* (Vitoria: Universidad del País Vasco, 1993); Irene López-Goñi, 'La figura de Juan Huarte de San Juan en la bibliografía sobre historia de la educación', in *Juan Huarte au XXIe siècle. Actes de Colloque International Juan Huarte au XXIe siècle (2003. Saint-Jean-Pied-de-Port, Basse Navarre)*, ed. by Véronique Duché-Gavet (Anglet: Atlantica, 2003), pp. 255–78. A 1983 agreement of the Spanish Deans of Psychology declared Huarte patron of Spanish psychology, and 23 February, the day in which the Examen was first published, its festivity.

[4] Chomsky mentioned Huarte first in a long footnote in *Cartesian Linguistics* and then in *Language and Mind*. For Chomsky, Huarte constitutes, together with Francisco Sánchez de las Brozas (El Brocense), a precedent for Port-Royal's *Grammaire generale et raisonnée* (1660). T. Miranda, 'Sánchez de las Brozas, Huarte de San Juan y la gramática generativa', *Alcántara*, 15 (1988), 7–31; A. Martín-Araguz and C. Bustamante-Martínez, 'Examen de ingenios, de Juan Huarte de San Juan, y los albores de la Neurobiología de la inteligencia en el Renacimiento español', *Revista de Neurología*, 38.12 (2004), 1176–85 (p. 1179); Malcolm K. Read, 'Ideologies of the Spanish Transition Revisited: Juan Huarte de San Juan, Juan Carlos Rodríguez, and Noam Chomsky', *Journal of Medieval and Early Modern Studies*, 34.2 (2004), 309–43 (pp. 311–13), and Javier Virués Ortega, 'Juan Huarte de San Juan in Cartesian and Modern Psycholinguistics: An Encounter with Noam Chomsky', *Psicothema*, 17.3 (2005), 436–40.

and to a specific kind of education in the arts, based on the ideal of *copia* and the ability of students to use quotable material. Erasmus's *De duplici copia verborum et rerum* (1477) or Meres's *Palladis Tamia: Wits Treasury* (1598) illustrate this phenomenon. In this regard, wit became synonymous with 'mental acumen', and at times connoted 'a flow of ideas and words ample for the development of any topic at length, along with quick comprehension of thought and readiness in answering'.[5] In Huarte wit denotes the totality of the psychological abilities of an individual; more precisely, an individual ability or predisposition dependent on temperament, linked to the qualities of the four basic elements (earth, air, water and fire), and organically connected to the brain and under the influence of other organs.[6] The starting point for Huarte's theory of wits is that the temperature of the four qualities (hot, cold, moist and dry) has an impact upon the function of the rational soul as well as upon that of the sensitive, and intemperate and ever-changing environmental conditions lead to a diversity of the wits. Wit is then subject to age, region of birth, sex, currents of air, weather, diet, physical exercise and lifestyle in general, as all sorts of circumstances have an impact upon the bodies of men, the natural combination of the four primary qualities, and the predominance in every individual of one of the three powers of the intellective soul: memory, imagination, or understanding.[7] Thus Huarte ultimately explains

---

[5] Willian G. Crane, *Wit and Rhetoric in the Renaissance* (New York: Columbia University, 1937), p. 9. The book concentrates on the relationship between wit, rhetoric, imitation, and processes of rhetorical amplification and ornamentation. Crane refers to Huarte only once and in passing, merely to say that Aristotle's *Rhetoric* influenced him (p. 83).

[6] David Pujante, 'La melancolía hispana, entre la enfermedad, el carácter nacional y la moda social', *Revista de la Asociación Española de Neuropsiquiatría*, 28.102 (2008), 401–18 (p. 403).

[7] The variable understanding of 'wit' in early modern England is parallel to that of the term 'humour'. Both notions, used consistently and accurately by Huarte throughout his treatise, were household words for his contemporaries in England, although the general understanding of them was for the most part vague. The term 'humour' appears in English in the fourteenth century within the medical context and within the framework of humoral theory. A century later, however, humoral theory 'passes into non-medical writings of the lay people', and while lay people discuss humours in a rather abstract manner, generally meaning people's mood, specialists in medicine employ more professional and technical definitions. Anu Lehto, Raisa Oinonen and Päivi

his theory of wits in medical terms drawing on the theory of humours, which at this time still provided the general framework for western medicine, and for the philosophical and literary representation of personality.[8] Huarte employs this theory in a personal way and applies it to very specific purposes, and in this regard his work is a mixture of tradition and innovation. His goal is to clearly delineate what makes a man capable of one science and incapable of another; to discover the number of differences of wits, the arts or sciences that correspond to each; and finally, and most importantly, to illustrate how all this can be known.

The question of the proper method of analysis and classification of reality and nature was certainly a shared concern among a generation of humanists. It is apparent in Juan Luis Vives's attempt to establish an encyclopaedia of the knowledge of his time (*De prima philosophia, De anima et vita, De disciplinis*), and it extends to Sir Francis Bacon's classificatory enterprise and, naturally, to Cartesian philosophy. The eighteenth century would witness a scientific context in which 'the authorities of antiquity competed with advocates of new claims to natural knowledge that justified themselves through their "method" instead of any classical precedent'.[9] 'Knowing how' had started 'to become as important as "knowing why"'.[10]

The first edition of the *Examen*, the so-called *princeps* edition, was produced in Baeza in 1575 by the printer Juan Bautista de

---

Pahta, 'Explorations through Early Modern English Medical Texts. Charting Changes in Medical Discourse and Scientific Thinking', in *Early Modern English Medical Texts: Corpus Description and Studies*, ed. by Irma Taavitsainen and Päivi Pahta (Amsterdam/Philadelphia: John Benjamins Publishing Company, 2010), pp. 151–66 (p. 155).

[8] Cristina Müller, *Ingenio y melancolía: una lectura de Huarte de San Juan*, trans. by Manuel Talens and María Pérez Harguindey (Madrid: Biblioteca Nueva, 2002), p. 31.

[9] Heikki Mikkeli and Ville Marttila, 'Change and Continuity in Early Modern Medicine (1500–1700)', in *Early Modern English Medical Texts*, ed. by Taavitsainen and Pahta, pp. 13–27 (p. 26).

[10] Peter Dear, *Revolutionizing the Sciences: European Knowledge and its Ambitions, 1500–1700* (Basingstoke: Palgrave, 2001), p. 170. See also p. 169. For a discussion of the philosophical birth of the modern scientific method see Manuel Garrido Palazón, 'El *Examen de ingenios para las ciencias* de Huarte de San Juan y el enciclopedismo retórico y didáctico de su tiempo', *Revista de Literatura*, 61.122 (1999), 349–73 (p. 349).

INTRODUCTION

Montoya. The printing expenses of the agreed 1,500 copies were paid by Huarte himself. The book, structured in fifteen chapters, was preceded by two prefaces, one addressing King Philip II, to whom it was dedicated, and the second addressing the reader. Huarte took his theories very seriously and truthfully believed that they could have practical repercussions upon the society of his time. The dedicatory epistle to King Philip II suggests in fact a law by which subjects exclusively performed the profession, art or science that corresponded to them by nature:

> To the end that artificers may attain the perfection requisite for the use of the commonwealth, me thinketh, Catholic royal Majesty, a law should be enacted that no carpenter should exercise himself in any work which appertained to the occupation of a husbandman, nor a tailor to that of an architect, and that the advocate should not minister physic, nor the physician play the advocate, but each one exercise only that art to which he beareth a natural inclination, and let pass the residue. (P2. 2–6)

Huarte envisioned appointing 'men of great wisdom and knowledge who might discover each one's wit in his tender age, and cause him perforce to study that science which is agreeable for him, not permitting him to make his own choice', 'to the end he may not err in choosing that which fitteth best with his own nature' (P2. 13–18). Huarte's reasoning was that if every man carried out the job that suited his natural capacities best, progress in the scientific, artistic and technological production of Spain would promptly follow, and a body of naturally accomplished and efficient professionals would ensue. Furthermore, by making all men take the career path most in accordance with their own nature, disasters of all sorts could be prevented, for a heterodox preacher could promote heresy, an incompetent physician may worsen the condition of the patient, an inept military leader may lose battles and conduct his soldiers to a sure death, and unqualified law-makers or judges may devise unfair laws or apply fair ones unjustly. In a nutshell, Huarte draws a correspondence between socio-political necessities and individual aptitudes to achieve a perfect integration of the individual microcosmos in the social macrocosmos. This leads to a system in which ideally only those individuals naturally able to be trusted with positions of responsibility or with academic professions actually — and exclusively — have access to them.[11] Huarte allows for a body of intellectuals defined by their merits, and not by the social class into which they are born; nature should then be made the key for social mobility. Nobility by birth is no guarantee of sophisticated wits, and Huarte remarks that precisely within the highest strata of society numerous witless children are born, whereas often poor families produce witty offspring. Huarte takes King Philip II as an example to illustrate his theories. King Philip II is king not only because he is the son and heir of Charles V but, most importantly, because nature had bestowed on him the perfect temperament for such position: a temperate one by which heat, cold, moisture and draught are in their exact proportion. For this, he is 'very wise, of great memory for things passed, of great imagination to foresee those to come, and of great understanding to find out the truth of all matters' (XIV. 349–351). And because this temperament that Philip II possesses is so rare and perfect, 'only the office of a king is in proportion answerable thereunto, and in ruling and governing ought the same solely to be employed' (XIV. 357–358).

The ambitious goals of the *Examen* turn it into a quintessential representative of the aspirations of medicine to have its own place within the politics of the Spanish monarchy in the context of the general crisis of the Spanish empire at the end of the sixteenth century. The political dimension of Huarte's treatise is clear, and has often been related to the proposals of the early modern *arbitristas*, or promoters of utopian schemes, and to what has been called 'political medicine'.[12] Medicine was then waking up to full consciousness of its political strength, and intended to establish itself as the third power after religion and politics, and to rule along with them. This explains the appearance of a peculiar medico-political literature in the second half of the

---

[11] Guillermo Serés, 'Introducción', in Juan Huarte de San Juan, *Examen de ingenios para las ciencias* (Madrid: Cátedra, 1989), pp. 21–129 (p. 51).

[12] For more on this notion, see Diego Gracia Guillén, 'Judaismo, medicina y mentalidad inquisitorial en la España del siglo XVI', in *Inquisición española y mentalidad inquisitorial*, ed. by Ángel Alcalá (Barcelona: Ariel, 1984), pp. 328–52 (pp. 342–43). See also José Antonio Maravall, *Estado moderno y mentalidad social: (siglos XV a XVII)* (Madrid: Revista de Occidente, 1972). Maravall mentions Huarte on numerous occasions in the two volumes of his work. He weaves Huarte's theories into an analysis of the general political thought of the time.

sixteenth century and the first decades of the seventeenth, illustrated by titles such as *República original sacada del cuerpo humano* (1587) by the physician Jerónimo Merola, *Retrato del perfecto médico* (1595) by Enrique Jorge Enríquez, and *Medicus politicus, sive de officiis medico-politicis tractatus* (1614) by the Portuguese Rodrigo de Castro.[13] Indeed, the physician was envisioned as a governor ruling over a microcosmos in parallel to the king's rule over his kingdom, and God's over the macrocosmos.[14] That medicine was intertwined with politics in early modern Spain was undeniable, if only because since 1501 it was compulsory to prove *limpieza de sangre* ('purity of blood') to work as a physician or a surgeon.

The national and international success of Huarte's *Examen* in the sixteenth and seventeenth centuries was enormous, and the book instantly became a bestseller in Spain and throughout western Europe. During those two centuries, the *Examen* was published in six different languages: in Spanish fifteen times, twenty-five in French, six in Italian, five in English, three in Latin, and one in Dutch. In total nine translators rendered this work into other tongues, and the book was printed in twenty different European cities: from Baeza to Leipzig, from London to Cremona, from Rouen to Amsterdam. Its influence was far from being limited to the arena of medicine, and it quickly extended into the realms of literature and philosophy, reaching authors as varied as Miguel de Cervantes, Baltasar Gracián, Francisco de Quevedo, Antonio Possevino, Sir Francis Bacon and G. E. Lessing. The *Examen*, nonetheless, had to face the Inquisition in Portugal, Spain and Italy, and was included in various inquisitorial indexes, censored, and forced to undergo revision, which led to the appearance of a revised and posthumously published version in 1594.

---

[13] Merola, Jorge Enríquez and de Castro openly acknowledge their debt to Huarte and refer to him in terms of praise.

[14] The idea of the human body as a microcosmos had become a commonplace throughout early modern Europe. It is Phineas Fletcher's allegorical poem *The Purple Island or The Isle of Man* (1633) where the analogy between the human body and the body politic reaches its climax in England. See Muriel Cunin, 'Corps humain, corps politique, corps poétique: *The Purple Island or The Isle of Man* de Phineas Fletcher (1633)', *Études Anglaises*, 63.3 (2010), 274–88.

In England the *Examen de ingenios para las ciencias* became *The Examination of Mens Wits* thanks to the inquisitiveness of Richard Carew (1555–1620), a gentleman of varied intellectual interests and curiosities, and the work's first translator into English. Carew was an antiquary, a poet and translator from Cornwall who entered public service in the 1580s, taught himself Greek, Italian, German, Spanish and French, translated several foreign works into English, and wrote some others of his own invention. Carew translated the *Examen* from the Italian version by Camillo Camilli (first published in 1582 and based on the 1578 Spanish edition of the book), *Essame de gl'ingegni de gli huomini per apprender scienze*, and Adam Islip printed it in London for the first time in 1594, and then again in 1596, 1604 and 1616. Later in the seventeenth century the book would be retranslated, this time directly from the Spanish and by the rather obscure figure of Edward Bellamy, who used as his model the 1594 expurgated edition of the book, and entitled it *The Tryal of Wits* (London, 1698). The differences between the two translations are numerous, and include the editorial success of Carew's work versus the single edition of Bellamy's.[15]

Carew's translation of the Spanish *ingenios* as 'wits' (through the Italian *ingegni*) is hardly surprising, for in the sixteenth century translators typically rendered the Latin term *ingenium* as the English 'wit'.[16] Early modern authors of dictionaries rely on the same translation: William Thomas translated *ingegno* as 'witte',[17] and so did Thomas Thomas and John Florio with

---

[15] *The Tryal of Wits* was Bellamy's first translation, which he called 'my Virgin-Essay', and it seems that it was also his only one. Edward Bellamy, *The Tryal of Wits: Discovering the Great Difference of Wits among Men, and What Sort of Learning Suits Best with Each Genius* (London: Richard Sare, 1698), a1^r.

[16] C. S. Lewis, 'Wit', *Studies in Words* (Cambridge: Cambridge University Press, 1960), pp. 86–110, offers a succint overview of the transformation of the meaning of wit from the Anglo-Saxon *wit* or *gewit* (understood as mind, reason or intelligence) to later understandings closer to the meaning of *ingenium*. In England 'wit' was not unfrequently related to Spaniards, and in the early modern period English translators of Spanish works on occasion included the term 'witty' in their titles even if it was absent in the original. Patricia Shaw, '*Wits, Fittes and Fancies*: Spanish *ingenio* in Renaissance England', *Estudios Ingleses de la Universidad Complutense*, 12 (2004), 131–48 (p. 133).

[17] William Thomas, *Principal Rules of the Italian Grammer with a Dictionarie for the Better Understandyng of Boccace, Petrarcha, and Dante* (London: Thomas Berthelet, 1550), R1^r.

ingĕnĭum,[18] and Richard Percyvall with *ingenio*.[19] Carew's introduction of the term 'men's' in the title of his translation is also appropriate and precise, for the wits that can be truly examined following Huarte's precepts are only men's. While there is variability in the wits of men, women are by definition unfailingly cold and moist, as that enables them to bear offspring. Because it is 'impossible that a woman can be temperate or hot' (XV. 252–253), differences between the wits of women are but slight; the only possible variation is their degree of coldness and moisture, which is given away by their abilities and outward appearance in general. While women are confined to coldness and moisture, men may be hot and dry, hot and moist, cold and dry, and even cold and moist, although in a much lesser degree than women. According to Huarte, the 'defect of wit in the first woman grew for that she was by God created cold and moist, which temperature is necessary to make a woman fruitful and apt for childbirth, but enemy to knowledge' (XV. 313–316). Only in the rare case of a 'gift from above' is a woman spared from her ineptitude to learn, but otherwise, 'whilst a woman abideth in her natural disposition, all sorts of learning and wisdom carrieth a kind of repugnancy to her wit', and consequently her 'sex admitteth neither wisdom nor discipline' (XV. 338–342). Huarte therefore does not address women in his treatise, nor the fathers of young girls, and he is most certainly not concerned with the education of the female sex:

> Those parents who seek the comfort of having wise children and such as are towards for learning, must endeavour that they may be born male, for the female, through the cold and moist of their sex, cannot be endowed with any profound judgement. Only we see that they talk with some appearance of knowledge in slight and easy matters with terms ordinary and long studied, but being set to learning, they reach no farther than to some smack of the Latin tongue, and this only through the help of memory. For which dullness themselves are not in blame, but

---

[18] Thomas Thomas, *Dictionarium linguae latinae et anglicanae* (London: Richard Boyle, 1587), Gg2r. John Florio, *A Worlde of Wordes, or Most Copious, and Exact Dictionarie in Italian and English* (London: Arnold Hatfield, 1598), Q1r.

[19] Richard Percyvall, *Bibliotheca Hispanica Containing a Grammar; with a Dictionarie in Spanish, English, and Latine* (London: John Jackson, 1591), O3v.

that cold and moist which made them women, and these self qualities [...] gainsay the wit and ability. (XV. 640–650)

Nothing can redeem women and help them escape their uneducated fate, and in the long history of humankind only a handful have overcome their natural abhorrence of learning. To Huarte's horror, 'for one man that is begotten, six or seven women are born to the world, ordinarily' (XV. 717–718).[20]

As will be seen in what follows, the story of the production, publication, distribution and reception of Huarte's *Examen* and of Carew's *Examination* is the story of early modern intellectual, scientific, political, social, and religious tensions and anxieties proper to early modern Europe. And the tracking of the motivations and concerns of Huarte and Carew, and the analysis of the transformation of the *Examen* into *The Examination* via the *Essame de gl'ingegni*, constitutes a textual adventure that takes the reader to the heart of restless Counter-Reformation Europe.

JUAN HUARTE AND RICHARD CAREW: BIOGRAPHICAL OVERVIEW

Huarte de San Juan and Richard Carew did not exactly have what one could call parallel lives. Identification with the author of the book could not have been an encouraging factor for Carew to carry out his translation, for neither Carew nor anybody else could have discovered much about the life of Huarte by reading his treatise. Even now little is known of his biography. Perhaps the mystery surrounding the *Examen* was for Carew part of the work's appeal, and the singularity of Huarte's ideas was at the roots of Carew's decision to translate the book. To this peculiarity the second translator of the *Examen* into English, Edward Bellamy, refers in his dedicatory preface when he states that he is 'pleased and edified with the Curious and Uncommon Notions of this *Spaniard*'.[21]

---

[20] See L. García Vega, 'El antifeminismo científico de Juan Huarte de San Juan, Patrón de la Psicología', *Revista de Psicología General y Aplicada*, 42 (1989), 533–42; and Gabriel-André Pérouse, 'Les Femmes et l'Examen des esprits du Dr Huarte (1575)', in *Les représentations de l'autre du Moyen Age au XVIIe siècle*, ed. by Evelyne Berriot-Salvadore (Saint-Etienne: Université de Saint-Etienne, 1995), pp. 273–83.
[21] Bellamy, *The Tryal of Wits*, a2ᵛ.

INTRODUCTION

Richard Carew (1555–1620) was an antiquary and poet born at Antony, Cornwall, into a family of landowners. His father, Thomas Carew (*c.* 1527–1564), had died when Richard was eight and, being the eldest of three, he inherited the estate that his family had had since the late fifteenth century.[22] Richard was sent to Christ Church, Oxford, at the early age of eleven, but seems not to have taken a degree. There he befriended William Camden and Sir Philip Sidney, with whom he was called to dispute *extempore* on one occasion before, among others, the Earls of Leicester and Warwick, Sidney's uncles. After Oxford, Carew sought legal training at Clement's Inn, an Inn of Chancery which students often joined before being admitted to an Inn of Court. In 1574 Carew entered the Middle Temple, where he spends three years studying law, just like his father before him, and like his brother George and his son Richard after him. Given the early death of his father, Carew could not undertake a foreign tour following his formal education. He eventually returned to Cornwall to solve issues related to his inheritance, and in 1577 married Juliana (1563–1629), of the wealthy and powerful family of the Arundells. She would become the mother of his ten children. Carew not only dedicated his time to managing his manors and enjoying the life of a country gentleman of multiple interests (from bee-keeping to fishing), but got also involved in government business. His long career in public service began in 1581, when he was appointed Justice of the Peace; in 1582, he was made Sheriff of Cornwall; burgess in Parliament for the town of Saltash in 1584, and of Mitchell's in 1597; in 1586, in the midst of the war with Spain, he was made deputy lieutenant with special responsibility for Cawsand Bay and the adjacent coast. The supreme command was in the hands of Sir Walter Raleigh, Lord Warden of the Stannaries and Lord Lieutenant.[23]

---

[22] For a detailed account of Richard Carew's family tree, see Frank Ernest Halliday, 'Introduction', in Richard Carew, *The Survey of Cornwall* (London: Adams & Dart, 1969), pp. 15–71 (pp. 15–22).

[23] S. Mendyk, 'Carew, Richard (1555–1620)', in *Oxford Dictionary of National Biography* (Oxford University Press, 2004) <http://www.oxforddnb.com/view/article/4635> [accessed 18 May 2013]. For an illustration of Richard Carew as a man involved in the affairs of Cornwall, see P. L. Hull, 'Richard Carew's Discourse about the Duchy Suit (1594)', *Journal of the Royal Institution of Cornwall*, 4 (1962), 181–251.

In 1598 Carew joined the Society of Antiquaries,[24] and in this context his name appears tied to his fellow antiquary Sir Henry Spelman, author of *De Non Temerandis Ecclesiis; a Tract of the Rights and Respect Due unto Churches* (1613). In 1615 Carew sent a letter to Spelman in which he manifested some doubts regarding this work; Spelman replied to it by means of an epistle in Latin in which he praised Carew's fertile talents.[25] Carew's intellectual curiosity made him learn Greek, Italian, German, Spanish and French 'only by Reading, without any other teaching', as his son, also named Richard Carew, explains.[26] Carew applied his knowledge of foreign tongues to translating not only *The Examination*, but also *Godfrey of Bulloigne, or The Recouerie of Hierusalem* (1594).[27] The fantasy tale in verse *A Herrings Tayle* (1598) was an original production of Carew's.

His interest in language led Carew to write his essay 'The Excellencie of the English Tongue' probably soon after April 1605.[28] 'The Excellencie', first published within the second edition

---

[24] Halliday, 'Introduction', in Carew, *Survey of Cornwall*, attributes to Carew a discourse dated 20 November 1599 entitled 'Of the Antiquity, Variety, and Etimology of Measuring Land in Cornwayl', published in 1720 in Thomas Hearne's *A Collection of Curious Discourses*, which gathers short dissertations by members of the Society of Antiquaries (pp. 39–40).

[25] Sir Henry Spelman, 'Henricus Spelmannus Richardo suo Careo viro praestanti Sal. P. D.', in *The English Works of Sir Henry Spelman Kt. Published in his Life-Time* (London: D. Browne, 1723), pp. 37–38.

[26] Richard Carew (son), 'The True and Ready Way to learn the Latine Tongue: Expressed in an Answer to a Quere, Whether the ordinary Way of teaching Latine by the Rules of Grammar, be the best Way for youths to learn it? By the late Learned and Judicious Gentleman Mr. Richard Carew of Anthony in Cornwall', in Samuel Hartlib, *The True and Readie Way to Learne the Latine Tongue* (London: R. and W. Laybourn, 1654), G3^r–H1^r (G4^r).

[27] For many years Richard Carew had been identified with the 'Carew' signing *A World of Wonders: Or An Introduction to a Treatise Touching the Conformitie of Ancient and Moderne Wonders* (1607). This was a translation of *Apologie pour Herodote*, by the Huguenot publisher and author Henri Estienne, friend of Sir Philip Sidney's. D. N. C. Wood's 'Ralph Cudworth the Elder and Henri Estienne's World of Wonders', *English Language Notes*, 11 (1973), 93–100, however, highly convincingly argues that Richard Carew could not have been the translator of this work. Wood suggests instead Ralph Cudworth the Elder, unquestionably a much more plausible candidate than Carew for biographical as well as for ideological reasons.

[28] D. N. C. Wood, 'Elizabethan English and Richard Carew', *Neophilologus*, 61.2 (1977), 304–15 (p. 304).

INTRODUCTION

of *Remaines, Concerning Britaine* (1614) by his friend from Oxford William Camden, was intended to rebut Richard Verstegan's *A Restitution of Decayed Intelligence of Antiquities* (1605). The essay shows that Carew became much involved in the intense debates of the time about the English language and the extent to which English should assimilate foreign words. Carew's stand was conciliatory: if Saxon was the 'natural language' of England, the contributions of foreign tongues to English should also be celebrated. Carew moreover authored *Survey of Cornwall* (1602), a detailed study of his home county which he dedicated to Sir Walter Raleigh.[29] The book provides a description of Cornwall, its landscape, climate and history, and a brief (and inaccurate) description of the Cornish language. It additionally discusses the affairs of the Cornish gentry and other varied matters such as etymology.[30] Carew had begun writing the *Survey* long before it was finally published. Camden, for example, already in 1586, at the end of the account of Cornwall in the first edition of his *Britannia*, mentions Carew's enterprise. Despite the fact that the *Survey* found a welcoming reception, Carew's project for a second edition was ruined by his decaying health:[31] around 1611 his eyesight began to fail, and for a period of two years, from 1613 to 1615, he went almost blind. In addition, Carew was often sick and suffered from afflictions such as a rupture and the stone. Richard Carew finally died at Antony in November 1620.

Juan Huarte de San Juan (1529–1588) was twenty-six years older than Carew, came from a lower social stratum, went to two universities and left with degrees from both. He was no politician but worked his entire life as a physician, and the only other

---

[29] A modern edition of the *Survey of Cornwall* was edited by Frank Ernest Halliday (London: Andrew Melrose, 1953; London: Adams & Dart, 1969). S. Nixon attributes to Carew a short poem entitled 'Epitaph on Brawn the Irishman but Cornish Beggar', which is not contained in the *Survey*. S. Nixon, 'Mr. Carew and Brawn the Beggar', *Notes and Queries*, 47.2 (2000), 180–81.

[30] For Alexandra Walsham, 'the *Survey of Cornwall* is at once a forerunner of the entertaining travel memoir and informative tour guide to memorable sights and a precursor of the amateur and professional scholarly impulses', 'Richard Carew and English Topography', *The Survey of Cornwall*, ed. by John Chynoweth, Nicholas Orme and Alexandra Walsham (Exeter: Devon and Cornwall Record Society, 2004), pp. 17–41 (p. 17).

[31] The second edition of the *Survey* would only appear in 1723, and the third in 1769.

language apart from Spanish that he truly seems to have known was Latin. He did not engage in translation and was not a prolific author. In fact, writing the *Examen* was a lifelong project, and the book was Huarte's only work. Not much is known about the friendships he had, or about his interests, and it is through an analysis of the intellectual milieu in which he moved that we can gain an insight into the mind of the man and an explanation for some of his anxieties.

It is certain that Huarte was originally from San Juan de Pie de Puerto, at the time part of Navarre.[32] Given the constant harrying of French troops in the area and the difficulties of its defence, Emperor Charles V left the castle and the fortifications unguarded and to their fate in 1530. Huarte's family, just like many other families of the region, decided to leave and moved south, and at some point between 1530 and 1540 finally settled in Baeza (Jaén, Andalusia). At the time Baeza had over 20,000 inhabitants and was enjoying a particular moment of prosperity as a result of the economic growth of the River Guadalquivir valley. Huarte would have studied in the Colegio de la Santísima Trinidad of Baeza, and later at the recently founded University of Baeza, achieving the degree of *licenciado* in Arts in 1553. The school for children of the Santísima Trinidad was founded in 1538, and to it was added in 1542 a Colegio Mayor able to award university degrees in Arts and Theology at the three levels of *bachillerato* (baccalaureate), *licenciatura* (graduate licentiate) and *doctorado* (doctorate).[33] The University of Baeza, which by the end of the sixteenth century had about 800 students enrolled, was to a certain extent an educational experiment: it was created by the professors Rodrigo López and Juan de Ávila (the so-called 'Apostle of Andalusia'), both of convert origin, considerably influenced by Erasmianism, concerned with the education of the lowest classes, and in trouble with the Inquisition.[34]

---

[32] San Juan de Pie de Puerto would officially pass to France in 1660, with King Philip IV and after the Peace of the Pyrenees.

[33] The University of Baeza had in common with that of Alcalá that it did not teach Law.

[34] María E. Álvarez, *La Universidad de Baeza y su tiempo: 1538–1824* (Jaén: Instituto de Estudios Jiennenses, 1958); Ricardo Sáez, 'La Baeza del siglo XVI y su imborrable presencia en la obra de Huarte de San Juan', *Huarte de San Juan*, 1 (1989), 81–95 (p. 92).

INTRODUCTION

Once in possession of a *licenciatura* in Arts from Baeza, Huarte enrolled in Medicine at the University of Alcalá.[35] The Faculty of Medicine at Alcalá was one of the most reputed in Spain in the sixteenth century, along with those of the Universities of Salamanca, Valladolid and Valencia.[36] In them medical humanism developed and flourished, and in the case of Alcalá, home of the grammarian Antonio de Nebrija, in addition to the humanist movement, Erasmian ideas were also prominent and perceptible. The thought and trajectories of physicians such as Juan Gil (also known as doctor Egidio), Constantino Ponce de la Fuente or Francisco de Vargas illustrate this.[37] In Alcalá medical humanism was represented by three professors and royal physicians: Fernando Mena, Cristóbal de Vega and Francisco Vallés (also known as the 'Spanish Galen').[38] The three launched translation enterprises to render anew texts by Galen and the most important books of the *Corpus hippocraticum*, which they enriched with scholarly commentaries. Their translations and commentaries converted Hippocrates' texts into landmarks of medical theory and practice without concurrently questioning the authority of Galen, and in their writings they critically assess medical tradition through their personal clinical experience. No doubt the work of Mena, Vega and Vallés left an indelible impression upon Huarte, who would make personal clinical observation the basis of his treatise while conceptually framing his theories within the hippocratic-galenic scheme. Certainly in

---

[35] At the time, a necessary requirement to enroll in medical studies was to be in possession of the degree of Bachelor of Arts.

[36] The teaching of Medicine at the University of Alcalá appears to have begun in the academic year 1509–1510. In sixteenth-century Spain, medicine was taught at many universities: Barcelona, Lérida, Zaragoza, Huesca, Seville, Granada, Santiago de Compostela, Alcalá, Valencia (notably influenced by Italian physicians), Salamanca and Valladolid. Also, a number of smaller, 'minor', universities such as Toledo, Sigüenza, Osuna, Gandía, Orihuela, Almagro, Irache, Estella and Oñate gave certificates that allowed the practice of medicine. Plus, in the American colonies, the universities of Santo Domingo, México and Lima offered medical studies.

[37] Juan Ramón Corpas Mauleón, 'Introducción', in Juan Huarte de San Juan, *Examen de ingenios para las ciencias* (Pamplona: Ediciones y Libros, 2003), pp. 9–40 (p. 14).

[38] Mena was a physician to Prince Don Carlos and King Philip II; Cristóbal de Vega, the successor of Reinoso at Alcalá, left university in 1557 to serve Prince Don Carlos, and Francisco Vallés left Alcalá in 1572 to work for Philip II.

Huarte's eyes the weight of authority is not enough in itself, and as he puts it when discussing medicine, 'experience beareth more sway than reason, and reason more than authority' (XI. 83–84).

In Alcalá Huarte became *bachiller* in May 1555, received the title of *licenciado* in October 1559, and that of *doctor* in December that same year. Learning in the Faculty of Medicine revolved around the reading of and commentary on medical texts, and around anatomical practices, introduced in the curriculum after the insistent demands of students from 1534. In 1559 a royal provision effectively established that the corpses of executed men or of those who died in the hospitals of Alcalá were given to the University to be dissected. The renowned Pedro Jimeno, disciple of Andreas Vesalius, moved from Valencia to Alcalá in the 1550s to explain the art of dissection, in which he was an expert.[39] In the process he introduced Vesalian anatomy in Alcalá, and given the dates of his teaching at its Faculty of Medicine, he might have had Huarte as a student.[40] Huarte is thus heir to one of the most splendid moments in the history of the University of Alcalá, and his seven years of medical studies at the institution decisively influenced his thought and shaped his method for the study of nature.[41]

After becoming *doctor* in 1559, Huarte moved to the small village of Tarancón, of only about 700 inhabitants. He bought a house and lived there six years. During this period, he married Doña Águeda de Velasco, also a native of San Juan de Pie de

---

[39] Because Jimeno left Valencia in the summer of 1550, he must have become the first anatomy professor of the University of Alcalá the next academic year. Jimeno stayed in Alcalá until his death. Ana Isabel Martín Ferreira, *El humanismo médico en la universidad de Alcalá (Siglo XVI)* (Alcalá de Henares: Universidad de Alcalá de Henares, 1995), p. 49.

[40] For more on medical humanism and the Faculty of Medicine of the University of Alcalá, see Luis Alonso Muñoyerro, *La Facultad de medicina en la Universidad de Alcalá de Henares* (Madrid: CSIC; Instituto Jerónimo Zurita, 1945); Luis S. Granjel, *Humanismo y Medicina* (Salamanca: Universidad de Salamanca, 1968) and *Los médicos humanistas españoles. Capítulos de la medicina española* (Salamanca: Universidad de Salamanca, 1971), pp. 13–29; José Luis Peset and Elena Hernández Sandoica, *Estudiantes de Alcalá* (Alcalá de Henares: Ayuntamiento de Alcalá de Henares, 1983), and Ana Isabel Martín Ferreira, *El humanismo médico en la universidad de Alcalá (Siglo XVI)* (Alcalá de Henares: Universidad de Alcalá de Henares, 1995).

[41] Jon Arrizabalaga, 'Juan Huarte en la medicina de su tiempo', in *Juan Huarte au XXIe siècle*, ed. by Duché-Gavet, pp. 65–98 (p. 70).

Puerto, and two of his children (Águeda and Luis) were born — the couple had a large family, three sons and four daughters. In 1566 they moved to Baeza, and in 1571 the town asked Huarte to become the city's physician. It has been speculated that Huarte might have practised medicine in Granada, or that he might have taught at the University of Huesca in the academic year of 1569–1570, or at the Colegio-Universidad of San Antonio de Portaceli in Sigüenza, which he would have given up in 1576.[42] It is known for certain that in 1571 the family had its residence in Linares, and that a year later King Philip II authorized the city of Baeza to hire Juan Huarte — a contract that stipulated an annual salary of 200 ducats and two tonnes of wheat. Although the contract was initially for two years, it was renewed up until Huarte's death, which occurred in 1588. Huarte died leaving few possessions and some debts behind, as well as an unfinished new (expurgated) version of the *Examen* adjusted to the demands of the Inquisition. His son Luis continued Huarte's revision enterprise and ensured that the new edition was published in 1594. In addition to the emotional motivation he might have had for so doing, he was no doubt partly moved by a lack of income and the hope to make some money out of the new edition. No matter how much he managed to make, surely the amount was not on a par with the unprecedented success that the book enjoyed throughout early modern western Europe.

INQUISITORIAL CENSORSHIP AND *SUB-PRINCEPS* EDITION (1594)

Edward Bellamy explains in the prologue to his translation that he carried it out from a new version of the Spanish text published after that which Camilli had taken as his model: '*Huartes* a few Years before he died, made some Additions to, and Retrenchments in several Places, entirely leaving out the Seventh, in the Old, and adddding the First, Second, and Fifth Chapters in the New Edition, with a large Supplement to the *Proem*'.[43]

---

[42] Francisco Javier Sanz Serrulla, 'El doctor Huarte de San Juan, médico y catedrático en Sigüenza. Aspectos biográficos inéditos', *Anales Seguntinos*, 1.3 (1986), 309–13.
[43] Bellamy, *The Tryal of Wits*, a1^v.

Bellamy omits to say that the only reason why the new version, first published in Baeza in 1594, was produced was to satisfy the demands of the Inquisition. Huarte's troubles with the Inquisition were fundamentally derived from the *Examen*'s postulates on the organic relations between the brain and the understanding. The idea that the understanding, the base of intellective life, had a material foundation made the Inquisition uneasy, for it implied that the soul and the understanding ultimately depended on the brain. Such an assertion could easily be regarded as a premise for the negation of the immortality of the soul: if such dependence existed, mortality of the brain would signify finiteness of the soul. Furthermore, it could suggest that the soul depended humorally on the body; this made the *Examen* look as if it advocated determinism and denied the dogma of free will. Nevertheless, even if Huarte affirms that the immortality of the soul cannot be demonstrated apodictically by reason, at no point does he deny or even question such immortality. For him, the issue is solely a matter of faith:

> it is a thing certain that the infallible certainty of our immortal soul is not gathered from human reasons or from arguments which prove that it is corruptible, for to the one and the other an answer may easily be shaped: it is only our faith which maketh us certain and assured that the same endureth forever. (VII. 35–39)

All this is at the core of the entirely censored Chapter VII, significantly entitled 'It is showed that though the reasonable soul have need of the temperature of the four first qualities, as well for his abiding in the body, as also to discourse and syllogize, yet for all this, it followeth not that the same is corruptible and mortal'.

Interestingly, the first edition of the *Examen* (1575) was printed with all the licences necessary for it to circulate in the kingdoms of Castile and Aragon, and even opened with most favourable criticism from the Augustine Fray Lorenzo de Villavicencio, who praised the originality of the doctrine as well as Huarte's fidelity to revealed truth:

> I have seen this book and its doctrine is wholly Catholic and healthy without anything against the faith of our holy Church of Rome. This is a doctrine of great and new wit, constructed upon and taken from the best philosophy that can be taught.

It discusses some very serious topics and explains them in a scholarly manner. Its main argument is so necessary for all heads of households to consider that, if they followed what this book recommends, the Church, the Republic and the families would have singular ministers and most important subjects.[44]

In February 1578, Diego Álvarez, a student of Divinity in Córdoba, sent Huarte a seventy-page manuscript entitled *Animadversión y enmienda de algunas cosas que se deben corregir en el libro que se intitula examen de ingenios del Dr. Juan Huarte de San Juan*. The purpose of it was to warn Huarte of the danger of the doctrine of the organicity of the understanding and its implications regarding the Thomistic argument of the immortality of the soul, which was based on the soul's immateriality. Diego Álvarez suggested modifications in Chapters III to VI, and the complete suppression of Chapter VII — also demanded by the Inquisition years later. Diego Álvarez's is friendly advice which Huarte disregarded: what Huarte actually revised was solely what the Inquisition would later on indicate in less friendly terms.

The *Examen* is first singled out by the Inquisition in Portugal: in 1581 it entered the catalogue of forbidden books of Lisbon, the *Católogo dos libros que se prohiben nestes Regnos e Senhorios de Portugal*. Two years before, in 1579, Alonso Pretel, Professor of Positive Divinity at the University of Baeza and regional *comisario* of the Holy Office in Baeza, wrote a virulent attack against the *Examen* and denounced it before the Inquisition of Córdoba, which then sent the denouncing document to Madrid. It is often stressed that in Pretel's virulent attack there was a factor of spite and revenge against Huarte, given that the *Examen* discredits the wit of positive divines. In addition, issues of a different nature may have been behind Pretel's report to the Inquisition.

---

[44] My translation. In Spanish it reads: 'He visto este libro, y su doctrina toda es catholica y sana sin cosa que sea contraria ala fee de nuestra madre la sancta Yglesia de Roma. Sin esto es doctrina de grande y nuevo ingenio, fundada y sacada de la mejor philosophia que puede enseñarse. Toca algunos lugares de scriptura muy grave y eruditamente declarados. Su principal argumento es tan necessario decōsiderar, de todos los padres de familias, que si siguiessen lo que este libro adviette, la Yglesia, la Republica, y las familias, terniā singulares ministros y subjetos importantissimos.' Juan Huarte de San Juan, *Examen de ingenios para las ciencias* (Baeza: Juan Bautista de Montoya, 1575), A2r.

At the time of the publication of the *Examen*, Baeza was a well-sized city of 20,000–25,000 inhabitants with a printing centre for literature of Erasmian tendencies, and a noticeable record of cases of psychopathies, possessions and revelations.[45] These phenomena were linked to a new form of spirituality that inquisitorial documents denominated *alumbradismo*, and that found many followers in the region of Jaén, particularly in Úbeda and Baeza. The term *alumbrado* appears for the first time in a 1524 Castilian text referring to a new sect or heresy, and *alumbradismo* becomes synonymous with a phenomenon of interior illuminism, a process of finding refuge in an ardent interior life.[46] The *alumbrados* encompassed mystics and reformists, and spiritualists with or without heretical tendencies whom sometimes might have shared a set of affinities with Erasmianism or Lutheran Protestantism.[47] In general terms, however, the *alumbrados* of Baeza remained within orthodoxy in high levels of asceticism, which often led to episodes of ecstasy, visions and prophecies. Inquisitorial repression was active against the *alumbrados*, for the Inquisition feared that the University of Baeza would become a centre of heterodoxy — after all, almost the entire faculty were

---

[45] María Dolores Higueras Quesada, 'Estudio sobre la evolución de la población de Baeza, 1550–1750', *Boletín del Instituto de Estudios Giennenses*, 176.1 (2000), 141–94; Claude Benoit, 'Présences de Juan Huarte de San Juan à Baeza', in *Juan Huarte au XXIe siècle*, ed. by Duché-Gavet, pp. 23–36 (pp. 30–31). José Luis Abellán, 'Los médicos-filósofos: Juan Huarte de San Juan, Miguel Sabuco y Francisco Vallés', in *Historia crítica del pensamiento español: 2. La edad de oro: Siglo XVI* (Madrid: Espasa-Calpe, 1986), pp. 207–22 (p. 208). Ricardo Sáez, 'La Baeza del siglo XVI', 1989, pp. 90–91. Felice Gambin, *Azabache. El debate sobre la melancolía en la España de los siglos de Oro*, trans. by Pilar Sánchez Otín (Madrid: Biblioteca Nueva, 2008), p. 148. This is a translation from the Italian *Azabache: Il dibattito sulla malinconia nella Spagna dei Secoli d'Oro* (Pisa: ETS, 2005). Álvaro Huerga Teruelo, 'El *Examen de Ingenios* y los fenómenos místicos', in *Los alumbrados de Baeza* (Jaén: Instituto de Estudios Giennenses. Diputación Provincial, 1978), pp. 103–30 (p. 123). Mystic spirituality, although of a different kind, finds a representative in Fray Juan de la Cruz, who arrived in Baeza in June 1579 to found the Colegio de San Basilio, of which he would become the first president. De la Cruz may have met Huarte, as in 1580 Baeza was struck by the flu and Fray Juan de la Cruz had almost all his friars sick in bed. Huerga Teruelo, 'Huarte de San Juan y San Juan de la Cruz', in *Los alumbrados de Baeza* (1978), pp. 126–30.

[46] Antonio Márquez, *Los alumbrados: orígenes y filosofía (1525–1559)* (Madrid: Taurus, 1980).

[47] Huerga Teruelo, *Los alumbrados de Baeza*, pp. 193, 197.

INTRODUCTION

new Christians.[48] If during the life of Juan de Ávila, the co-founder of the University of Baeza, there had been warnings on the part of the Inquisition, after his death in 1569 surveillance intensified, as did the inquisitorial processes against the *alumbrados*.[49] In Baeza three visits of agents of the Inquisition occurred between 1571 and 1575, which might also explain why Professor Alonso Pretel was especially alert.[50]

The *Examen* was eventually included in the two great indexes of the Spanish Inquisition of the sixteenth century, those published in 1583 and 1584 by the Inquisitor General Gaspar de Quiroga. The first is the *Index et catalogus librorum prohibitorum*, a list of forbidden books. The second, the *Index librorum expurgatorum*, which specifies the passages that had to be amended and corrected; these run for five pages in the case of the *Examen*. Only three chapters are free from inquisitorial objections (VIII, IX and XI); the rest suffered mutilations in various degrees. Then, at the beginning of the seventeenth century, the Vatican Indexes (1604–1605) prohibited the *princeps* edition of the *Examen*. The Spanish inquisitorial indexes had autonomy from those of Rome, and it was not unusual for Roman and Spanish indexes to disagree, and for books banned in one place not to be banned in the other. Also, because the *Expurgatorum* did not have to be enforced in the Low Countries, the 1575 version kept on being published there even after the 1594 edition was released in Spain. Hence, the texts of the editions published in Leyden in 1591 and 1652, in Antwerp in 1593 and 1603, in Amsterdam in 1662, and in Brussels in 1702, were pre-inquisitorial.[51]

The Spanish index of 1584 authorized the distribution of Huarte's work provided that the censored passages were amended. The Inquisition expurgated forty-four fragments which vary in length: from a single word or a short sentence to a chapter in its entirety (in the text the censored fragments are indicated in

---

[48] Huerga Teruelo, *Los alumbrados de Baeza*, p. 17.

[49] Huerga Teruelo, *Los alumbrados de Baeza*, p. 37.

[50] It was in the 1570s when the Inquisition accused Hernando de Herrera and Diego Pérez de Valdivia, two clergymen of the circle of Juan de Ávila, of being *alumbrados*. Years before, Juan de Ávila had himself been imprisoned by the Inquisition in Seville (1531–1533). Luis S. Granjel, *Juan Huarte y su "Examen de ingenios"* (Salamanca: Real Academia de Medicina, 1988), p. 26.

[51] Serés, 'Introducción', in Huarte, *Examen de ingenios*, p. 123, includes a visually clarifying family tree of all the editions.

21

footnotes). The reasons for censoring some of the extracts seem rather arbitrary, while on other occasions, fairly evident. For example, certain fragments are censored because they seem to presuppose in animals powers of the rational soul specific to mankind; others, because they presuppose in God an election when giving out gifts to men. Sometimes it is because Huarte appears to debase the wit of preachers and positive divines, or because he attempts to explain the temperament of Christ. On one occasion, regarding biblical exegesis, Huarte seems to defend the literal sense (i.e. the closest to the letter), and disregard the tropological sense, which according to the Church is in many cases the true one:

> Of many Catholic senses which the Holy Scripture may receive, I hold that ever better which taketh the letter than that which reaveth the terms and words of their natural signification. (XV. 1935-1937)

This may suggest that an individual of fine wit is able to grasp revealed truth on his own, and therefore discern the true sense of the Scriptures without anyone else's help. Such a statement dangerously draws Huarte towards the defenders of the *libre examen*, who questioned the exclusive right of the Church to interpret the Holy Scriptures as established at the Council of Trent. Yet it is chiefly his thesis on the organicism of the understanding that is most problematic, for it seems to question the immortality of the soul.

Huarte spent the rest of his life amending his book, but died in 1588 before seeing the new version published. His son Luis undertook the continuation of the editing process. The amended edition — known as *sub-princeps* — was published in Baeza in 1594 by the same printer, Juan Bautista de Montoya, and without much variation in the title: *Examen de ingenios para las ciencias, en el cual el lector hallará la manera de su ingenio para escoger la ciencia en que más ha de aprovechar, y la diferencia de habilidades que hay en los hombres, y el género de letras y artes que a cada uno responde en particular*. Huarte wrote three new chapters,[52] significantly enlarged the second prologue, divided the previous Chapter XV into five independent chapters, and added a second

---

[52] Chapters I, II and V in the reformed edition, which means that the 1575 Chapter I becomes Chapter III in 1594.

prologue to the reader. Omissions, amendments, and added fragments of, at times, considerable length abound in the posthumously published work, also filled with typographical errors and repetitions no doubt due to the disorder in which Huarte left his notes when dying. Despite all modifications, the 1594 edition retains the ideological core of the 1575 book, to the extent that the content is substantially the same in the two versions. Thus, in the 1594 edition Huarte reaffirms, even if more subtly, what he had been forced to erase, and so preserves his major thesis, namely, the organicity of the understanding.

The International Dimension: Translation and Reception

Edward Bellamy's dedicatory preface to *The Tryal of Wits* (1698), the second translation of Huarte's book into English, comments on the international reception of Huarte's *Examen*:

> Nor need I offer one Word in Behalf of this Excellent Book, because it speaks sufficiently for itself, it is well known among the Learned, and was well received when first Writ, and is yet no less in Esteem amongst most Men of Letters.
>
> There have been no less than Five or Six several Editions of the Original, Three of the *Italian*, Ten or Eleven of the *French*, into which it was at two several times Translated; as also Once into *Latin*, and as often into *Dutch*. If all this Proclaims not its Merit, at least it speaks its good Fortune, in the kind Reception this Book has met with in the World.[53]

The reception of Huarte's book was truthfully more than kind. If in England Carew's *The Examination* was first published in 1594, reprinted in 1596, 1604 and 1616, and then retranslated by Bellamy as *The Tryal of Wits* (1698), Continental Europe was no less welcoming. The *Examen* became a bestseller in Spain during the sixteenth and seventeenth centuries: the book, first published in Baeza in 1575, was subsequently published in Spanish in Pamplona, 1578;[54] Bilbao, 1580; Valencia, 1580; Huesca, 1581;

---

[53] Bellamy, *The Tryal of Wits*, a1r–v.
[54] The Pamplona edition was the basis of all the rest, both Spanish as well as foreign, until the 1594 expurgated edition was published.

Leiden, 1591; Antwerp, 1593; Baeza, 1594; Antwerp, 1603; Medina del Campo, 1603; Barcelona, 1607; Alcalá, 1640; Leiden, 1652; Amsterdam, 1662; and Madrid, 1668. During those two centuries the treatise was rendered into French by three different translators (Gabriel Chappuys, Charles Vion Dalibray, and François-Savinien d'Alquié) and published in French in Lyon in 1580, 1597, 1608, 1609, 1668 and 1672; in Paris in 1588, 1614, 1618, 1619, 1631, 1633, 1634, 1645, 1650, 1655, 1661, 1675 and 1698; in Rouen in 1598, 1602, 1607, 1613 and 1619; and in Amsterdam in 1672. Gabriel Chappuys rendered the title as *Examen des esprits propres et naiz aux sciences*, and his translation was used until 1633, and after that only for the 1698 Paris edition. From then on, and until 1675, Charles Vion Dalibray's rendering of the book as *L'examen des esprits pour les sciences* was used by publishers. François-Savinien d'Alquié preserved Dalibray's title and his translation was employed in the 1672 printing of the book in Amsterdam.

The *Examen* was translated into Italian twice: first by Camillo Camilli under the title of *Essame de gl'ingegni de gli huomini, per apprender le scienze*, and secondly by Salustio Gratii, who entitled his translation *Essamina de gl'ingegni de gli huomini accomodati ad apprendere qual si voglia scienza*. Camillo Camilli's translation was the one employed in all sixteenth-century editions of the book, while Salustio Gratii's was the one published during the seventeenth century. All the Italian printings of Huarte's work save one (Cremona, 1588) appeared in Venice (in 1582, 1586, 1590, 1600 and 1603). In 1605 the book was banned by the Inquisition in Italy, which accounts for the sudden cessation of its printings in the country after having achieved considerable editorial success.[55] In the seventeenth century, the *Examen* was rendered into Latin by Aescatius Major (Joachim Caesar), who entitled his translation *Scrutinum ingeniorum pro iis, qui excellere cupiunt*.[56] In Latin, the book was published in Leipzig in 1622,

---

[55] For a discussion of Camilli's (and Gratii's) rendering of the *Examen* into Italian see, Felice Gambin, 'Il gesuita e il medico: le annotazioni alla traduzione italiana dell'*Examen de ingenios para las ciencias* di Juan Huarte de San Juan', in *Malattia e scrittura. Saperi medici, malattie e cure nelle letterature iberiche*, ed. by Silvia Monti (Verona: Cierre Grafica, 2012), pp. 147–83.

[56] For more on the Latin translation see Martin Franzbach, *Lessings Huarte-Übersetzung (1752): Die Rezeption und Wirkungsgeschichte des "Examen de ingenios para las ciencias" (1575)' in Deutschland* (Hamburg: Cram, de Gruyter,

and in Jena in 1637 and 1663.[57] Additionally, in 1659 the *Examen* was printed in Dutch in Amsterdam under the title of *Onderzoek der byzondere Vernuftens*, in a translation by Henryk Takama.

In the history of the reception of Huarte's work diversity is key: there is variety in terms of the countries in which the *Examen* was known and read, variety in the languages into which it was translated, and in regard to the cities where it was printed. For instance, in Spain alone, it appeared published in Baeza, Pamplona, Bilbao, Valencia, Huesca, Medina del Campo, Barcelona, Alcalá and Madrid. What is striking about this list is that it covers the great publishing hubs in Spain (Alcalá, Madrid, Valencia, and to a lesser extent Medina del Campo), and cities that were not particularly renowned within the publishing business (Baeza, Pamplona, Bilbao and Huesca).[58] Such frequent reprintings of a medical book were unusual at a national level (in the Spanish context at least), let alone at an international one. Consider that in Spain the publication of medical books begins in 1475, and from that year until 1599, 350 out of the 541 titles published correspond to first editions, the majority written by Spaniards, many of whom only published a single book.[59] This is not to say that Huarte was alone in writing a successful medical

---

1965). This book was translated into Spanishby Luis Ruiz Hernández as *La traducción de Huarte por Lessing (1752): recepción e historia de la influencia del "Examen de ingenios para las ciencias" (1575) en Alemania* (Pamplona: Diputación Foral de Navarra, Institución Príncipe de Viana, 1978), pp. 40–58.

[57] A translation into German only appeared in the eighteenth century; it was carried out by Gotthold Ephraim Lessing as part of his doctoral thesis at Wittenberg (*Johann Huarts Prüfung der Köpfe zu den Wissenschaften*), and published in 1752 and 1785.

[58] In Castile, Salamanca and Alcalá were the main printing hubs due to the importance of their universities; Madrid joined the group after it was made capital of Spain in 1561. Seville was another important centre of publication, followed by Barcelona, Zaragoza and Valencia. Both Valladolid and Medina del Campo had certain relevance within the publishing business too. See Juan Riera, 'La literatura científica en el Renacimiento', in *Ciencia, medicina y sociedad en el Renacimiento castellano*, ed. by Juan Riera, *et al.* (Valladolid: Universidad de Valladolid, 1989), pp. 5–17 (p. 10).

[59] Granjel, *La medicina española renacentista*, pp. 54–60. For more on the publication of Spanish scientific texts, see José María López Piñero, *Ciencia y técnica en la sociedad española de los siglos XVI y XVII* (Barcelona: Labor, 1979), pp. 118–20. According to this scholar, the number of first editions of Spanish scientific works from 1475 to 1600 was 366 in the field of medicine, and 69 in natural philosophy.

work: titles by other contemporary Spanish physicians such as Cristóbal de Vega, Francisco Vallés and Luis Mercado, for instance, were printed outside Spanish borders, and the works by Juan Valverde de Amusco, Nicolás Monardes, Cristóbal de Acosta and several surgeons were also rendered into other languages. However, what is exceptional in the case of the *Examen* is that its impact was not narrowed to a specific field of knowledge, for it became well-known outside medical and natural philosophy circles. Thus, in addition to physicians and natural philosophers, poets and literary authors, and men of letters and science in general became acquainted with the treatise and took from it what could be of use to them in their respective fields.[60]

Miguel de Cervantes's *La Galatea* (1585), *El licenciado Vidriera* (1613), *Viaje del Parnaso* (1614), *La elección de los alcaldes de Daganzo* (1615), *Trabajos de Persiles y Segismunda* (1616) and his masterpiece *El ingenioso hidalgo Don Quijote de La Mancha* (1605) betray the influence of Huarte's postulates.[61] Indeed, in *Don*

---

[60] Knowledge of the *Examen* does not imply partial or total agreement with its theories. The physician Andrés Velásquez, a former colleague of Huarte at Alcalá, critically responds to Huarte in his *Libro de la melancolía* (1585), the first work in a vernacular language specifically and entirely devoted to melancholy. Indeed, *Libro de la melancolía* appeared a year before Timothy Bright's *A Treatise of Melancholy* (1586), and some years before *Discours de la conservation de la veue; des maladies melancholiques; des catarrhes et de la vieillesse* (1597), by the French physician André Du Laurens. Likewise, the physician Francisco Villarino disagreed with Huarte on several topics. His *Advertimientos sobre el libro intitulado Examen de ingenios del Doctor Juan Huarte* rigorously criticizes the *Examen* chapter by chapter, at times almost page by page. In *Advertimientos* Villarino refers to the Jesuit Antonio Possevino's previous analysis of Huarte's work, which appeared in some chapters of the first book of his *Bibliotheca Selecta* (1593) — later published independently, first in Italian as *Coltura degl'Ingegni* (1598), and then in Latin as *Cultura ingeniorum ... Examen ingeniorum J. Huartis expenditur* (1604). See, Antonio de la Granda, 'Juan Huarte de San Juan y Francisco Villarino', in *Estudios de Historia Social de España*, IV vols, ed. by Carmelo Viñas y Mey (Madrid: CSIC; Instituto Balmes de Sociología, 1949), I, pp. 655–69; and, for a discussion in detail of Andrés Velásquez's *Libro de la melancolía*, Roger Bartra, *Cultura y melancolía: las enfermedades del alma en la España del Siglo de Oro* (Barcelona: Anagrama, 2001), pp. 19–150; trans. by Christopher Follet, *Melancholy and Culture: Essays on the Diseases of the Soul in Golden Age Spain* (Cardiff: University of Wales Press, 2008) pp. 41–164.

[61] Rafael Salillas, *Un gran inspirador de Cervantes: El Doctor Juan Huarte y su 'Examen de ingenios'* (Madrid: Imprenta de Eduardo Arias, 1905); Mauricio de Iriarte, *El ingenioso hidalgo y El examen de ingenios: (qué debe Cervantes al*

Quixote the term *ingenioso* directly points at Huarte's theories.⁶² Cervantes surely knew the *Examen* and no doubt read the book, and yet not once does he mention Huarte or the *Examen*. Even Francisco de Quevedo in *La España defendida y los tiempos de ahora* at one point asks '¿Qual philosopho exçedio ni igualo el *Examen de Injenios* nuestro?' ('Which philosopher exceeded or equalled our *Examination of Men's Wits*?').⁶³ Huarte's notions are moreover present in Baltasar Gracián's emphasis on the natural gifts and genius of the poet; like Cervantes, Gracián fails to mention Huarte explicitly.⁶⁴

Dr. Huarte de San Juan) (San Sebastián: *Separata de Revista Internacional de los Estudios Vascos*, 24.4, 1933); Fernando Escobar, *Huarte de San Juan y Cervantes en la locura de D. Quijote de la Mancha. Breve estudio clínico psico-somático* (Granada: Universidad de Granada, 1949); Otis H. Green, 'El Licenciado Vidriera: Its Relation to the *Viaje del Parnaso* and the *Examen de Ingenios* de Huarte', *Linguistic and Literary Studies in Honor of Helmut A. Hatzfeld*, ed. by Alessandro S. Crisafulli (Washington: Catholic University Press, 1964), pp. 213–20; Harry Sieber, 'On Juan Huarte de San Juan and Anselmo's "locura" in "El curioso impertinente"', *Revista Hispánica Moderna*, 36.1–2 (1970–1971), 1–8; Chester S. Halka, 'Don Quijote in the Light of Huarte's *Examen de ingenios*: A Reexamination', *Anales Cervantinos*, 19 (1981), 3–13; Esteban Torre, *Sobre lengua y literatura en el pensamiento científico español de la segunda mitad del siglo XVI: Las aportaciones de G. Pereira, J. Huarte de San Juan y F. Sánchez el Escéptico* (Sevilla: Universidad de Sevilla, 1984); Brian McCrea, 'Madness and Community: Don Quixote, Huarte de San Juan's *Examen de ingenios* and Michel Foucault's *History of Insanity*', *Indiana Journal of Hispanic Literatures* 5 (1994), 213–24; David F. Arranz Lago, 'Sobre la influencia del *Examen de ingenios* en Cervantes. Un tema revisitado', *Castilla: Estudios de Literatura*, 21 (1996), 19–38; and José Luis Peset, *Las melancolías de Sancho: humores y pasiones entre Huarte y Pinel* (Madrid: Asociación Española de Neuropsiquiatría, 2010).
⁶² On the temperament and wit of Don Quixote, see Bartra, *Cultura y melancolía*, pp. 151–96; trans. by Follet, *Melancholy and Culture*, pp. 165–205.
⁶³ Francisco de Quevedo, *España defendida y los tiempos de ahora: de la calumnias de los noveleros y sediciosos*, ed. by R. Selden Rose (Madrid: Fortanet, 1916), p. 70.
⁶⁴ C. M. Hutchings, 'The *Examen de ingenios* and the Doctrine of Original Genius', *Hispania: A Journal Devoted to the Teaching of Spanish and Portuguese*, 19.2 (1936), 273–82; Hennig Mehnert, 'Der Begriff "ingenio" bei Juan Huarte und B. Gracián', *Romanische Forschungen*, 91.3 (1979), 270–80; and Guillermo Serés, 'El ingenio de Huarte y el de Gracián: Fundamentos teóricos', *Insula: Revista de Letras y Ciencias Humanas*, 655–56 (2001), 51–53. Additionally, the application of Huarte's theories to literary criticism is illustrated by Juan Díaz Rengifo, Alonso López Pinciano, Luis Alfonso de Carballo, Juan de la Cueva and by the Sevillian school of poetry of the late sixteenth century. Esteban Torre, *Ideas lingüísticas y literarias del doctor Huarte de San Juan* (Sevilla: Publicaciones

In Spain the influence of the *Examen* decreased after the end of the seventeenth century. In the mid-eighteenth century, father Benito Jerónimo Feijoo complains that it was only through foreign authors that he came to learn of the existence of the *Examen*. This stands as proof of the phenomenal success of Huarte's treatise abroad, of the gradual decline of the book in Spain, and of the complex system of tides and flows of influence by which knowledge of a Spanish work is brought back to its native land via commentaries and translations of it into foreign tongues:

> I believe that many very good books by Spanish authors would have gone lost if foreigners had not preserved them; this is the reach of our, I will not say negligence, but literary drowsiness. Some are listed in his *Biblioteca* by Don Nicolás Antonio,[65] of which he did not have notice but for foreign authors. Not long ago, reading the third volume of the *Spectador Anglicano*, in Discourse number 49,[66] I found quoted a book whose title is *Examen de ingenios para las ciencias*, and its author Juan Huarte, a Spanish physician. From what the English writer says about this book I judged the idea excellent and the matter important. [...]
>
> As before finding the news about the physician Huarte in the *Spectador* I had not read nor heard his name, I could not but marvel [...] that it took so long for me to receive the first news of a Spanish author of such merit, and so much so because this first piece of news came to me from an Anglican writer....] This Spanish author, while widely acknowledged among foreigners, is almost entirely forgotten amongst Spaniards.[67]

---

de la Universidad de Sevilla, 1977), pp. 129–31; Serés, 'Introducción', in Huarte, *Examen de ingenios*, pp. 71–78.

[65] Nicolás Antonio was a Spanish bibliographer of the seventeenth century renowned for his works *Bibliotheca hispana vetus* (1672) and *Bibliotheca hispana nova* (posthumously published in 1696), focused on Spanish authors from 1500 to 1647.

[66] *The Spectator* was a daily publication founded by Joseph Addison and Richard Steele in England and published from 1711 to 1712. Feijoo refers here to *The Spectator* of 21 February, 1712.

[67] My translation. In Spanish: 'Creo que no pocos Libros muy buenos de Autores Españoles se hubieran perdido, si no los hubieran conservado los Estrangeros, que es a quanto puede llegar nuestra, no diré ya negligencia, sino modorra literaria. Algunos nombra en su Biblioteca Don Nicolas Antonio, de

Father Feijoo stresses the impact of the book in France by reproducing the opinion of a Mr. Berteud (or Bertaud), according to whom Spaniards, by comparison with the French, were not generally acquainted with Huarte or with his treatise:

> In the second volume of the *Menagiana* [*Menagiana ou les bons mots et remarques critiques … de Monsieur Menage*] of the Paris edition of the year 1729, on page 18, where in the name of Mr. [Gilles] Menage Spaniards are branded as not exceedingly scholarly, towards the end of the page there is the following note in little font, placed there by the adder: *Mr. Berteud in his travels says that in Spain Doctor Huarte is not known, neither his book the Examen de los Ingenios.*[68]

In France the *Examen* was introduced in 1580 via the translation of Gabriel Chappuys, which first appeared in Lyon the same year as the first *Essais* of Montaigne.[69] It enjoyed phenomenal success

---

los quales no tuvo noticia, sino por Autores Estrangeros. No ha mucho tiempo, que leyendo el tercer Tomo del Spectador Anglicano, en el Discurs. 49 hallé citado un Libro, cuyo título es: *Examen de Ingenios para las Ciencias*, y su Autor Juan Huarte, Médico Español. Por lo que dice de este Libro el Escritor Inglés hice juicio de la excelencia de la idea, y de la importancia del asunto [...].

Como yo, antes de ver la noticia del Medico Huarte en el Spectador, no había leído, ni oído su nombre, no dexé de estrañar [...] que tan tarde llegase a mi la primera noticia de un Autor Español de tanto mérito; y aun esa primera noticia derivada a mí de un Escritor Anglicano. [...] este Autor Español, al paso que muy famoso entre los Estrangeros, casi está enteramente olvidado de los Españoles.' Benito Jerónimo Feijoo, 'Carta XXVIII. Del descubrimiento de la circulación de la sangre, hecho por un albeytar español', in *Cartas eruditas, y curiosas*, IV vols (Madrid: Imprenta Real de la Gazeta, 1774), III, pp. 314–23 (pp. 320–21).

[68] My translation. In Spanish: 'En el segundo Tomo de la Menagiana de la edición de París del año de 1729, a la página 18, donde en nombre de Mr. Menage son censurados de poco eruditos los Españoles, hay al fin de la página la nota siguiente de letra menuda, puesta por el Addicionador: *Mr. Berteud en su viage dice, que en España no es conocido el Doctor Huarte, ni su Libro del Examen de los Ingenios.*' Benito Jerónimo Feijoo, 'Carta XXVIII', 1774) pp. 314–23 (p. 321).

[69] Montaigne must have heard of the *Examen*, as during his visit to Lyon in 1581 Chappuys's translation would have been a topic for discussion. It is highly unlikely that the first edition of Montaigne's *Essais* was influenced by the *Examen*, as Montaigne most likely did not read the treatise before that year. Gabriel-André Pérouse, *L'Examen des esprits du docteur Juan Huarte de San Juan. Sa diffusion et son influence en France aux XVIe et XVIIe siècles* (Paris: Les Belles lettres, 1970), p. 69; Harald Weinrich, 'Prólogo', in Huarte *Examen*

until 1637, when Huarte started to be eclipsed by Descartes, who might have heard of Huarte's theses too.[70] In France Pierre de Deimier, Guillaume Bouchet, Pierre Charron, Jourdain Guibelet, Jean de Silhon and Montesquieu, among others, seem to have known about the *Examen* in different degrees. In other countries such as Italy or Germany, the list of physicians, philsophers and authors is not shorter or less impressive. Among the Italians, we find Antonio Possevino, Giovan Battista della Porta, Antonio Zara, the physician Domenico Gagliardelli, Alessandro Tassoni, Giorgio Baglivi, and Juan Imperial; among the Germans, Christian Thomasen, Diezinger, Hamann and Herder, who mentions Huarte in his essay *On the Cognition and Sensation of the Human Soul* (1778). Kant's trascedental idealism as well as the thought of J. F. Buddeus, Mendelssohn, Garve, Johann Justus von Einem, J. C. Lavater, Goethe, and Schopenhauer have also been connected at points with some of Huarte's theories.[71]

---

de ingenios para las ciencias, ed. by Serés, pp. 9–16 (p. 15). Weinrich admits that despite the chronological difficulty, he often wonders whether the term *essais* used by Montaigne does not owe something to *examen*, as it suggests connotations that match Huarte's use of the word. See also Gabriel-André Pérouse, 'Montaigne et le Dr Huarte: Avec un mot sur Pierre Charron', *Bulletin de la Société des Amis de Montaigne*, 8.13–14 (1999), 11–22.

[70] Pérouse, *L'Examen des esprits du docteur Juan Huarte*, p. 145.

[71] Malcolm K. Read, *Juan Huarte de San Juan* (Boston: Twayne Publishers, 1981), analyzes the influence of the *Examen* in Spain and abroad, pp. 106–22. See also Pierre Mauriac, 'Montesquieu connaissait Jean Huarte ...', *Figaro Littéraire*, 8 (1959), 8; Martin Franzbach, 'La influencia del *Examen de ingenios para las ciencias* (1575) de Juan Huarte de San Juan en Alemania', *Medicina Española*, 62 (1969), 450–56; Malcolm K. Read and J. Trethewey, 'Juan Huarte and Pierre de Deimier: Two Views of Progress and Creativity', *Revue de Littérature Comparée*, 51.1 (1977), 40–54; Felice Gambin, 'Sobre la recepción y la difusión del Examen de ingenios para las ciencias de Huarte de San Juan en Italia', in *Filosofía y literatura en el mundo hispánico. Actas del IX Seminario de Historia de la Filosofía Española e Iberoamericana*, ed. by Antonio Heredia Soriano and Roberto Albares Albares (Salamanca: Universidad de Salamanca, 1997), pp. 409–25; José Biedma López, 'Poder de la imaginación y fecundidad del entendimiento en el *Examen de ingenios para las ciencias* (Sobre el origen hispano de la filosofía crítica)', in *Juan Huarte au XXIe siècle*, ed. by Duché-Gavet, pp. 213–36; Eduardo Gil Bera, 'Faire le point: el concepto de verdad en Huarte y su visión de la condición humana', in *Juan Huarte au XXIe siècle*, ed. by Duché-Gavet, pp. 207–12; E. García García and A. Miguel Alonso, 'El *Examen de ingenios* de Huarte de San Juan en la *Bibliotheca selecta* de Antonio Possevino', *Revista de Historia de la Psicología*, 3–4 (2003), 387–997, and 'El Examen de Ingenios de Huarte en Italia. La anatomia ingeniorum de Antonio

INTRODUCTION

RICHARD CAREW: FROM THE *GERUSALEMME* TO *THE EXAMINATION*

In 1594 Richard Carew published translations of two works in Italian: his incomplete translation of Torquato Tasso's *Gerusalemme liberata*, which he entitled *Godfrey of Bulloigne or, The Recoverie of Hierusalem*, and his *The Examination of Mens Wits*. The reasons for carrying out these translations remain unknown: his *Godfrey of Bulloigne* was published without his consent and hence without a preface by him, and the laconic dedication 'To the right worshipful Sir Francis Godolphin Knight, one of the deputy lieutenants of Cornwall' that precedes *The Examination* is not illuminating in this respect:

> Good Sir, your book returneth unto you clad in a Cornish gaberdine, which if it become him not well, the fault is not in the stuff, but in the botching tailor, who never bound prentice to the occupation, and working only for his pastime, could hardly observe the precise rules of measure. But such as it is, yours it is, and yours is the workman, entirely addicted to reverence you for your virtues, to love you for your kindness, and so more ready in desire than able in power to testify the same, do with my dewiest remembrance take leave, resting at your disposition,
> R. C. (Pl. 3–12)

This was the dedication published in 1594 and in all subsequent reprintings of the book (all done by Adam Islip) in 1596, 1604 and 1616, and to it not a word was added. What the dedicatory letter reports is that the copy from which Carew carried out the translation belonged to his fellow Cornishman Sir Francis Godolphin, father-in-law of his brother George, and that it was lent to him so that he would render it into English.[72] It is certain

---

Zara', *Revista de historia de la psicología*, 25.4 (2004), 83–94; Guillermo Serés, 'Possevino entre Huarte y Gracián: el cultivo del ingenio y la imaginación creativa', in *La traduzione della letteratura italiana in Spagna (1300–1939)*, ed. by M. de las Nieves Muñiz Muñiz (Florence: Franco Cesati, 2007), pp. 429–42; Miguel Ángel González Manjarrés, 'Introducción', Giovan Battista della Porta, *Fisiognomía*, II vols (Madrid: Asociación Española de Neuropsiquiatría, 2007–2008), I, pp. 7–19; and Stephen Pender, 'Introduction: Reading Physicians', in *Rhetoric and Medicine in Early Modern Europe*, ed. by Stephen Pender and Nancy S. Struever (Farnham & Burlington: Ashgate, 2012), p. 17.

that Godolphin's copy was the *Essame de gl'ingegni de gli huomini per apprender scienze*, the Italian rendering of Huarte's work by Camillo Camilli. Nonetheless, the circumstances by which Sir Francis Godolphin came to own this book in the first place, the context in which Sir Francis Godolphin and Richard Carew would begin a conversation about the book in question, the reasons why Sir Francis Godolphin might have asked Carew to translate the book or, for that matter, the reasons why Carew accepted (or offered) to render the book into English are unknown. We are thus ignorant of what made the *Examen* particularly appealing to Carew, what made him want to invest his time in translating such a long treatise. A door opens to speculation on this matter when taking into consideration the other translation by Carew also published in 1594, his *Godfrey of Bulloigne, Or The Recouerie of Hierusalem*, a rendering in *ottava rima* (an eight-line stanza previously used by Harington in the *Orlando Furioso*) of the first five cantos of Torquato Tasso's monumental work. Arranged as a bilingual edition with the English on the verso and the Italian on the recto pages,[73] and printed in quarto by John Windet for Christopher Hunt, it lacks a prologue by Richard Carew. There is only a foreword, 'To the Reader', by Christopher Hunt, who explains that the absence of a single word by Carew is due to the fact that, in his own haste to see the work in print, he had it published without the consent of Carew, whom he allegedly did not really know:

> Gentlemen, let it be lawful for me with your leaves to trouble you a little. It was my good hap of late to get into my hands an English translated copy of *Seig. Tasso's Hierusalem*, done (as I

---

[72] Sir Francis Godolphin (*c.* 1534–1608) served as JP for Cornwall from the mid-1570s; in 1580 he was knighted, and sat as knight for Cornwall in 1589 and MP for Lostwithiel in 1593. J. P. D. Cooper, 'Godolphin, Sir William (*c.* 1518–1570)', in *Oxford Dictionary of National Biography* <http://www.oxforddnb.com/view/article/67867> [accessed 22 June 2013].

[73] The Italian text facing Carew's translation was the Osanna edition of 1584, not the one Carew used. As Walter L. Bullock explains, 'Richard Carew used as the Italian original of his *Godfrey of Bulloigne or the Recoverie of Hierusalem* ... a copy of the edition of *Il Goffredo* printed at Venice, 1582, by Gratioso Perchacino; either in its original form, or with the altered title-page imprint "... presso Francesco de' Franceschi Senese 1583"', 'Carew's Text of the *Gerusalemme Liberata*', *PMLA*, 45.1 (1930), 330–35 (p. 334).

was informed) by a Gentleman of good sort and quality, and many ways commended unto me for a work of singular worth and excellency. Wherupon, by the advise, or rather at the instance of some of my best friends, I determined to send it to the Press. Wherin if my forwardness have fore-run the Gentleman's good liking, yet let me win you to make me happy with the sweet possession of your favours, for whose sakes I have done whatsoever herein is done. When first I sent it to the Printer, I did not certainly know whose work it was, and so rested deprived of all means to gain his assent and good liking thereunto. And yet notwithstanding the persuasions of some that would fain have prevailed with me, I resolved (at the motion, no doubt of some rare excellent spirit that knew and foresaw this to be the readiest means to draw him to publish some of his many most excellent labours) to go on with what I had begun, ever assuring my selfe, and never doubting, but that you would like of it yourselves, and entertain it with such dear affection as it doth worthily merit. Now if it shall not in each part lively resemble the absolute perfection of the doer thereof, yet is he blameless, and the fault as it is mine, so I will acknowledge it for mine, for by my haste it proves his untimely birth, and doubtless miserably wanteth of that glorious beauty wherewith it otherwise would, and hereafter happily may be richly honoured withal. (¶2^{r-v})

It is fitting that this incomplete translation of Tasso's *Gerusalemme liberata* should appear without the consent of the translator, as Tasso's *Gerusalemme* itself was also first printed (by Domenico Cavalcalupo at the request of Marcantonio Malaspina) without the author's knowledge, and appeared incomplete under the title of *Il Goffredo di m. Torquato Tasso. Nuouamente dato in luce* (Venice, 1580).[74] Interestingly, an analysis of the translation reveals that 'there are no signs of Carew's *Godfrey* having been written in haste or of its being in an unfinished state'.[75] On that account, one wonders whether the

---

[74] A year later the work was printed in its entirety by Erasmo Viotti as *La Gierusalemme liberata, ouero il Goffredo* (Parma, 1581), by Vittorio Baldini in Ferrara and in Lyon by Alessandro Marsili. Soko Tomita, *A Bibliographical Catalogue of Italian Books Printed in England, 1558–1603* (Farnham: Ashgate, 2009), p. 257.

final outcome would have been any different had Carew translated the work in full or had he had the opportunity to revise his translation before its publication — if in fact he had not done so. After all, there was no real haste to print the translation, for it was not as if a craze for Tasso's *Gerusalemme* had invaded England overnight: in England, before the rendering by Richard Carew, only Book 1 of *La Gerusalemme liberata* had been translated (and that was into Latin) by Scipione Gentili and printed by John Wolfe as *Torquati Tassi Solymeidos* in 1584. It would not be until 1600 that Tasso's complete poem appeared in English. This was Edward Fairfax's translation, printed in folio by Arnold Hatfield for John Jaggard and Matthew Lownes under the title of *Godfrey of Bulloigne, or The Recouerie of Ierusalem*.[76] In any case, true or false, the excuse provided by Hunt explains why there are no remarks by Carew upon the process of translating Tasso or the outcome of his translation venture.

Quite unexpectedly Christopher Hunt's 'To the Reader' preceding *Godfrey of Bulloigne* sheds light upon Carew and *The Examination*. Consider the following fragment:

> Now whereas I thought you should have had all together, I must pray you to accept of the five first Songs, for it hath pleased the excellent doer of them (for certain causes to himself best known) to command a stay of the rest till the summer. (¶2ᵛ)

'A stay of the rest' may mean that Carew had indeed translated the other cantos but decided to hold their publication 'till the summer'. It is more plausible, however, that it signifies that Carew would wait until the summer to resume his translation of Tasso. Carew would not translate the rest of Tasso's poem — or if he did there is no record of it — and, most likely, the reason why he postponed the translation of the rest of Tasso was that by then he was already engaged in translating Huarte. *The Examination* was by all appearances published months after the *Godfrey*:

---

[75] Massimiliano Morini, *Tudor Translation in Theory and Practice* (Aldershot: Ashgate Publishing, 2006), p. 121; on Carew's translation see, pp. 121–28.

[76] For a discussion of Carew's translation and a contrast with Fairfax's version (1600) see Ralph Nash, 'On the Indebtedness of Fairfax's Tasso to Carew', *Italica*, 34.1 (1957), 14–19. See also R. E. N. Dodge, 'The Text of the *Gerusalemme Liberata* in the Versions of Carew and Fairfax', *PMLA*, 44.3 (1929), 681–95, and Bullock, 'Carew's Text of the *Gerusalemme Liberata*'.

INTRODUCTION

Hunt's preface to the *Godfrey* is signed 'From Exceter the last of February, 1594' (¶2ᵛ), and although there is no evidence regarding the month of publication of *The Examination*, it was certainly published after February that year. That *Godfrey* was published before *The Examination* is corroborated by the fact that, of the three editions of *The Examination* that Adam Islip published in London in 1594, one was for 'C. Hunt of Excester',[77] an indication that by the time of the publication of *The Examination*, Christopher Hunt had grown well acquainted with Carew and with his work as a translator.

There exists yet another connection between Carew's *Godfrey* and *The Examination*, or rather, between Tasso's *Gerusalemme* and Huarte's *Examen*, one that may ultimately explain why Carew decided to translate Huarte in the first place. This link is a third man: the enigmatic Camillo Camilli. Little is known for certain about Camilli: his date of birth remains undetermined; he has been related to Siena, or more probably to Monte San Savino, within the region of Arezzo, and from 1600 until his death in 1615 he seems to have lived in Ragusa. Camilli published in Venice an original work entitled *Le imprese illustri di diversi, coi discorsi di C. Camilli* (1586), dedicated to Cardinal Ferdinando de Medici, and also wrote a religious poem, *Le lagrime di santa Maria Maddalena* (the first edition was published in Perugia, s.d., although probably 1592; it was reprinted in Palermo in 1597). He was furthermore a prolific translator: from Latin he rendered Ovid's *Heroides* and *Fasti* (Venice, 1587), and from Spanish *Prima parte dell'oratione et meditatione* (Venice, 1580–82) by Luis de Granada, Juan de Ávila's *Trattato spirituale sopra il verso 'Audi filia'* (Venice, 1581), Martín de Azpilcueta's *Manuale de confessori et penitenti* (Venice, 1592), Luis de Granada's *Trattato primo coll'aggiunta del Memoriale della vita cristiana* (Venice, 1594), as well as Agostino Agostini's *I Sette Salmi penitenziali mutati in rima* (Antwerp, 1595).[78] More importantly, at least for our purposes, Camillo Camilli translated Huarte's *Examen* as

---

[77] Another was for Richard Watkins (the one used in this edition), and a third for Thomas Man.
[78] Renato Pastore, 'Camilli, Camillo', in *Dizionario Biografico degli Italiani* (Rome: Istituto Enciclopedia Italiana, 1974), pp. 210–12. See also Tommaso Chersa, *Degli illustri Toscani stati in diversi tempi a Ragusa* (Padua: tip. della Minerva, 1828), pp. 16–18.

Essame de gl'ingegni de gli huomini per apprender scienze, which was first issued in Venice by Aldus in 1582, and which, by 1590, had reached four impressions in two editions. It was dedicated to the distinguished philosopher Federico Pendasio, Professor of Natural Philosophy first in Padua and later at Bologna, teacher of Federico Borromeo, Scipione Gonzaga, and of the illustrious Antonio Possevino.

But Camilli's most popular work was *I Cinque Canti di Camillo Camilli aggiunti al Goffredo del signor Torquato Tasso* (Venice, Francesco de Franceschi, 1583). Camilli's five cantos were intended as a continuation of Tasso's *Il Goffredo*, and so were published appended to it at least thirteen times until 1823.[79] In the sixteenth century, *I Cinque Canti* were reprinted by the press of Altobello Salicato in Venice in 1584, 1585, 1588 and 1593; by Osanna in Mantua in 1584, by Cagnacini in Ferrara in 1585 and by the heirs of Francesco de Franceschi in Venice in 1600.[80] Although Carew never translated Camilli's continuation of Tasso's poem — he never even finished translating Tasso in the first place — he must have read Camilli's addendum, for it was in the same volume, and probably liked it.[81] Thus, because in translating Tasso Carew came across Camilli and read his cantos, when he discovered that Camilli was also behind the translation of a volume into Italian owned by his friend Sir Francis Godolphin he must have celebrated the happy coincidence. This, together with his trust in the Italian's literary tastes and his friendship with the owner of the copy must have led Carew to read the *Essame* and subsequently to undertake the translation of

---

[79] Carmine Jannaco, *Storia letteraria d'Italia, Vol. 8, Il Seicento* (Milan: Vallardi, 1986), p. 566.

[80] Trustees of the British Museum, eds, *Short-title Catalogue of Books Printed in Italy and of Italian Books Printed in Other Countries from 1465 to 1600 Now in the British Library* (London: British Library, 1986), p. 140, pp. 659–61.

[81] According to Walter L. Bullock, 'Carew's Italian original, *Il Goffredo del S. Torquato Tasso*, may have borne the imprint either "In Venetia, Appresso Gratioso Perchacino. MDLXXXII", or "In Venetia, presso Francesco de' Franceschi Senese 1583"'. He explains that 'the latter "edition" consists merely of the remainder sheets of the former, with a new title-page at the beginning, and the *Cinque Canti* of Camillo Camilli added at the end. The text, therefore, is identical; and it is impossible to say which was the imprint on the copy Carew used.' 'Carew's Text of the *Gerusalemme Liberata*", p. 334. I claim that it was the 1583 edition with the *Canti* attached the one that Carew used.

it, presumably encouraged by Godolphin. This was not to be the end of the influence of Camilli upon Carew. Four years after *Godfrey* and *The Examination*, a fantasy tale in *ottava rima* blending Arthurian romance with natural history and entitled *A Herrings Tayle* (1598) was anonymously published; the influence of Camilli's *L'Epistole d'Ovidio* (Venice, 1587) has been remarked upon it,[82] and the text is nowadays invariably attributed to Carew.

RICHARD CAREW: LANGUAGE AND TRANSLATION

In his prologue to *The Examination* Carew refrains from commenting on the process of translating Huarte via Camilli or on his ideas about translation. Instead, we can get a glimpse of Carew's views about language in general and the English tongue in particular in his essay 'The Excellencie of the English Tongue'. 'The Excellencie', probably composed soon after April 1605, was first published in 1614 in the second edition of William Camden's *Remaines, Concerning Britaine*.[83] It rebutted *A Restitution of Decayed Intelligence of Antiquities* (Antwerp, 1605) by Richard Verstegan, a purist Saxonist who, unlike Carew, rejected foreign borrowings (particularly if they were from Romance languages), praised the monosyllabic trait of English, and tried to put back in circulation numerous archaisms. Carew, who never mentions Verstegan in his essay, discusses the Saxon quality of English, the process of linguistic borrowing, the ability of English to form compounds, and the enriching influence of other languages upon English: 'yea even we seek to make our good of our late Spanish enemy,and fear as little the hurt of his tongue as the dint of his sword'.[84] His attitude is not of opposition to linguistic import, but of rejoice in lexical abundance: 'Seeing then we borrow (and that not shamefully) from the *Dutch*, the *Britain*, the *Roman*, the *Dane*, the *French*, the *Italian*, and *Spaniard*; how can our stock be other

---

[82] Carmen Rogers, 'Introduction', in Juan Huarte de San Juan, *Examen de Ingenios: The Examination of Men's Wits (1594). A Facsimile Reproduction* (Gainesville: Scholars' Facsimiles & Reprints, 1959), p. x.

[83] Wood, 'Elizabethan English and Richard Carew', p. 304.

[84] Richard Carew, 'The Excellencie of the English Tongue', in William Camden, *Remaines, Concerning Britaine* (London: John Legatt, 1614), F2ᵛ–G2ᵛ (F4ᵛ). Wood, 'Elizabethan English and Richard Carew', 1977, pp. 304–15.

than exceeding plentiful?'.[85] Nevertheless, Carew is not unaware of the voices that warn against the negative effects of mass linguistic importation, and of the resulting linguistic confusion: 'It may be objected that such patching maketh *Littletons* hotchpot of our tongue, and in effect brings the same rather to a Babellish confusion, than any one entire language'.[86] From Carew's perspective, English is more pleasant than other tongues partly because it has managed to absorb all the positive qualities of many foreign languages while successfully disregarding their drawbacks. The result is a filtered language that unites the perfections of many others; English speakers '(like Bees) gather the honey of their good properties and leave the dregs to themselves'.[87] Carew himself, according to the OED, is accountable for the creation of over fifty new terms in English in which Saxon and Romanic roots and affixes combine. On occasion he adopts foreign terms directly when he considers it appropriate.[88]

Carew's interest in the English language is linked to his general curiosity in the past, which moves him to become a member of the Society of Antiquaries. In a letter to his fellow antiquary Sir Robert Cotton dated 7 April, 1605, Carew mentions Camden's *Remaines* and describes the research in history of the language — the 'derivation of the English names' — as 'both a profitable and pleasant labour'.[89] In that 1605 letter to Sir Robert Cotton, Carew draws on Huarte's ideas when touching upon the fact that England had no Academy of the language after the manner of some countries in the Continent. 'It imports no little disgrace to our Nation', Carew affirms, 'that others have so many Academies, and we none at all, especially seeing we want not choice of wits every way matchable with theirs, both for number and sufficiency'.[90] So it seems that Carew in fact added the Huartean notion of wit to his personal stock of tools to

---

[85] Carew, 'The Excellencie', F4^v.
[86] Carew, 'The Excellencie', F4^v.
[87] Carew, 'The Excellencie', G2^r.
[88] See the Neologisms section for a full list of Carew's coinages.
[89] Richard Carew, 'XXVII. Richard Carew of Anthony to Sir Robert Cotton', in *Original Letters of Eminent Literary Men: Of the Sixteenth, Seventeenth, and Eighteenth Centuries*, ed. by Sir Henry Ellis (London: John Bowyer Nichols and Son, 1843), pp. 99–100.
[90] Carew, 'XXVII. Richard Carew of Anthony to Sir Robert Cotton', p. 99.

conceptualize men. Furthermore, such a reference suggests that Cotton, the addresee of the letter, was also acquainted with the Huartean theory of wits (no doubt thanks to Carew's translation), and therefore fully understood Carew's comment.[91]

Carew's intention in 'The Excellencie' to 'prove that our English language, for all, or the most, is matchable, if not preferable before any other in use at this day',[92] is related to Huarte's choice to write the *Examen* in the vernacular at a time when it was the norm to discuss medicine in Latin. Choosing Spanish instead amounted to making a social and political statement.[93] The fact that the *Examen* was in Spanish is consistent with the book's purpose to enable a large amount of readers to discover their own wit and that of their sons so as to orient their professional careers accordingly. Writing the *Examen* in a language unknown to the majority of Huarte's contemporaries would sabotage the main reason for writing the treatise in the first place. With time Spanish gained force as a language for medicine due to ignorance of Latin particularly among surgeons, which explains why many treatises on anatomy were in Spanish. The greatest resistance to Spanish seems to have come from university physicians, who opposed the practice of medicine or related fields without proper qualifications.[94]

Similar reactions proliferated in the English medical context: the physicians of the Royal College did not approve of vernacular translations of landmark medical texts in Latin on the basis that they would reveal the secrets of the profession over to laymen.[95] As a result, translators such as Thomas Gale, a renowned surgeon of the sixteenth century, had to defend their translation activity incessantly.[96] The fact that Gale was a surgeon is relevant, for in

---

[91] More will be said about Sir Robert Cotton in the following pages, particularly when discussing Ben Jonson's appropriation of Huarte's theory of wits via Carew's translation.

[92] Carew, 'The Excellencie', F3^r.

[93] Similarly, the interest of antiquaries like Carew in the past had a component of 'patriotic defence' which may indicate that there was something more than sheer curiosity in Carew's willingness to look into the past of the English tongue. Wood, 'Elizabethan English and Richard Carew', p. 306.

[94] Riera, 'La literatura científica en el Renacimiento', p. 37.

[95] Elizabeth Lane Furdell, *Publishing and Medicine in Early Modern England* (Rochester: University of Rochester Press, 2002), p. 37.

England, as in Spain, the education of surgeons was based on practical apprenticeship in the context of guilds. In other words, it did not take place at university and it did not require knowledge of Latin. For this reason, only the elite of surgeons could read Latin and translate from it.[97] Nonetheless, as the seventeenth century progressed, the number of scientific texts in English grew: if to the middle of the seventeenth century Latin dominated within learned medical writing in England, between 1640 and 1660, 207 out of the 238 medical works published were in English.[98]

Doubtless the most defining feature of Carew as a translator is his adherence to the original work. Carew's rendering of Tasso has been described as exceptionally 'conscientious' and remarkable because of its 'steady faithfulness'.[99] When six years later Fairfax translated Tasso's poem in full, his version looked as if it had been 'translated loosely' by comparison with Carew's,[100] for Fairfax adds 'allusions, explanations, moralistic asides, misogynist comments, and proverbs of his own coinage'[101] which produce 'a modern version in which Tasso's versification and diction were adapted to the habits of English readers'.[102] By contrast, Carew's willingness to reproduce Tasso's complicated syntax, in some cases almost word by word, rendered his translation 'often unintelligible',[103] and 'so "foreignizing" as to

---

[96] Irma Taavitsainen and Jukka Tyrkkö, 'The Field of Medical Writing with Fuzzy Edges', in *Early Modern English Medical Texts*, ed. by Taavitsainen and Pahta, p. 69.

[97] Jukka Tyrkkö, 'Surgical and Anatomical Treatises', in *Early Modern English Medical Texts*, ed. by Taavitsainen and Pahta, p. 120. See also Andrew Wear, *Knowledge and Practice in English Medicine, 1550–1680* (Cambridge: Cambridge University Press, 2000), pp. 216–18.

[98] C. Webster, *The Great Instauration: Science, Medicine and Reform 1626–1660* (London: Duckworth, 1975), p. 276, and Irma Taavitsainen and Päivi Pahta, 'Vernacularisation of Medical Writing in English: A Corpus-Based Study of Scholasticism', *Early Science and Medicine*, 3.2 (1998), 157–85.

[99] Dodge, 'The Text of the *Gerusalemme Liberata*', p. 681. Alberto Castelli singles out Carew's 'fedeltà di fronte al testo italiano' and affirms that, when compared to the source text, in his translation 'Le differenze sono poche, e non vitali' (p. 76). Alberto Castelli, *La Gerusalemme liberata nella Inghilterra di Spenser* (Milan: Vita e Pensiero, 1936); particularly on the translations by Carew and Fairfax, see pp. 71–112.

[100] Dodge, 'The Text of the *Gerusalemme Liberata*', p. 687.

[101] Morini, *Tudor Translation*, p. 124.

[102] Morini, *Tudor Translation*, p. 101.

sound almost uncanny at times'.[104] Nonetheless, *The Examination* is in no manner linguistically challenging to the reader, even if, as in his translation of Tasso, Carew follows his model extremely closely — with the exception of the omission of two fragments against Protestant preachers and beliefs.

From the *Examen* to *The Examination*: Textual Differences

Camillo Camilli had been meticulous when translating Huarte,[105] and perhaps the most perceptible difference between the Italian and the Spanish texts is that Camilli translates into Italian all the quotations that Huarte keeps in Latin. Camilli's motivation for translating all Latin quotations into Italian was to make the treatise accessible to a large readership. Indeed, in Italy the book was not intended to address an intellectual minority, no matter what the dedication of the editor Manassi to Federico Pendasio might seem to suggest.[106] Camilli also translates the Latin quotations without indicating their source, which the Spanish edition provides in the margins of the text.[107] Because notes in the margins are left out by Camilli, they are absent from Carew's version too, which renders into English Camilli's quotations

---

[103] Morini, *Tudor Translation*, p. 105.

[104] Morini, *Tudor Translation*, p. 121. Other comparative studies have reached similar conclusions: D. N. C. Wood, '*Gerusalemme liberata* — Englished by Richard Carew', *Cahiers Elisabethains: Etudes sur la pre-Renaissance et la Renaissance anglaises*, 13 (1978), 1–13; Roberto Weiss, 'Introduction', *Jerusalem Delivered. The Edward Fairfax Translation* (Carbondale: Southern Illinois University Press, 1962); Halliday, 'Introduction' in Carew, *Survey of Cornwall*, p. 32. For more on the translations by Carew and Fairfax see Charles Peter Brand, *Torquato Tasso, A Study of the Poet and of his Contribution to English Literature* (Cambridge: Cambridge University Press, 1965), pp. 238–46; for Fairfax's only, see *Godfrey of Bulloigne: A Critical Edition of Edward Fairfax's Translation of Tasso's Gerusalemme Liberata*, ed. by Kathleen M. Lea and T. M. Gang (Oxford: Clarendon Press, 1981).

[105] Mauricio Iriarte, *El doctor Huarte de San Juan y su 'Examen de ingenios': contribución a la historia de la psicología diferencial* (Madrid: CSIC, 1948), p. 76.

[106] Gambin, 'Il gesuita e il medico', p. 155.

[107] The subsequent translation of Gratii however retains almost all the quotes in Latin along with their accompanying references.

already in Italian. As a result, Latin is absent from Carew's rendering as well.

Readers of the book in Spanish would get a quotation in Latin and immediately afterwards a translation or a paraphrase of it into Spanish. This would please both educated readers proficient in Latin, and less educated ones who would not know it at all. By contrast, readers of the Italian or the English translations would have the impression that Huarte unnecessarily, tiresomely, reworded his ideas time and again. For example, in extracts such as the following, what appears underlined would be in Latin in the Spanish text. Then, Huarte would translate it or paraphrase it into Spanish after the expression 'as if he should say/have said':

> But there is no man who hath better verified this than the good Marcus Cicero, who through grief of seeing his son such a donought, with whom none of the means could prevail that he had procured to breed him wisdom, said in the end after this sort: '<u>What else is it, after the manner of the giants, to fight with the gods, than to resist against nature.</u>' As if he should have said: 'What thing is there which better resembles the battle which the giants undertook against the gods than that a man who wanteth capacity should set himself to study?' (I. 320–328)

> Galen said: '<u>Coldness is apparently noisome to all the offices of the soul.</u>' As if he should say: 'Cold is the ruin of all the operations of the soul', only it serves in the body to temper the natural heat and to procure that it burn not overmuch. (V. 138–142)

A close comparative analysis of the *Examen*, the *Essame*, and *The Examination* shows that the divergences between the Spanish text and the translations are minimal. Leaving aside Carew's deliberate omission of two fragments against Protestantism, both Camilli's and Carew's renderings prove highly precise. The 'Textual Notes' section gathers all the differences between the three texts; as can be seen, this is a rather short list, particularly when taking into consideration the length of the treatise. Most differences are merely minor changes and omissions, and simple translation mistakes, such as accidentally omitting a word or confusing a figure. Sometimes, the differences are due to an attempt to adapt the text to a new target readership, as when Spanish city names are deleted to create a more imprecise and hence universal environment for the application of the book's theories. On other occasions, names of small cities (Simancas) are not erased but replaced by similar-sounding names of well-known cities (Salamanca, XIII. 857); the same happens with proper names of Spanish personalities fairly unknown in England (the poet Juan Boscán), by others of established authorities (Boccaccio, VIII. 232).

In addition to these rather minor changes, there are two relevant differences between the English and the Spanish texts which indicate that Richard Carew might have had access to Huarte's work in Spanish even if his translation was on the whole carried out from Camilli's rendering. If we did not admit that Carew did have access to a copy of the Spanish text, two terminological coincidences would be uncanny. The first is Carew's inclusion of the Latin term *solertia* (i.e. skill and cunning to do or go about something) in the sentence 'This property to attain suddenly the means is *solertia* (quickness)' (XIII. 115–116). In the Spanish original (1575, fol. 220^r), it only reads *solercia*, and the word is not followed by any clarifying noun in parentheses. In Italian, the word translating the Spanish *solercia* is *vivacità* (1586, O8^v), which comes without an explanatory note and without any allusion to *solertia* or *solercia*. The fact that Carew decides to use precisely the term *solertia* accompanied by an English synonym in brackets may indicate that, on this occasion, Carew checked a copy of the book in Spanish and decided to keep the Latin noun, closer to the Spanish term, rather than the Italian.

The second terminological match has to do with the English phrase 'the book of *Bezerro*' (XIII. 856), which corresponds to the Spanish 'libro d[e]l Bezerro' (1575, fol. 247^r). In Italian, however, the expression is 'libro del Gioverico' (1586, Q7^r). Literally, 'libro d[e]l Bezerro' means 'book of the calf', because of the calfskin with which it was covered. It was the book which in the Middle Ages recorded the privileges and possessions of churches and monasteries, and it often became synonymous with 'libro de apeo' (book of surveying), to an extent similar in contents, as it registered land and property, rights and privileges, renowned historical events and jurisdictions of a place or of a political or religious entity. The translation into Italian of 'libro dl Bezerro' as 'libro del Gioverico' ('giovenco' means young ox in Italian) is a strictly literal one, for there was nothing in sixteenth-century Italy known as 'libro del Gioverico' or 'giovenco'.[108] Still, given the context, Italian readers of the time might have inferred that the expression alluded to a book or a registry of privileges. The fact that Carew uses just the Spanish word *Bezerro*, entirely absent from Camilli's translation, instead of the Italian term or a translation of it into English may indicate that he did possess or have access to a copy of the work in Spanish. Carew might have opted for reproducing the original expression in Spanish out of the feeling that the translation into Italian was a word-for-word rendering void of true meaning and lacking an actual referent that the Spanish, by contrast, did have.

MENTAL FACULTIES, THEORY OF THE WITS AND RELIGION

Contrary to the view of Aristotle and following Plato, Hippocrates and Galen instead, Huarte argues that 'the brain is the principal seat of the reasonable soul' (III. 22–23). In his view, in order for the reasonable soul to discourse and philosophize, the brain 'should be tempered with measurable heat and without excess of the other qualities' (III. 127–128), and divided into four ventricles, 'distinct and severed, each duly bestowed in his seat and place' (III. 42–43). Huarte describes the ventricles of the brain as four little hollows of 'one self composition and figure without anything coming between which may breed a difference' (V. 29–30). The three ventricles in the forepart of the head are used to 'discourse and philosophize' (V. 73), while the fourth ventricle deals with the least noble operations, as it 'hath the office of digesting and altering the vital spirits and to convert them into animal' (V. 65–67). The conviction that the three mental powers (understanding, imagination and memory) necessarily work in collaboration with each other — to the extent that without one the rest would malfunction — makes Huarte conclude that 'in every ventricle are all the three powers' (V. 123). By asserting so he rejects allocating a single faculty to one ventricle only:

> all the powers are united in every several ventricle and […] the understanding is not solely in the one, nor the memory solely

---

[108] Professor Felice Gambin, from the University of Verona, has confirmed this to me.

in the other, nor the imagination in the third, as the vulgar philosophers have imagined[.] (V. 91–95)

That the three powers are in the three ventricles explains why even if a ventricle is seriously damaged, one of the faculties does not simply vanish. This explanation is along the lines of the thought of Galen, who criticized rigid departmentalization theories.

According to Huarte, the brain temperament and the predominance in it of one of the primary qualities accounts for the differences in every man. In other words, the variety of wits in men does not result from divergences in their souls, which are identical, always 'of equal perfection (as well that of the wiser, as that of the foolish)' (II. 211–212). Rather, from heat, moisture and dryness all the differences of wit originate. Dryness fosters a better working of the understanding, memory is born from moisture and imagination from heat. In contrast, cold is 'unprofitable to any operation of the reasonable soul', as in excess 'all the powers of man do badly perform their operations' (V. 132–135). Of all possible combinations, cold and dry 'is most appropriate to the operations of the reasonable soul' (V. 423–424), unlike moisture, which is 'contrary to the reasonable faculty' (V. 246) but an essential requirement for memory. Consequently, blood and phlegm, which share a fair degree of moisture, 'cause an impairing of the reasonable faculty' (V. 253–254). Great understanding and equally good memory are therefore incompatible in the same person, considering that understanding is based on drought while memory on moisture, and so 'it is impossible that the brain should of his own nature be at one self time dry and moist' (V. 323–324). For this reason, Huarte states, 'it is a miracle to find a man of great imagination who hath a good understanding and a sound memory' (V. 351–352).

Huarte distinguishes three general differences of wits — 'for there are no more but three qualities whence they may grow' (V. 361–362) — and many subcategories depending on the multiple sub-degrees of heat, moisture and dryness. Given that every difference of wit corresponds to an art or a science, Huarte aims to explain 'what difference of science is answerable in particular to what difference of wit' (VIII. 24–25). He establishes that 'Latin, grammar, or of whatsoever other language, the theory of the laws, divinity positive, cosmography, and arithmetic' (VIII. 28–30) are the arts and sciences of the memory. To the understanding correspond 'school divinity, the theory of physic, logic, natural and moral philosophy, and the practice of the laws, which we term pleading' (VIII. 31–33). To the imagination, 'poetry, eloquence, music, and the skill of preaching, the practice of physic, the mathematics, astrology, and the governing of a commonwealth, the art of warfare, painting, drawing, writing, reading, [...] all the engines and devices which artificers make' (VIII. 36–40), and 'music and the stage' (X. 424). In the *Examen* there is no detailed analysis of all the previously enumerated arts, sciences and occupations; instead Huarte focuses on utilitarian professions that in his day enjoyed social recognition within the fields of jurisprudence (including government), medicine, warfare and divinity.[109] The idea is to apply the method to a sample of profiles and careers in order to show readers how it works, and by so doing to enable them to apply it independently to their own interests.

In his examination of divinity, Huarte makes an initial distinction between positive divinity and school divinity. Positive divinity he says 'appertaineth to the memory and is nought else save a mass of words and Catholic sentences taken out of the holy doctors and the divine Scripture, and preserved in this power, as the grammarian doth with the flowers of the poets' (IX. 120–123). That is, positive divines remember and repeat the quotations they have learned by heart from the Scriptures. Huarte does not envision positive divines in a favourable light, as he believes that 'a man of many words ordinarily wanteth understanding and wisdom' (IX. 141–142), which makes it difficult for him to 'search out the bottom of the truth' (IX. 139). Because positive divines are good at languages and 'understand well the Hebrew, Greek and Latin tongues' by virtue of their memory, they think that they are qualified to interpret the divine Scriptures. However, in reality in so doing 'they ruinate themselves', because since 'their understanding is defective' (IX. 147–151) they cannot well discern the true signification of the text. In contrast, school divines are men of great understanding able to dive into the textual depths of the Scriptures. The problem with school divines is that while their understanding is strong, their memory is poor, and so is their Latin.

---

[109] Huarte only mentions in passing non-utilitarian arts such as poetry; after all, he primarily aspires to provide solutions to the problems of the administration, the universities, the practice of medicine and the army.

INTRODUCTION

It is evident for Huarte that Spaniards are men of understanding, and consequently excel in school divinity while failing in the learning of languages on account of their bad memory. By contrast, northern Europeans seem generally to possess worse understandings but better memories: 'For which cause, the one can [acquire] no skill of Latin, and the other easily learn the same' (VIII. 413–415). And so, Huarte affirms, even if they do not excel in communicating their knowledge, Spaniards interpret the Holy Scriptures correctly due to their greater understanding.

If the northern nations excel in anything within the realm of divinity, Huarte affirms, it is in producing eloquent and persuasive preachers, as a result of their memory and imagination. Because a preacher is not a man of understanding and hence 'can pierce no farther than into the upper skin of things',

> it falleth out a dangerous matter that the preacher enjoyeth an office and authority to instruct Christian people in the truth, and that their auditory is bound to believe them, and yet they want that power through which the truth is digged up from the root. (X. 400–404)

It is precisely for this reason that Huarte would not entrust northern men with preaching at all, as they, with good eloquence but meagre understanding, 'draw the auditory after them and hold them in suspense and well pleased, but when they least misdoubt it, they fetch a turn to the Holy House' (X. 652–654). For Huarte this undoubtedly accounts for the spread of Protestantism in the north of Europe.

Carew's Censorship and Criticism of Huarte's Text

Huarte's criticism of the men of the north in the context of divinity functions as an attack on Protestantism. Yet Huarte's most explicit criticism of Protestantism did not reach readers of *The Examination*, as Carew decided to leave out the two most critical paragraphs against the countries which joined the Protestant Reformation. The omission was deliberate on Carew's part, as the two extracts are present in Camilli's Italian text; indeed, for Camilli and his Catholic readership they were no offence. The first extract censored by Carew (in my translation below), openly refers to the loquaciousness of the English and their shared responsibility in the tragic fragmentation of Christianity:

> The verbosity and garrulity of German, English, Flemish and French divines, and of the rest of them that live in the North, ruined the Christian audience with such skill of languages, with so much ornament and grace in preaching, for having no understanding to reach truth. And that these lack understanding we have already proved by opinion of Aristotle as well as of many other reasons and experiences brought for the case. But if the English and German audience were warned in what St Paul wrote to the Romans (pressed by false preachers too) perchance they would not be so easily deceived. *Rogo autem vos fratres, ut observetis eos, qui dissensiones & offendicula praeter doctrinā quā vos didiscistis faciunt & declinate ab illis huiusmodi enim Christo domino nostro non serviūt sed suo vētri; & perdulces sermones et benedictiones seducunt corda inoscentium* [I beseech you, my brethren, to observe those who are in disagreement and put up obstacles because of the teachings that you have given, and move away from them, for in that manner they do not serve Christ, our Lord, but their own belly, and, through sweet conversation and praise, seduce the hearts of the innocent]. As if he said: 'My brethren, for the love of God I beseech you beware those that teach you a different doctrine from the one that you have learned, and walk away from them, for they do not serve our lord Jesus Christ but their own vices and sensuality, and they are so well-spoken and eloquent that, with the sweetness of their words and reasons, fool those who know little.[110]

---

[110] This should have appeared in X. 405. In the original in Spanish: 'La vaniloquencia y parlería, delos theologos Alemanes, Ingleses, Flamencos, Franceses, y de los demas, que abitan el septentrion, echo aperder el auditorio christiano: con tanta pericia de lenguas, con tanto ornamento y gracia en el predicar: por no tener entendimiento, para alcançar la verdad. Y que estos sean faltos de entendimiento, ya lo dexamos provado atras, de opinion de Arist. aliende de otras muchas razones y experiencias, q truximos para ello. Pero si el auditorio ingles, y aleman, estuviera advertido, en lo que S. Pablo escrivio a los romanos (estando tābiena ellos apretados de otros falsos predicadores) por ventura, no se engañaran tan presto. *Rogo autem vos fratres, ut observetis eos, qui dissensiones & offendicula praeter doctrinā quā vos didiscistis faciunt & declinate ab illis huiusmodi enim Christo domino nostro non serviūt sed suo vētri;*

INTRODUCTION

The second extract censored by Carew erases from *The Examination* a list of beliefs proper to Protestantism and, again, an overt criticism of eloquent Protestant preachers:

> And thus they work in interpreting the divine Scripture, in agreement with their natural inclination, making those who know little believe that priests can marry, and that Lent is not needed, nor fastings, and that letting the confessor know our offences to God is not advisable. And with these wiles (with wrongly brought Scripture) they disguise as virtues their wrong deeds and vices, and make people think of them saints.[111]

Not even in a note in the margins does Carew comment on the censored extracts; in this case, Carew works effectively but silently. Nonetheless, Carew does speak his mind regarding some of Huarte's views in the form of short sentences placed in the margins of his translation. In total there are eleven instances of such responses, many on religious and political issues. They become more abundant towards the end of the book, probably because by then Carew had grown annoyed at translating beliefs that he did not share. Consider the following sentence of *The Examination*:

> In matters of faith propounded by the Church, there can befall none error: for God, best weeting how uncertain men's reasons are, and with how great facility they run headlong to the deceived, consenteth not that matters so high and of so weighty importance should rest upon our only determination. (XI. 452–455)

& per dulces sermones et benedictiones seducunt corda inoscentium. Como si dixera: hermanos mios, por amor de Dios os ruego q tengays cuēta particular con essos que os enseñā otra doctrina, fuera de la q aveys aprendido: y apartaos dellos: por que no sirven a nuestro señor Jesuchristo sino a sus vicios, y sensualidad: y sö tambien hablados y eloquentes, que con la dulçura de sus palabras y razones, engañan a los q poco sabē.' Huarte, *Examen de ingenios*, fol. 153[r-v].

[111] This should have appeared in X. 445. In the original in Spanish: 'Y assi trabajan de interpretar la escritura divina, de manera que venga bien con su inclinacion natural: dando a entender a los que poco saben que los sacerdotes se pueden casar: y q no es menester que aya cuaresma, ni ayunos ni conviene manifestar al confessor, los delictos q contra Dios cometemos. Y usando d esta maña (con escriptura mal trayda) hazen parecer virtudes, a sus malas obras y vicios; y que las gētes los tengā por sanctos.' Huarte, *Examen de ingenios*, fol. 155[r-v].

To this an irritated Carew responds: 'Take heed you receive no hurt for leaving out the Pope' (XI. Footnote 109). On other occasions, the object of Carew's scorn is King Philip II and Huarte's flattering comments on his regal, flawless nature:

> No sort of knowledge is found so distinctly and severed from another but that the skill in the one much aideth to the others' perfection. But how shall we do if having sought for this difference of wit with great diligence in all Spain, I can find but one such? (XIV. 50–53)

'No doubt your own king' (XIV. Footnote 160), Carew replies. Shortly after this, we read a veiled allusion of Huarte to King Philip II:

> Wherethrough I may now conclude that the man who is auburn haired, fair, of mean stature, virtuous, healthful, and long lived, must necessarily be very wise and endowed with a wit requisite for the sceptre royal. (XIV. 671–674)

Carew makes explicit what had remained implicit via a taunting comment: 'And such a one if you mistake not is your king Philip' (XIV. Footnote 181). Carew also replies to Huarte's complaint that the union of understanding and imagination is highly unusual:

> We have also (as by the way) disclosed in what sort great understanding may be united with much imagination and much memory, albeit this may also come to pass and yet the man not be temperate. But nature shapeth so few after this model that I could never find but two amongst all the wits that I have tried. (XIV. 675–679)

This finds a playful retort in Carew: 'Your king and yourself' (XIV. Footnote 182). Also, Carew genuinely disagrees with some of Huarte's interpretations of the Sacred Scriptures. Interestingly, his disagreement is not so much based on official discrepancies between Catholic and Protestant readings, as on personal interpretations of the text. Consider, for example, Huarte's analysis of kings David and Saul:

> [S]eeing in a people so large (as that of Israel) God could not find one to choose for a king, but it behooved him to tarry till David was grown up and the whiles made choice of Saul. For

INTRODUCTION

> the Text saith that he was the best of Israel, but verily it seemed he had more good nature than wisdom, and that was not sufficient to rule and govern. (XIV. 522–528)

To this Carew notes in the margin: 'A weak reason, rather God chose Saul as a carnal man fit for the Jews obstinate asking, and David as a spiritual man, the instrument of his mercy' (XIV. Footnote 177). While Huarte gives credit to the account that Publius Lentulus provides of Christ (XIV. 553–603), Carew drastically disregards it: 'And I hold it untrue, because the phrase utterly differeth from the Latin tongue as *speciosus valde interfilios bomimum* [very fair amongst the son of men]' (XIV. Footnote 179). In reality, the extract that Huarte mentions of Publius Lentulus, predecessor of Pontius Pilate, is considered an apocryphal text.[112] In addition, Carew undermines Huarte's statement that the first token of a temperate man was to have

> his hair and beard of the colour of a nut fully ripe, which to him that considereth it well appeareth to be a brown auburn, which colour God commanded they [*sic*] heifer should have which was to be sacrificed as a figure of Christ. (XIV. 579–582)

Carew dismisses this remark with a succint 'Unwritten varities' (XIV. Footnote 180). In the like manner, he qualifies as 'A high speculation' (XIV.Footnote 184) Huarte's reflections on the first man's reaction when comparing his nature to that of angels:

> And principally he [the first man] shamed seeing that the angels, with whom he had competence, were immortal and stood not in need of eating, drinking, or sleeping for preservation of their life; neither had the instruments of generation but were created all at once without matter and without fear of corrupting. (XIV. 93–98)

While still engaged in the description of the nature of angels, Huarte affirms the following in regard to the human soul:

> And that this is a well fitting answer we evidently perceive, for God to content the soul after the universal judgement, and to bestow upon him entire glory, will cause that his body shall

---

[112] Carew was definitely not the only reader of Huarte to hold it false. Pierre Bayle, *Dictonnaire historique et critique*, VIII (Paris: Desoer, 1820), p. 292.

partake the properties of an angel, bestowing thereupon subtleness, lightness, immortality, and brightness, for which reason he shall not stand in need to eat or drink as the brute beasts. (XIV. 102–108)

Carew notes that this may be a proof to the immortality of the soul: 'Note here a sign which showeth the immortality of the soul' (XIV. Footnote 185). This may indicate that Carew was glad to discover a contradiction in Huarte in this respect, and that he firmly believed in the immortality of the human soul and in the possibility of its demonstration.

The final two comments by Carew have to do with the contents of Chapter XV, that is, with the treatise on eugenesis.[113] The first appears by this sentence shortly after the beginning of the chapter:

And that is that man (though it seem otherwise in the composition which we see) is different from a woman in nought else, saith Galen, than only in having his genital members without his body. (XIV. 151–154)

By these words Carew writes the following warning: 'This is no chapter for maids to read in sight of others' (XIV. Footnote 187). Funnily, it does not so much oppose young women reading it as much as their doing so in front of others out of social decorum. Carew might even have thought it advantageous for women to have a knowledge of physiological issues related to reproduction and the begetting of wise children. Finally, Carew corrects Huarte when he asserts that 'in the regions towards the north (as England, Flanders, and Almaine)', 'no married woman was ever childless, neither can they there tell what barrenness meaneth, but are all fruitful and breed children through their abundance of coldness and moisture' (XIV. 211–217). 'You are much mistaken' (XIV. Footnote 189), Carew retorts in the margin, as if he addressed Huarte directly and not a readership made up of his fellow Englishmen, who would have no doubt appreciated the comic relief provided by Carew's humorous remarks.

---

[113] José María Gondra Rezola, 'Juan Huarte de San Juan y la eugenesia', in *Criminología y derecho penal al servicio de la persona: libro-homenaje al profesor Antonio Beristain*, ed. by J. L. Cuesta, I. Dendaluze, and E. Echeburua (San Sebastián: Instituto Vasco de Criminología, 1989), pp. 199–210.

INTRODUCTION

THE RECEPTION OF *THE EXAMINATION* IN ENGLAND

Early modern England witnessed the publication of many works on the nature of man and the workings of the mind, either originally written in English, or translated from the classical languages or from other vernaculars into English. Richard Carew's translation of the *Examen* is related to a series of works expressing similar concerns. By the time of the publication of *The Examination* (1594), English readers were already acquainted with Sir Thomas Elyot's *The Castel of Helth* (1539), Thomas Rogers's *Anatomie of the Minde* (1576) or Timothy Bright's *Of Melancholy* (1586). Also they could read in English translations Levinus Lemnius's *The Touchstone of Complexions* (1576), Pierre Boaistuau's *Theatrum Mundi* (1566?), Pierre de la Primaudaye's *The French Academie* (1586) and Thomas Gale's translation of Galen (1586), among others. The first printing of *The Examination* was followed by works such as Sir Richard Barckley's *A Discourse of the Felicitie of Man* (1598), Robert Allott's *Wits Theater of the Little World* (1599), Thomas Wright's *The Passions of the Minde in Generall* (1601), Thomas Walkington's *The Optick Glasse of Humors* (1607), Robert Burton's *Anatomy of Melancholy* (1621) and Edward Reynolds's *A Treatise of the Passions and Faculties of the Soul of Man* (1640). None of these titles explicitly acknowledge acquaintance with Huarte and *The Examen* or with Carew and *The Examination*.[114] Yet it does not follow from the lack of

---

[114] Poems such as Sir John Davies's *Nosce Teipsum* (1599), John Davies of Hereford's *Mirum in Modum* (1602) and *Microcosmos* (1603), Phineas Fletcher's *The Purple Island* (1633), and plays such as Tomkis's *Lingua* (1607) or Peter Heylyn's *Microcosmus* (1621) deal with the nature of man and human psychology, and do not mention Huarte either. It is worth remarking here that nowhere is Huarte mentioned in *Early Modern English Medical Texts*, ed. by Taavitsainen and Pahta, and that *The Examination* is not one of the texts chosen to be included in the EMEMT corpus, which does include translations of foreign works into English. EMEMT is a collected corpus of medical texts printed between 1500 and 1700 which gathers circa 450 full texts or samples — shorter texts are included in their entirety while longer ones are represented by extracts of approximately 10,000 words. The reason for not including *The Examination* may be that, despite their flexible understanding of the category of 'early modern medical writing', the authors considered *The Examination* not so much a medical text as a treatise in natural philosophy. Searches in the corpus of 'Huarte', 'Carew' and the title of Huarte's treatise either in Spanish or in English (in full or abridged forms) resulted in no hits.

explicit references that their authors were oblivious to Huarte's treatise, by then not only circulating in English, but also in the form of French and Italian renderings (as Sir Francis Godolphin's Italian copy demonstrates). After all, this was a time when classical authorities were profusely quoted while recognition of indebtedness to contemporary authors was either silenced or kept to a minimum. Huarte was no exception to this trend, as he also chose not to name any contemporary author or title, which undoubtedly makes establishing a clear map of influences more challenging.[115]

Some English authors demonstrate that they had read *The Examination* not by means of direct allusions to Carew or Huarte, but through the inclusion of textual references that unmistakably have it as their source. This is the case of Carew's friend William Camden, who, in the first edition of *Remaines of a Greater Worke, Concerning Britaine* (1605), in a veiled way alluded to Carew's translation by slipping in the name of an imagined giant *Traquitantos*. *Traquitantos* was the subject of an Huartean anecdote on words in general and proper names in particular:

> [A] Spanish gentleman who made it his pastime to write books of chivalry [...] had a certain kind of imagination which enticeth men to feigning and leasings. Of him it is reported that being to bring into his works a furious giant, he went many days devising a name which might in all points be answerable to his fierceness. Neither could he light upon any until playing one day at cards in his friend's house, he heard the owner of the house say: 'Ho, sirrah, boy, *tra qui tantos*.'[116] The gentleman so soon as he heard this name *traquitantos*, suddenly he took the same for a word of full sound in the ear, and without any longer looking arose saying: 'Gentlemen, I will play no more,

---

[115] Even though Huarte extensively refers to traditional authorities such as Plato, Aristotle, Cicero or Galen, he nonetheless never quotes nor refers explicitly to any contemporary author. Thus, he makes not the slightest reference to the abundant literature produced in his own time on similar subjects by, for instance, Juan Luis Vives, Pedro Simón Abril, Gómez Pereira, Francisco Vallés, Rodrigo Sánchez de Arévalo, Luis Lobera de Ávila, Alonso López de Corella, Pedro Mexía, Alonso de Fuentes or Pedro Mercado.

[116] '*Tra qui tantos*' is how 'trae aquí tantos' ('bring here this many', followed by a noun possibly indicating objects symbolizing money) sounds when spoken quickly.

for many days are past sithence I have gone seeking out a name which might fit well with a furious giant whom I bring into those volumes which I now am making, and I could not find the same until I came to this house, where ever I receive all courtesy.' (VIII. 459–474)

Camden's attentive reading of what for him was a particularly appealing section of *The Examination* resulted in him using precisely this episode of the name of the giant *Traquitantos* to close a glossary of names of women:

> Neither do I think in this comparison of Names that any will prove like the Gentleman who, distasting our names, preferred King *Arthurs* age before ours, for the gallant, brave, and stately names then used; as sir *Orson*, sir *Tor*, sir *Quadragan*, sir *Dinadan*, sir *Launcelot*, *etc*. which came out of that forge, out of the which the Spaniard forged the haughty and lofty name *Traquitantos* for his Giant, which he so highly admired, when he had studied many days and odd hours, before he could hammer out a name so conformable to such a person as he in imagination then conceited.[117]

Textual evidence in Ben Jonson's works seems to indicate that he was acquainted with *The Examination* as well. *Every Man in His Humour* (1601), *Every Man Out of His Humour* (1600) and *The Magnetic Lady* (1641) are the central pieces of Ben Jonson's comedy of humours. Jonson in fact develops a theory of comedy based on correspondences between the humours and various kinds of human temperaments, by which, consequently, all characters act in accordance with their particular humours. In the opening of *Every Man Out of His Humor*, Jonson explains what he exactly means by 'humor':

> Is *Humour*: so in every human body
> The choller, melancholy, flegme, and blood,
> By reason that they flow continually
> In some one part, and are not continent,
> Receive the name of Humours. Now thus far
> It may by Metaphor apply itselfe

---

[117] William Camden, *Remaines of a Greater Worke, Concerning Britaine* (London: G[eorge] E[ld], 1605), M4^v.

Unto the general disposition,
As when some one peculiar quality
Doth so possess a man that it doth draw
All his affects, his spirits, and his powers
In their confluctions all to run one way,
This may be truly said to be a Humour[.][118]

Jonson elaborates further on the varieties of wits in his posthumously published *Timber* (1640), where his words flawlessly match Huarte's:

> *In the difference of wits*, I have observed there are many notes, and it is a little *Mastery* to know them, to discern what every nature, every disposition will bear; for before we sow our land, we should plough it. There are no fewer forms of minds than of bodies amongst us. The variety is incredible, and therefore we must search. Some are fit to make *Divines*, some *Poets*, some *Lawyers*, some *Physicians*; some to be sent to the plough and trades.
>
> There is no doctrine will do good where nature is wanting. Some wits are swelling and high; others low and still. Some hot and fiery; others cold and dull. One must have a bridle, the other a spur.[119]

The scholar Harry Levin was probably the first to note that Jonson must have read *The Examination*, and that this book decisively shaped his understanding of the concept of humour.[120] Indeed, Jonson could have known Carew through the Cotton Circle, a literary circle that revolved around the antiquarian Sir Robert Cotton and that, in addition to Jonson and Carew, included, among others, John Selden, Lancelot Andrewes and Samuel Purchas. Through Cotton, not only could Jonson have

---

[118] Ben Jonson, *Every Man out of his Humor* (London: Printed [by P. Short], 1600), B1ʳ.

[119] Ben Jonson, *The Workes of Benjamin Jonson. The Second Volume* (London: Printed [by John Beale, James Dawson, Bernard Alsop and Thomas Fawcet], 1640 [i.e. 1641]), N3ᵛ.

[120] *Veins of Humor*, ed. by Harry Levin (Cambridge: Harvard University Press, 1972), p. 8. See also, Yumiko Yamada, 'Ben Jonson and Cervantes: The Influence of Huarte de San Juan on Their Comic Theory', in *Shakespeare and the Mediterranean*, ed. by Tom Clayton, Susan Brock, Vicente Forés and Jill Levenson (Newark: University of Delaware Press, 2004), pp. 425–36.

learned about the existence of Carew's translation, but also gained access to a copy of it, for, as has been remarked, 'there was little serious scholarship in the first quarter of the seventeenth century that was not indebted to Sir Robert Bruce Cotton and the pleasure he took in making his library available to his friends'; indeed, 'the accessibility of Cotton's house in Westminster, with its garden running down to the river, soon made Old Palace Yard the natural haunt of scholars and antiquaries'.[121] Both Cotton and Jonson had been educated at Westminster under William Camden, who, as we already know, was a dear friend of Richard Carew's too. Yet, like many other authors, Jonson does not mention either Carew or Huarte in his works.

The Examination furthermore connected with the literature produced by leading English educators such as Roger Ascham and Richard Mulcaster, whose works on the education of children appeared in print before Carew's translation: Ascham's *The Scholemaster or Plaine and Perfite Way of Teachyng Children* was published in 1570, and Mulcaster's *Positions vvherin those Primitiue Circumstances be Examined, which are Necessarie for the Training vp of Children*, in 1581. Since neither were translated into Spanish, they could not have been a direct influence upon Huarte, and yet, it is possible to speculate that Ascham, Mulcaster and Huarte interconnect thanks to the pivotal figure of Juan Luis Vives. As has been stressed, Huarte believed that education should begin by discovering the proper wit of every child in order to train him accordingly:

> ([A]t least if I were a teacher) before I received any scholar into my school, I would grow to many trials and experiments with him, until I might discover the quality of his wit, and if I found it by nature directed to that science whereof I made profession, I would willingly receive him […]. And, if not, I would counsel him to study that science which were most agreeable with his wit. But if I saw that he had no disposition or capacity for any sort of learning, I would friendly and with gentle words tell him: 'Brother, you have no means to prove a man of that possession which you have undertaken, take care not to lose your time and

---

[121] David Sandler Berkowitz, *John Selden's Formative Years: Politics and Society in Early Seventeenth-century England* (Washington and London: Folger Shakespeare Library; Associated University Presses, 1988), p. 25.

your labour, and provide you some other trade of living which requires not so great an hability as appertaineth to learning.' (I. 90–103)

The implication here is that school curricula cannot be the same for every child. Instead, a personalized array of subjects would be most advantageous to students — that is if their wit is appropriate to studying at all. The title of the first chapter makes this point clearly: 'if a child have not the disposition and hability which is requisite for that science whereunto he will addict himself, it is a superfluous labour to be instructed therein by good schoolmasters, to have store of books, and continually to study it' (I. 4–8).

If primary education is crucial, university education is no less important. Huarte was notably worried about the situation of the Spanish universities of his time, and this concern probably constitutes one of his main motivations for writing his treatise.[122] The fact that many students who enrolled in university studies eventually abandoned them without a degree particularly alarmed Huarte. Indeed, despite the large number of students,[123] the figure of graduates was relatively low, as in 'the late sixteenth century less than one-third of the students who began their course took a degree'.[124] Richard Carew was one of those students that would have concerned Huarte, as Carew did go to Christ Church, Oxford, but appears not to have taken a degree, and not because of any financial hardship. It is impossible to know whether this was the case of Edward Bellamy too, for his educational background remains a mystery. Nonetheless, Bellamy does express an interest in an educational reform that would only allow

---

[122] Elvira Arquiola, 'Consecuencias de la obra de Huarte de San Juan en la Europa moderna', *Huarte de San Juan*, 1 (1989), 15–28, p. 55.

[123] At the time of the publication of the *Examen*, the rates of university enrolment in Spanish universities were without precedent, and many of them had a number of students only equalled again at the end of the nineteenth century. During the last quarter of the sixteenth century, the high point of Castile's universities, there were about 20,000 students anually out of a population of 6.9 million. Richard L. Kagan, *Students and Society in Early Modern Spain* (London: Johns Hopkins University Press, 1974), p. 199. See also, Jon Arrizabalaga, 'Filosofía natural, psicología de las profesiones y selección de estudiantes universitarios en la Castilla de Felipe II: la obra y el perfil intelectual de Juan Huarte de San Juan', *Huarte de San Juan*, 1 (1989), 29–58 (p. 56), and Arrizabalaga, 'Juan Huarte en la medicina de su tiempo', p. 94.

[124] Kagan, *Students and Society*, p. 201.

the best students to enter university and to take only the courses towards which their natural wits were inclined: 'It would be no small Advantage to this Kingdom (in Particular) and to the Commonwealth of Learning (in General) if this Reformation were attempted, and put in Practice'.[125]

Before Huarte, Juan Luis Vives's long treatise *De disciplinis* (1531) suggested a radical reform of education which in England must have inspired Ascham and Mulcaster, among others. After all, Vives was well-known in early modern England, where his dialogues were commonly used in grammar schools.[126] In *De disciplinis* Vives dwells on the various dispositions of pupils and the method for judging their character and aptitudes, devises specific training for them, and ponders over the diversity of wits. In more than one aspect, then, Huarte's concerns and recommendations mirror Vives's, many of whose assertions bring Huarte to mind.[127] This occurs, for instance, when Vives states that only the boy naturally prepared for it should be sent to a preparatory school where 'his disposition may be investigated', where teachers 'discuss together the natures of their pupils' and 'apply to each boy to that study for which he seems more fit'.[128] However, if a boy 'is not apt at his letters but trifles with the school tasks, and what is more serious, wastes his time', he should be 'early transferred to that work for which he seems fitted, in which he will occupy himself with more fruitfulness'.[129] Also, Vives affirms that 'determining the instruction to be given to each person' is a matter that 'belongs to psychological inquiry', and therefore the 'natural powers of the mind' ('sharpness in observing, capacity for comprehending, and power in comparing and judging', which he explores in *De anima*), ought to be

---

[125] Edward Bellamy, *The Tryal of Wits*, a1^v.

[126] Joan Simon, *Education and Society in Tudor England* (Cambridge: Cambridge University Press, 1967), p. 105.

[127] See Mª Luisa Ruiz Gil, 'Juan Luis Vives y Juan Huarte de San Juan. Esquema comparativo de su doctrina psicológico-pedagógica', *Perspectivas pedagógicas*, 6.16 (1965), 64–84, and B. del. Moral, 'Estudio comparativo del "Ingenio" en Luis Vives y Huarte de San Juan', *Analecta Calasantiana*, 35–36 (1976), 65–143.

[128] Juan Luis Vives, *On Education: A Translation of the* De tradendis disciplinis *of Juan Luis Vives*, ed. by Foster Watson (Cambridge: Cambridge University Press, 1913), p. 62.

[129] Vives, *On Education*, pp. 70–71.

considered.¹³⁰ In other words, Vives delves into the matter by discussing issues of natural philosophy.

Both Ascham and Mulcaster dwell on similar subjects too. Ascham, for one, is concerned with the 'bringing up of children: of the nature of quick and hard wits: of the right choice of a good wit'.¹³¹ The twist in Ascham is of course his distrust of quick wits, as he regards them typically not prone to scholarship: 'those which be commonly the wisest, the best learned and best men also when they be old, were never commonly the quickest of wit when they were young'.¹³² For Ascham 'the quickest wits commonly may prove [...] ready of tongue to speak boldly, not deep of judgement'.¹³³ The shared concern of Vives and Mulcaster with, among other issues, the admission of students, has made some scholars affirm that Mulcaster's vision of education coincides more with Vives's than with that of his fellow Englishmen Sir Thomas Elyot or Roger Ascham.¹³⁴ In fact, on one occasion Mulcaster even mentions Vives:

> Among many if only *Vives*, the learned *Spaniard*, were called to be witness, he would crave pardon for his own person as not able to come for the gout, but he would substitute for his deputy his whole twenty books of disciplines, wherein he entreateth, how they [pupils] come to spoil, and how they may be recovered.¹³⁵

It is typically assumed, although impossible to fully prove, for not once does Ascham refer to Vives, that Ascham's 'indebtedness to

---

¹³⁰ Vives, *On Education*, p. 73.
¹³¹ Roger Ascham, *The Scholemaster or Plaine and Perfite Way of Teachyng Children, to Vnderstand, Write, and Speake, the Latin Tong* (London: by Iohn Daye, [1570]), B2ᵛ.
¹³² Ascham, *The Scholemaster*, C4ᵛ.
¹³³ Ascham, *The Scholemaster*, C4ᵛ.
¹³⁴ María O'Neill, 'Juan Luis Vives and Richard Mulcaster: A Humanist View of Language', *Estudios Ingleses de la Universidad Complutense*, 6 (1998), 161–75 (p. 163). The impact of Vives upon Mulcaster in the conception of a public elementary school is acknowledged in Richard L. DeMolen, 'Richard Mulcaster's Philosophy of Education', *Journal of Medieval and Renaissance Studies*, 2 (1972), 69–91 (p. 74).
¹³⁵ Richard Mulcaster, *Positions Concerning the Training up of Children*, ed. by William Barker (Toronto: University of Toronto Press, 1994), p. 256. Barker connects Mulcaster with Vives also; see 'Introduction', pp. xiii–lxxxvi.

Vives in matters of education was as great as was that of Mulcaster'.[136]

The preoccupation in early modern England with the proper natural wit of children is not exclusive to specialists in topics of education. John Lyly, for example, in his best-seller *Euphues the Anatomie of Wit* (1578) admits the importance of parents getting involved in the education of their offspring and feels it appropriate to show 'what wit can and will do if it be well employed'.[137] Lyly has a word for those parents who, 'when they see a sharp wit in their son', 'load him with continual exercise' and force him to study to exhaustion in order to 'outrun his fellows'. This Lyly suspects is a counter-productive approach to a healthy education, for 'the mind with indifferent labour waxeth more perfect, with over-much study it is made fruitless'.[138] Lyly's ideas on this subject are relevant if only because his *Euphues* proved a tremendous success: by the year of the publication of *The Examination*, *Euphues*, published for the first time in 1578, had been reprinted seven times, and after 1594 and until 1650, was reprinted another eight times.[139]

The Examination moreover interacts with numerous medical treatises on midwifery and procreation published in England in the early modern period. Indeed, Chapter XV of the book is a short treatise on eugenesis (i.e. the application of the biological laws of inheritance to the perfection of mankind) particularly concerned with four issues: (1) 'to show the natural qualities and temperature which man and woman ought to possess to the end they may use generation'; (2) to discuss 'what diligence the parents ought to employ that their children may be male and not female', and (3) 'how they may become wise and not fools', and finally, (4) 'how they are to be dealt withal after their birth for preservation of their wit' (XV. 124–130). This final chapter is of utmost importance to Huarte, who is of the opinion that 'parents

---

[136] Foster Watson, 'Introduction', in Vives, *Vives on Education*, pp. xvii–clvii (p. xxxv).
[137] John Lyly, *Euphues the Anatomie of Wit* (London: William Leake, 1606), M1^v.
[138] Lyly, *Euphues*, O3^{r-v}.
[139] For a discussion of the particular meaning of 'wit' in *Euphues*, see Richard McCabe, 'Wit, Eloquence, and Wisdom in *Euphues: The Anatomy of Wit*', *Studies in Philology*, 81.3 (1984), 299–324.

apply not themselves to the act of generation with that order and concert which is by nature established, neither know the conditions which ought to be observed to the end their children may prove of wisdom and judgement' (XV. 9–13). In other words, *The Examination* aims to prevent parents from engendering witless children out of ignorance, and by so doing to remedy the problems of society prior even to the moment of conception: 'if by art we may procure a remedy for this [begetting witless children], we shall have brought to the commonwealth the greatest benefit that she can receive' (XV. 16–18).

Debates on similar issues had been going on in England for years before the publication of *The Examination*. For instance, Richard Jonas's translation *The Byrth of Mankynde* (1540), the English rendering of the original in Latin by Eucharius Roesslin, includes a final book on 'the Conception of mankynde, and howe manye wayes it may be letted or furtheryd'. Jacques Guillemeau's *Child-Birth* (1612), translated from the French by an unknown translator, John Sadler's *The Sicke Womans Private Looking-Glasse* (1636), Nicholas Culpeper's *A Directory for Midwives* (1651), Jane Sharp's *The Midwives Book* (1671), or the anonymous *Aristoteles Master-Piece Or The Secrets of Generation* (1684) tackle similar matters. Consider, for instance, that the last title includes chapters on 'The Signs of Barrenness', 'The way of getting a Boy or a Girl', or 'the likeness of Children to Parents', which give an idea of the extent to which these preoccupations were still alive at the end of the seventeenth century.

Nonetheless, many of these treatises suggest explanations that differ greatly from Huarte's more scientific and systematic approach to procreation. This is the case with Stefano Guazzo's *The Court of Good Counsell Wherein is Set Downe the True Rules, How a Man Should Choose a Good Wife from a Bad, and a Woman a Good Husband from a Bad* (1607). Guazzo is concerned with topics such as 'How that many times, Wise Parents may have foolish Children, and foolish Parents wise Children & how that oftentimes, Rich mens Children (being left rich) become poore, and poore mens Children become rich'.[140] Huarte, intrigued by this question too, denies that wise fathers tend to have the wisest

---

[140] Stefano Guazzo, *The Court of Good Counsell Wherein is Set Downe the True Rules, How a Man Should Choose a Good Wife from a Bad, and a Woman a Good Husband from a Bad* (London: Ralph Blower, 1607), B1^v.

children, for the very same 'qualities which make him wise, namely, coldness and dryness' (XV. 1387–1388) also make man 'a coward, of small strength of body, a spare seeder, and not very able for procreation' (XV. 1385–1386). Thereupon, 'wise men's children are well-near always formed of their mother's seed, for that of the father's [...] is not fruitful for generation and in engendering serveth only for aliment' (XV. 1499–1502). The child fashioned after the cold and moist female seed is doomed to witlessness because of those same qualities. This being so, 'when the child proveth discreet and prompt, the same yieldeth an infallible token that he was formed of his father's seed' (XV. 1505–1506). Stefano Guazzo, in replying to the same question, puts forward different arguments and stresses nurture over nature:

> those men whom nature causeth their Children to be fooles, in my Judgement hapneth not by the birth, but by the bringing up [....] I would come to this, that the father, who through much travell and trouble, both of body and minde, hath gotten wealth and honour, though he getteth children of great wit, yet he is so over-gone in fatherly affection towards them, that knowing he hath provided for them sufficiently to live by at their ease, he cannot abide to see them travell and labour as he hath done, so that vanquished with a certaine tender affection, he suffereth them to be brought up dillicately, and wantonly, and is the cause, that by this idlenes their naturall force decayeth, and by Custome is quite chaunged into an other nature.[141]

Huarte's tips on procreation also found their way into English literary works such as John Marston's comedy *Parasitaster, or The Fawn* (1606). The protagonist of *Parasitaster*, Hercules, at one point digresses on the prerequisites for procreation by reflecting on the living conditions of the Israelites under the domination of the King of Egypt:

> *Pharoah* and his councell were mistaken, and their devise to hinder the encrease of procreation in the Israelites, with inforcing them to much labour of bodie, and to feed hard, with beetes, garlike, and onions (meats that make the orriginall of

---

[141] Guazzo, *The Court of Good Counsell*, B2ʳ⁻ᵛ.

man most sharpe, and taking) was absurd. No hee should have given barlie bread, lettice, mellones, cucumers [*sic*], huge store of veale, and fresh beefe, blown up their fleshe, held them from exercise, rould them in feathers, & most severely seene them drunke once a day, then would they at their best have begotten but wenches, and in short time their generation infeebled to nothing.[142]

In reality, his digression is a paraphrase of *The Examination*'s consideration on the same matter (XV. 795–900). As usual, neither Huarte's name nor the title of his treatise are mentioned.

Despite the publishing success of Carew's translation of the *Examen* and its various reprintings, it would seem that *The Examination* was not the sole vehicle for the circulation of Huarte's ideas in England. Translations of it into French and Italian, and maybe even copies of it in Spanish, were probably sold and bought on English soil not only before but after Carew's rendering. Given that Huarte became so well-known in France and Italy by the end of the sixteenth century, French and Italian works which reproduced Huarte's theories — with or without explicitly mentioning Huarte, and either in their original tongues or in English translation — became in their turn vessels for the expansion of Huartean postulates in England. This is the case with Pierre Charron's *De la sagesse* (Bordeaux, 1601), immensely successful in France and the 'most important conduit of Huarte's ideas' in that country.[143] The influence of Huarte upon Charron is well established, and it is generally acknowledged that the *Examen* 'appealed to Charron and [that] he copied its key classifications into his own work':

> Charron incorporated Huarte's scheme of matching temperament to mental ability into *Sagesse* and he copied the classifications of vocations according to brain types with a few minor changes. Charron was attracted also by Huarte's directions for guaranteeing the birth of strong, male children, such as eating warm meat before conception took place, and

---

[142] John Marston, *Parasitaster, or The Fawne* (London: Thomas Purfoot, 1606), C3ʳ. See also, Jürgen Schäfer, 'Huarte: A Marston Source', *Notes and Queries*, 18 (1971), 16–17.

[143] Henry C. Clark, *La Rochefoucauld and the Language of Unmasking in Seventeenth-Century France* (Geneva: Librairie Droz, 1994), p. 42.

he copied some of this advice.[144]

De la sagesse bore the title *Of Wisdome* in England, where it attained great success thanks to the translations by Samson Lennard (published in London eight times from 1608 to 1670), and George Stanhope, whose translation was published in 1697.[145]

Another French author that can claim to have introduced in England some of Huarte's ideas is the hispanist César Oudin, language instructor and the first translator into French of the first part of the *Quixote* (1614). César Oudin was moreover the author of the popular *Grammaire Espagnolle expliquée en François* (1597), which underwent over twenty editions in the seventeenth century and was translated into Latin (1607) and later into English as *A Grammar Spanish and English* (1622).[146] The language manual reproduces a long passage from the *Examen* on the etymology of the term *hidalgo*: 'The Etymology of *hidálgo* or *hijo dálgo*, is amply declared in *Examen de los ingenios*, where the studious may see it at their pleasure'.[147]

Finally, whether Sir Francis Bacon knew *The Examination* and used it for his tripartite classification of the sciences is a matter of controversy, and one that still merits thorough study. Bacon, considered one of the first spokesmen of modern science along with Galileo, Descartes and Harvey, and the father of the inductive method,[148] believed that traditional philosophy had to be refuted, and that an active science ought to replace

---

[144] Renée Kogel, *Pierre Charron* (Geneva: Librairie Droz, 1972), p. 43.

[145] F. Charles-Daubert, 'Charron et l'Angleterre', *Recherches sur le XVIIe Siècle*, 5 (1982), 53–56.

[146] Marc Zuili, 'Nuevas aportaciones sobre el hispanista francés César Oudin (1560?-1625)', *Thélème: Revista Complutense de Estudios Franceses*, 20 (2005), 203–211 (p. 208).

[147] César Oudin, *A Grammar Spanish and English: Or A Briefe and Compendious Method, Teaching to Reade, Write, Speake, and Pronounce the Spanish Tongue* (London: John Haviland, 1622), N6^r.

[148] 'For some present-day epistemologists, Bacon was a spokesman for a hopelessly naïve induction by enumeration, and had thus nothing to do with the development of modern science', Markku Peltonen, 'Introduction', in *The Cambridge Companion to Bacon*, ed. by Markku Peltonen (Cambridge: Cambridge University Press, 1996), pp. 1–24 (p. 1). See also Paolo Rossi, 'Bacon's Idea of Science', in *The Cambridge Companion to Bacon*, ed. by Peltonen, pp. 25–46 (p. 25).

contemplative science. In this scheme, a new classification of knowledge as well as new methods of acquiring knowledge were imperative. Bacon admits that particular sciences are different branches of one single science of nature: natural philosophy, the mother of all sciences. Indeed, Bacon talks about a *philosophia prima*, a universal science from which all individual sciences stem.[149] Still, in the *Advancement of Learning* (1605) and the *De dignitate et augmentis scientiarum* (1623), Bacon divides all knowledge into three parts: history, poesy, and philosophy. He pairs each with a different mental faculty: history with memory, poesy with imagination, and philosophy with reason.[150] Huarte, as deviser of a classification of the sciences based on the powers of the rational soul and the epistemological subject, has been seen as a forerunner of Bacon. Chronologically, Huarte's influence upon Bacon is possible; nonetheless, not all scholars admit this intellectual debt.[151]

The Examination was certainly one of the most influential Spanish scientific books of the early modern period in England, even if explicit acknowledgement of Huarte or Carew is unusual amongst early modern English authors. Carew's rendering of Huarte was no doubt the main vehicle for the dissemination of

---

[149] Robert McRae, 'The Unity of the Sciences: Bacon, Descartes, and Leibniz', *Journal of the History of Ideas*, 18.1 (1957), 27–48.

[150] Sachiko Kushukawa, 'Bacon's Classification of Knowledge', in *The Cambridge Companion to Bacon*, ed. by Peltonen, pp. 47–74.

[151] Among those who acknowledge it: Gaston Sortais, *La philosophie moderne depuis Bacon jusqu'à Leibniz, Vol. 1* (Paris: P. Lethielleux, 1920–1922), p. 351; Marcelino Menéndez y Pelayo, *La ciencia española* (Madrid: Librería General de Victoriano Suárez, 1933); Herman Jean de Vleeschauwer, *Autour de la classification psychologique des sciences: Juan Huarte de San Juan, Francis Bacon, Pierre Charron, D'Alembert* (Pretoria: University of South Africa, 1958), pp. 41–42; Garrido Palazón, 'El *Examen de ingenios para las ciencias* de Huarte de San Juan y el enciclopedismo retórico'. Among those who deny such influence: José Mallart, 'Huarte y las modernas corrientes de ordenación profesional y social', in *Estudios de Historia Social de España*, IV vols, ed. by Carmelo Viñas y Mey (Madrid: CSIC; Instituto Balmes de Sociología, 1952), II, pp. 113–151; Grazia Tonelli Olivieri, 'Galen and Francis Bacon: Faculties of the Soul and the Classification of Knowledge', in *The Shapes of Knowledge: From the Renaissance to the Enlightenment*, ed. by Donald R. Kelley and Richard Henry Popkin (Dordrech: Kluwer Academic, 1991), pp. 61–81 (pp. 69, 71–73). Among the skeptical, Sachiko Kushukawa, 'Bacon's Classification of Knowledge', p. 52.

INTRODUCTION

Huarte's novel theory of wits in England. That was certainly so even if the *Examen* circulated in England in Spanish or in translations into other languages, or, as has been discussed, indirectly within foreign works (often rendered into English) that had absorbed its principles. The metamorphosis of the *Examen* into *The Examination* illustrates the complex workings of the early modern book trade, including the social and political tensions derived from national and religious confrontations, and their repercussion upon translation and printing. In the midst of these anxieties, Carew follows Huarte's treatise closely, rarely alters the text, and makes himself visible as a translator by means of short but meaningful comments in the margins which give a sense not only of his intellectual standpoint before the text he was rendering, but also of his vivid humorous wit.

FURTHER READING

Beecher, Donald and Grant Williams, eds, *Ars Reminiscendi: Mind and Memory in Renaissance Culture* (Toronto: Centre for Reformation and Renaissance Studies, 2009)

Brann, Eva T. H., *The World of the Imagination: Sum and Substance* (Savage: Rowman & Littlefield, 1991)

Craig, Hardin, *The Enchanted Glass: The Elizabethan Mind in Literature* (Oxford: Basis Blackwell, 1950)

Craik, Katharine A., *Reading Sensations in Early Modern England. Early Modern Literature in History* (New York: Palgrave Macmillan, 2007)

Crawford, Patricia, *Blood, Bodies, and Families in Early Modern England* (Harlow: Pearson/Longman, 2004)

Cunningham, Andrew, *The Anatomical Renaissance: The Resurrection of the Anatomical Projects of the Ancients* (Aldershot: Scolar Press, 1997)

—— and Ole Grell, *Medicine and the Reformation* (London: Routledge, 1993)

Eccles, Audrey, *Obstetrics and Gynaecology in Tudor and Stuart England* (Kent: Kent State University Press, 1982)

Fissell, Mary E., *Vernacular Bodies: The Politics of Reproduction in Early Modern England* (Oxford: Oxford University Press, 2004)

Gordon, Andrew and Thomas Rist, eds., *The Arts of Remembrance in Early Modern England: Memorial Cultures of the Post Reformation* (Farnham: Ashgate, 2013)

Harrison, Peter, *The Bible, Protestantism and the Rise of Natural Science* (Cambridge: Cambridge University Press, 1998)

Harvey, E. Ruth, *The Inward Wits: Psychological Theory in the Middle Ages and the Renaissance* (London: Warburg Institute, 1975)

Katritzky, M. A., *Women, Medicine and Theatre, 1500–1750: Literary Mountebanks and Performing Quacks. Studies in*

Performance and Early Modern Drama (Aldershot: Ashgate, 2007)

Kerwin, William, *Beyond the Body: The Boundaries of Medicine and English Renaissance Drama* (Amherst and Boston: University of Massachusetts Press, 2005)

Lindemann, Mary, *Medicine and Society in Early Modern Europe* (Cambridge: Cambridge University Press, 2010)

Macfarlane, Alan, *Marriage and Love in England: Modes of Reproduction, 1300–1840* (New York: Basil Blackwell, 1986)

Maclean, Ian, *The Renaissance Notion of Woman: A Study in the Fortunes of Scholasticism and Medical Science in European Intellectual Life* (Cambridge: Cambridge University Press, 1980)

Miller, Jon, ed., *Topics in Early Modern Philosophy of Mind* (Dordrecht: Kluwer Academic Publishers, 2009)

Nutton, Vivian, ed., *Medicine at the Courts of Europe, 1500–1837* (London and New York: Routledge, 1990)

—— and Roy Porter, eds., *The History of Medical Education in Britain* (Amsterdam: Rodopi, 1995)

Pardo Tomás, José, *Ciencia y censura: la Inquisición española y los libros científicos en los siglos XVI y XVII* (Madrid: Consejo Superior de Investigaciones Científicas, 1991)

Porter, Dorothy and Roy Porter, *In Sickness and in Health: The British Experience, 1650–1850* (London: Fourth Estate, 1988)

Sawday, Jonathan, *The Body Emblazoned: Dissection and the Human Body in Renaissance Culture* (London and New York: Routledge, 1995)

Shapiro, Barbara J., *Probability and Certainty in Seventeenth Century England: A Study of the Relationships between National Science, Religion, History, Law and Literature* (Princeton: Princeton University Press, 1983)

Siraisi, Nancy, *Medieval and Early Renaissance Medicine: An Introduction to Knowledge and Practice* (Chicago: University of Chicago Press, 1990)

Stone, Lawrence, *Family, Sex, and Marriage in England, 1500–1800* (London and New York: Weidenfeld and Nicolson, 1977)

Sugg, Richard, *Murder after Death: Literature and Anatomy in Early Modern England* (Ithaca: Cornell University Press, 2007)

Wear, Andrew and I. M. Lonie, eds., *The Medical Renaissance of the Sixteenth Century* (Cambridge: Cambridge University. Press, 1985)

Webster, Charles, ed., *Health, Medicine and Mortality in the Sixteenth Century* (Cambridge: Cambridge University Press, 1979)

Wilson, Adrian, *Making of Man-midwifery: Childbirth in England, 1660–1770* (London and Cambridge MA: UCL Press and Harvard University Press, 1994)

THE EXAMINATION OF MEN'S WITS

The Examination of Men's Wits. In which, by discovering the variety of natures, is shown for what profession each one is apt and how far he shall profit therein. By John Huarte. Translated out of the Spanish tongue by M. Camillo Camili. Englished out of his Italian by R[ichard]. C[arew]. Esquire. London, Printed by Adam Islip for Richard Watkins. 1594.[1]

---

[1] This transcription of the title page is from the copy at the Bodleian Library, STC 13890.

To the right worshipful Sir Francis Godolphin Knight, one of the deputy lieutenants of Cornwall.

Good Sir, your book returneth unto you clad in a Cornish gaberdine°, which if it become him not well, the fault is not in the stuff, but in the botching tailor, who never bound prentice to the occupation, and working only for his pastime, could hardly observe the precise rules of measure. But such as it is, yours it is, and yours is the workman, entirely addicted to reverence you for your virtues, to love you for your kindness, and so more ready in desire than able in power to testify the same, do with my dewiest remembrance take leave, resting at your disposition,

R. C.

To the Majesty of Don Philip, our Sovereign.

To the end that artificers may attain the perfection requisite for the use of the commonwealth, me thinketh, Catholic royal Majesty, a law should be enacted that no carpenter should exercise himself in any work which appertained to the occupation of a husbandman, nor a tailor to that of an architect, and that the advocate should not minister physic, nor the physician play the advocate, but each one exercise only that art to which he beareth a natural inclination, and let pass the residue. For considering
10 how base and narrowly bounded a man's wit is for one thing and no more, I have always held it for a matter certain that no man can be perfectly seen in two arts without failing in one of them. Now, to the end he may not err in choosing that which fitteth best with his own nature, there should be deputed in the commonwealth men of great wisdom and knowledge who might discover each one's wit in his tender age, and cause him perforce to study that science which is agreeable for him, not permitting him to make his own choice. Whence this good would ensue to your states and signories that in them should reside the rarest artificers
20 of the world, and their works should be of the greatest perfection, for nought else than because they united art with nature. The like would I that the universities of our kingdoms did put in practice, for seeing they allow not that a scholar should pass to another faculty unless he perfectly understand the Latin tongue, they should have also examiners to try whether he who purposeth to study logic, philosophy, divinity, or the laws, have such a wit as is requisite for every of these sciences, for otherwise, besides the damage that such a one shall work afterwards to the commonwealth by using an art wherein he is not skilled, it is a grief to see
30 that a man should take pains and beat his brains about a matter wherein he cannot reap any advantage. For that at this day such a diligence is not used, those who had not a wit fit for divinity have destroyed the Christian religion. So do those who are untoward for physic shorten many a man's days. Neither possesseth the legal science that perfection which it might receive, because it is not made known to what reasonable power the use and interpretation of the laws appertaineth. All the ancient philosophers found by experience that where nature doth not dispose a man to knowledge, it falleth out a superfluous labour
40 to toil in the rules of art. But none hath clearly and distinctly

delivered what that nature is which maketh a man able for one science, and uncapable of another, nor how many differences of wits there are found in mankind, nor what arts or sciences do answer each in particular, nor by what tokens this may be known, which is the thing that most importeth.

These four points, though they seem impossible, contain the matter whereof I am to entreat, besides many others appurtenant to the purpose of this doctrine, with intention that curious parents may have an art and manner to discover the wit of their children, and may weet° how to set each of them in hand with that science wherein he shall principally profit. And this is an advice which Galen sayeth was given his father, namely that he should set him to study physic because for that science he had a singular wit.[2] By which your majesty shall understand how much it importeth the commonwealth that there be established in the same a choice and examination of wits for the sciences, seeing from the study which Galen bestowed in physic, there ensued so great good to the diseased of his time, and he left so many remedies in writing for the posterity. Even as Baldus (a notable man in profession of the laws) when he studied and practised physic, if he had passed further therein, would have proved but an ordinary physician, as he was not better, for that he wanted the difference of wit requisite for this science, but the laws should have lost one of the greatest helps that might be found amongst men for expounding them.[3]

When I therefore purposed to reduce this new manner of philosophy to art, and to prove the same in some wits, I remembered myself of your Majesty as the best known, and one at whom the whole world wondereth, beholding a Prince of so great knowledge and wisdom, of whom here we cannot conveniently entreat, the last chapter save one is your convenient place, where your Majesty shall see the purport of your own wit, and the art and learning wherewith you would have benefited your commonwealth if you had been a private person, as by nature you are our king and sovereign.

---

[2] 1. *Index*: 'And this is an advice [51] [...] a singular wit' [censored].
[3] Baldus] Baldus de Ubaldis (Baldo degli Ubaldi, in Italian) (1327–1400) was a renowned Italian jurist from Perugia. He taught at the Universities of Bologna, Perugia, Pisa, Florence, Padua and Pavia.

The second proem to the Reader.

When Plato would teach any doctrine grave, subtle, and divided from the vulgar opinion, he made choice amongst his scholars of such as he reputed best witted, and to those only he imparted his mind, knowing by experience that to teach delicate matters to persons of base understanding was loss of time, loss of pains, and loss of learning. The second thing which he did after this choice made was to prevent them with certain pre-suppositions, clear and true, which should not be wide from his conclusion, for the speeches and sentences which unlooked for are delivered against that which the vulgar believeth at the beginning serve for nought else (such prevention not being made) than to put in a confusion him that listeneth, and to breed such a loathing in men's minds as it causeth them to lose their good affection, and to abhor and detest this doctrine. This manner of proceeding would I that I might observe with thee, curious Reader, if means could be used that I might first treat with thee, and discover between thee and me, the disposition of thy wit. For if it be such as is requisite for this doctrine, and estranged from the ordinary capacities, I would in secret tell thee such new and special conceits as thou wouldest never have thought could fall within the compass of a man's imagination. But inasmuch as this will not be, and this work must issue in public for all sorts, I could not but set thy brains somewhat at work: for if thy wit be of the common and vulgar alloy°, I know right well thou art already persuaded that the number of the sciences, and their perfection, hath been accomplished many days ago. And hereto thou art moved by a vain reason, that they having found out no more what to add, it is a token that now there is in nothing any more novelties. Now if by hap thou art possessed of such an opinion, go no further, nor read thou any longer on, for thou wilt be much aggrieved to see how miserable a difference of wit possesseth thee. But if thou be discreet, well compounded, and sufferant, I will deliver unto thee three conclusions very true, albeit for their novelty they are worthy of great marvel.

The first is that of many differences of wit which are in mankind, one only with pre-eminence can fall to thy lot, if already nature, as very mighty, at such time as she framed it for thee did not bestow all her endeavour in uniting two only, or three, or (in that she could not effect the same) left thee a dolt, and deprived of them all.

The second, that to every difference of wit there answereth in pre-eminence but one only science and no more of that condition. So as if thou divine not to choose that which answereth thy natural hability, thou shalt be very remiss in the rest, though thou ply them night and day.

The third, that after thou hast known which the science is that most answereth thy wit, there resteth yet (that thou mayst not be deceived) another greater difficulty, which is whether thine hability be more appliable to the practic than the theoric, for these two parts (be it what science it will) are so opposite betwixt themselves, and require wits so different, that they may be placed one against the other, as if they were contraries. Hard are these sentences, but yet they have greater difficulty and hardness, viz., that we cannot appeal from them, nor pretend that we have received wrong. For God being the author of nature, and seeing that she gave not to each man more than one difference of wit (as I have said before), through the opposition or difficulty which cumbereth us in uniting them, He applied Himself to her, and of the sciences which are distributed amongst men by grace, it is a miracle if in an eminent degree He give more than one. But there are, sayth St Paul, divisions of graces, and the same spirit; there are divisions of ministries, and the same Lord; there are divisions of operations, but the same God, who worketh all things in all persons. To everyone is given the ministry of the spirit for profit, and to one is given by the spirit the word of wisdom, to another that of knowledge, after the same spirit, to another faith in the same spirit, and to another the grace of healing, in the same spirit, to another the working of virtues, to another prophesying and the description of spirits, to others the variety of tongues, to another the interpretation of words. But one self spirit, which divideth to every one as him pleaseth, worketh all these things.

This bestowing of sciences (I doubt not) God useth having regard to the wit and natural disposition of every person. For the talents which He distributed in St Matthew, the same Evangelist sayth that He gave them unto every one according to his proper virtue.

And to think that these supernatural sciences require not some dispositions in the subject before they be infused is an error very great,[4] for when God formed Adam and Eve, it is certain that

---

[4] 2. *Index*: 'This bestowing of sciences [71] [...] an error very great' [censored].

before He filled them with wisdom, He instrumentalized⁵ their brain in such sort as they might receive it with ease, and serve as a commodious instrument therewith to be able to discourse and to form reasons. And therefore the divine Scripture sayth God gave them a heart to think, and filled them with the discipline of understanding, and that according to the difference of which every one partaketh, one science is infused, and not another, or more or less of each of them is a thing which may be understood by this example of our first parents, for God filling them both with wisdom, it is a verified conclusion that He infused the lesser portion into her, for which reason the divines say that the devil took hardness to beguile her, and durst not tempt the man, as fearing his much wisdom. The reason hereof (as hereafter we will prove) is that the natural composition which the woman had in her brain is not capable of much wit nor much wisdom. In the angelical substances we shall find also the like count and reason, for God, to give an angel more degrees of glory and higher gifts, first giveth him a more delicate nature. And if you inquire of the divines whereto this delicate nature serveth, they answer that the angel who hath the deepest understanding and the best nature, with most facility converteth himself unto God, and useth his gift with the more efficacy, and that the like betideth in men.⁶ Hence we clearly infer that there being an election of wit⁷ for sciences supernatural, and that, not whatsoever difference of hability is their commodious instrument, human learning (with more reason) requireth the same, because it is to be learned by men with the force of their wit.

To be able then to distinguish and discern these natural differences of man's wit, and to apply to each by art that science wherein he may profit, is the intention of this my work. If I bring the same to end (as I have purposed) we will yield the glory to God, seeing from His hand proceedeth whatsoever is good and certain. And if not, thou knowest well, discreet reader, that it is impossible both to devise an art and to reduce the same to perfection. For so long and large are human sciences that a man's life sufficeth not to find them out and to give them that perfection which is requisite.

---

⁵ instrumentalize] to make or render instrumental to some end; to organize.
⁶ 3. *Index*: 'and that the like betideth in men' [censored].
⁷ 4. *Index*: 'of wit' [censored].

The first inventor performeth very much if he discover some notable principles, to the end that such as come after may with this seed take an occasion to amplify the art, and to bring it into that estimation and account which is due thereunto. Aristotle, alluding hereunto, sayth that the errors of those who first began to handle matters of philosophy are to be held in great reverence, for it proving a matter so difficult to devise new things, and so easy to add unto that which hath been already spoken and treated of, the defects of the first deserve not (by this reason) to be much reproved, neither he who addeth ought meriteth any great commendation. I confess that this my work cannot be excused from some errors, seeing the matter is so delicate and no way fore-opened to entreat thereof. But if the same be in a matter where the understanding hath place to think, in this case I pray thee, witty reader, that before thou give sentence thou read over the whole work, and assure thyself what the difference of thine own wit is, and if in the work thou find ought which in thine opinion is not well said, consider well of the reasons which sway the most against it, and if thou canst not resolve, then turn to read the eleventh chapter, for in that shalt thou find the answer which they may receive.

THE EXAMINATION OR TRIAL OF MEN'S WITS AND DISPOSITIONS

CHAP. I.

He proveth by an example that if a child have not the disposition and hability which is requisite for that science whereunto he will addict himself, it is a superfluous labour to be instructed therein by good schoolmasters, to have store of books, and continually to study it.

The opinion of Cicero was good, who, that his son Mark might prove such a one in that kind of learning, which himself had made choice of, as he desired, judged that it sufficed to send him to a place of study so renowned and famous in the world as that of Athens, and to give him Cratippus[8] for his schoolmaster, who was the greatest philosopher of those days, bringing him up in a city so populous where, through the great concourse of people which thither assembled, he should of necessity have many examples and profitings of strangers, fit to teach him by experience those things which appertained to the knowledge that himself was to learn. But, notwithstanding all this diligence, and much more besides, which (as a good father) he used, providing him books and writing some unto him of his own head, the historians report that he proved but a cod's-head° with little eloquence and less philosophy – a matter usual amongst men that the son abies the much wisdom of the father. Verily Cicero greatly beguiled himself imagining that albeit his son were not issued out of nature's hands with that wit and hability which is requisite for eloquence and philosophy, yet by means of the good industry of such a teacher, and the many books and examples of Athens, together with the young man's continual endeavour, and process of time, the defects of his understanding would be amended. But we see that finally he deceived himself, neither do I marvel thereat, for he had many examples to this purpose which encouraged him to believe that the same might also befall in the person of his son.

---

[8] Cratippus] Cratippus of Pergamon (first century BC) was a leading Peripatetic philosopher. Cicero, a contemporary of his, records the only aspects of his teaching known today.

For the same Cicero reports in his book of Destiny that Zenocrates[9] had a wit very untoward for the study of natural and moral philosophy, of whom Plato said that he had a scholar who stood in need of a spur, and yet notwithstanding, through the good industry of such a master, and the continual travail of Zenocrates himself, he became a very great philosopher. And he writes the like also of Cleantes,[10] who was so doltish and void of understanding that no teacher would receive him into his school, whereat the young man, aggrieved and ashamed, endured so great toil in studying that he came afterwards to be called a second Hercules for wisdom. No less untoward for matters of eloquence seemed the wit of Demosthenes, of whom it is said that when he was now grown big, he could not yet speak plain, but labouring and applying the art by hearing of good teachers, he proved the best orator of the world. And specially, as Cicero recounts, he could not pronounce the letter 'r', for that he did somewhat stammer, and yet by practice he grew to articulate it so well as if he had never had that way any defect. Hence took that proverb his original, which saith that man's wit in matters of science is like a player at dice, for if anyone prove unlucky in throwing his chance, by artificial practice he comes to amend his evil fortune.

But none of these examples produced by Cicero remains without a convenient answer in my doctrine: for (as we will hereafter prove) there is in young men a certain dullness which argues a greater wit in another age, than if the same had been sharp from their childhood. Nay, it is a judgement that they will prove loutish men, when they begin very soon to discourse and be quick of conceit. Wherefore, if Cicero had known the true tokens by which wits are in their first age to be discovered, he would have held it a good sign that Demosthenes was rude and slow of speech, and that Zenocrates had need of a spur whilst he learned. I take not from a good instructor art and industry, their virtue and force to manure wits, as well rude as pliant. But that

---

[9] Zenocrates] Xenocrates (*c.* 396/5–314/3 BC) of Chalcedon was a Greek philosopher and mathematician whose teachings followed those of Plato. He presided over the Platonic Academy from 339/8 until his death.

[10] Cleantes] Cleanthes (*c.* 330 BC–*c.* 230 BC), originally a boxer, became a Stoic philosopher by attending Zeno's lectures in Athens. After the latter's death, he became his successor as the head of the Stoic school in Athens. Chrysippus was the most renowned of Cleanthes' pupils.

which I will say is that if a young man have not of himself an understanding capable of precepts and rules which properly belong to the art he would learn, and to none other, that the diligence used by Cicero with his son was as vain as that which any other parent shall use with his son will be in the like. Those who have read Plato shall easily know that this doctrine is true, who reports that Socrates was the son (as he also reported himself) of a midwife, and that as his mother (albeit she were much praised in the art) could not make a woman to be delivered, that before her coming to her was not with child, so he (performing the like office as his mother) could not make his scholars bring forth any science if of themselves they had not their understanding conceived therewith. He was of opinion that sciences were (as it were) natural to those men only who had their wits appliable thereunto, and that in such it befell, as we see by experience in those who have forgotten somewhat which they first knew, who if we put them in mind but of one word, gather from that all the residue.

Masters (for ought that I can gather) have none other office with their scholars than to bring learning to their remembrance, for if they have a fruitful wit, they make them with this only to bring forth wonderful conceits, otherwise they do but afflict themselves and those whom they instruct, nor ever obtain their desires.[11] And (at least if I were a teacher) before I received any scholar into my school, I would grow to many trials and experiments with him, until I might discover the quality of his wit, and if I found it by nature directed to that science whereof I made profession, I would willingly receive him, for it breeds a great contentment in the teacher to instruct one of good towardliness. And, if not, I would counsel him to study that science which were most agreeable with his wit. But if I saw that he had no disposition or capacity for any sort of learning, I would friendly and with gentle words tell him: 'Brother, you have no means to prove a man of that possession which you have undertaken, take care not to lose your time and your labour, and provide you some other trade of living which requires not so great an hability as appertaineth to learning.' Hereof is seen very plain experience, for we behold a great number of scholars enter the course of

---

[11] 5. *Index*: 'so he (performing the like [76] [...] nor ever obtain their desires' [censored].

whatsoever science and (be the teacher very good or very bad) finally every day some prove of great skill, some of mean, and some in their whole course have done nought else than lose their time, spend their goods, and beat their brains without any manner of profit.

I wot never whence this effect may spring, they all hearing one self teacher, and with equal diligence and care, and perhaps the dull take more pain than the witty, and this difficulty grows the greater by seeing that those who are untoward for one science are very apt to another, and the toward in one sort of learning, passing to another sort, can understand nothing. But myself am at least a good witness in this truth, for there were three companions of us who entered together to study the Latin tongue, and one of us learned the same with great facility, the rest could never make any commendable composition. But all passing on to logic, one of those who could not learn grammar proved in that art a principal eagle, and the other two, in the whole, never learned one ready point. Then all three coming to hear astrology, it was a matter worthy of consideration that he, who could no skill of Latin or logic, in few days knew more in astrology than his master that taught them, and the rest could never learn it. I then marvelling hereat began forthwith to make discourses and play the philosopher hereon, and so I found that every science required a special and particular wit, which reaved from that, was little worth in other sorts of learning. And if this be true (as verily it is, and we will so prove it hereafter) he that at this day should enter into the schools of our times making proof and assay of the scholars' wits, how many would he change from one science to another, and how many would he send into the fields for dolts and unable to learn? And how many would he call back[12] of those who, for want of hability, are occupied in base exercises and yet their wits were by nature created only for learning? But sithence this cannot be brought about nor remedied, it behoves to stay no longer hereon, but to pass forward.

It cannot be denied but that (as I have said) there are wits found capable of one science which are untoward for another. And therefore it behoves, before the child be set to study, to discover the manner of his wit, and to see what science agreeth with his

---

[12] call back] restore.

THE EXAMINATION OF MEN'S WITS

capacity, and then to provide that he may apply the same. But it is necessary also to consider that this which hath been said sufficeth not to make a man prove sufficiently learned, but we must have regard of other conditions no less requisite than is this of towardliness. For Hippocrates sayth that man's wit holds the like proportion with knowledge as the earth doth with seed, which though of herself she be fruitful and fat, yet it behoves to manure her and use advisement to what sort of seed her natural disposition inclineth; for every sort of earth cannot without distinction produce every sort of seed. Some better brings forth wheat than barley, and some barley better than wheat. And of wheats some bring a plentiful increase of good Lammas-wheat,[13] and cannot away with the basest sort.

Neither doth the good husbandman content himself to make this only distinction, but after he hath manured the earth in due season, he looks for convenient time to sow it, for it cannot be done at all times of the year, and after that the grain is sprung up, he cleanseth and weedeth it that it may increase and grow, giving the fruit which of the seed is expected. After this sort, it is necessary that the science being known, which best fitteth with the person, he begin to study from his first age, for this, sayth Aristotle, is the most pliant of all others to learning. Moreover, man's life is very short, and the arts long and toilsome, for which it behoves that there be time sufficient to know them, and space to exercise them, and therewith to profit the commonwealth. Children's memory, saith Aristotle, is a table without any picture, because it was but a little while since they were born, and so they receive anything whatsoever with facility — and not as the memory of old men, which, full of those many things they have seen in the long course of their life, is not capable of more. And therefore Plato sayth that in the presence of youth, we should recount honest tales and actions, which may incite them to virtuous doings, for what they learn in that age abides still in their minds and not, as Galen sayth, that then it behoves to learn the arts, when our nature hath accrued all the forces that she can have, which point is void of reason if you admit no distinction. He that is to learn the Latin tongue or any other language, ought to do it in his childhood, for if he tarry till the body be hardened, and take

---

[13] Lammas-wheat] winter wheat.

the perfection that it ought to have, he shall never reap available profit. In his second age, namely boy's state, it is requisite that he travail in the art of syllogisms, for then the understanding begins to display his forces, which hath the same proportion with logic as shackles have with the feet of mules not yet trained, who going some days therewith, take afterward a certain grace in their pace, so our understanding shackled with the rules and precepts of logic, takes afterwards a graceful kind of discoursing and arguing in sciences and disputations. Then follows youth, in which all the sciences appertaining to the understanding may be learned, for that hath a ripened knowledge.

True it is that Aristotle excepteth natural philosophy saying a young man is not of fit disposition for this kind of doctrine, wherein it seemeth he hath reason, for that it is a science of deeper consideration and wisdom than any other.

Now the age thus known in which sciences are to be learned, it behoves to search out a commodious place for the same, where nothing else save learning may be handled, and such are the universities. But the youth must forgo his father's house, for the dandling of the mother, brethren, kindred, and friends which are not of his profession, do greatly hinder his profiting. This is plainly seen in the scholars who are native of the cities and places where universities are seated, none of which (save by great miracle) ever become learned. And this may easily be remedied by changing universities, and the native of one city going to study in another. This faring° that a man takes from his own country to make himself of worth and discretion is of so great importance that there is no master in the world who can teach him more, and especially, when a man sees himself sometimes abandoned of the favour and delights of his country. Depart out of thy land (said God to Abraham) and sever thyself from amidst thy kindred and thy father's house, and come to the place where I will show thee, in which thou shalt make thy name great, and I will give thee my blessing. The like says God to all men who desire to prove of value and wisdom, for albeit He can bless them in their native country, yet He will that men dispose themselves by this mean which He hath ordained, and that wisdom be not attained by them with idleness. All this is meant with a foregoing presupposal that a man have a good wit and be apt, for otherwise, he that goes a beast to Rome, returns a beast again. Little avails it that a dullard go to learn in the famous places of study, where there is no chair of understanding, nor wisdom, nor a man to teach it.

The third point of diligences is to seek out a master who hath a direction and method in teaching, whose doctrine is sound and firm, not sophistical nor of vain considerations. For all that the scholar doth whilst he is alearning is to credit all that which his master propounds unto him, for he hath no sound judgement or discretion to discern or separate falsehood from truth, albeit this is a chanceful[14] case, and not placed in the choice of such as learn that the scholars come in due time to study, and that the universities have good or unfit instructors. As it befell certain physicians of whom Galen reports that, having convinced them by many reasons and experiments, and showed them that the practice which they used was false and prejudicial to men's health, the tears fell from their eyes and in his presence they began to curse their hard hap in lighting on such bad masters as bare sway during the time that they were learners. True it is that there are found some scholars of so ripe wit as they straightway look into the condition of the teachers and the learning which he teacheth, and if it be vicious, they know how to confute the same, and to give allowance to such as deliver soundly. These at the year's end teach their master much more than their master taught them, for doubting and demanding wittily, they make him to understand and answer things so exquisite as he himself never knew nor should have known if the scholar with the felicity of his wit had not brought them to his mind. But those who can do this are one or two at the most, and the dullards are infinite, through which it would do well (seeing this choice and examination of wits for every science is not had) that the universities always made provision of good teachers endued with sound learning and a clear discerning wit to the end they may not instruct the ignorant in errors and false propositions.

The fourth diligence requisite to be used is to study every science with order, beginning at his principles and passing through the midst to the end without having matter that may presuppose another thing before. For which cause, I have always held it an error to hear many lessons of diverse matters and to carry them all home fardelled up° together. By this means there

---

[14] chanceful] dependent on chance

is made a mass of things in the understanding which afterwards, when they come to practice, a man knows not how to turn to use the precepts of his art, nor to assign them a place convenient. And it is much better to bestow labour in every matter by itself, and with that natural order which it holds in his composition, for in the self manner as it is learned, so is it also preserved in the memory. And more in particular, it is necessary that they do this who of their own nature have a confused wit. And this may easily be remedied by hearing one matter by itself, and that being ended, to enter into the next following, till the whole art be achieved. Galen well understanding of how great importance it was to study matters with order and conceit, wrote a book to teach the manner that was to be held in reading his works to the end that the physician might not be tangled in confusion.[15] Others add hereunto that the scholar, whilst he learneth, have but one book which may plainly contain the points of his learning, and that he attend to study that only and no more least he grow into a garboil[16] and confusion, and herein they are warranted by great reason.

The last thing which makes a man prove of rare learning is to consume much time at his book and to expect that knowledge have his due digestion and take deep root. For as the body is not maintained by the much which we eat and drink in one day, but by that which the stomach digesteth and turneth, so our understanding is not filled by the much which we read in little time, but by that which by little and little it proceeds to conceive and chew upon. Our wit day by day disposeth itself better and better, and comes (by process of time) to light on things which before it could neither understand nor conceive. Understanding hath his beginning, his increase, his standing, and his declining, as hath a man, and other creatures and plants. It begins in boyage, hath his increase in youth, his standing in middle or man's age, and in old age it begins to decline. Who so therefore would know at what time his understanding enjoyeth all the forces which it may partake, let him weet° that it is from the age of thirty and three until fifty, little more or less, within which compass we may best give credit to grave authors, if in the discourse of their life they

---

[15] a book] *De ordine librorum suorum* [On the Order of his own Books]. In this book Galen explains the order in which his main treatises ought to be read.
[16] garboil] disturbance

have held contrary opinions. And he that will write books, let him do it about this age, and not before nor after, if he mean not to unsay again, or change opinion.

But man's age hath not in all people a like measure and reason, for in some, childhood ends in twelve years, in some at fourteen, some have sixteen, and some eighteen. Such live very long because their youth arrives to little less than forty years, and their ripe or firm age to three-score, and they have afterward twenty years of old age, wherethrough their life amounts to four-score, and this is the term of those who are very strong. The first sort, who finish their childhood at twelve years, are very short lived, and begin speedily to discourse, their beard soon sprouteth out, and their wit lasteth but a small time, these at thirty five years begin to decline, and at forty and eight finish their life.

Of all the conditions above specified, there is not any one which is not very necessary, profitable, and helpful in practice for a young man to receive notice of, but to have a good and answerable nature to the science which he pretendeth to study is the matter which most makes for the purpose. For with this, we have seen that diverse men have begun to study, after their youth was expired, and were instructed by bad teachers with evil order, and in their own birthplaces, and yet for all that have proved great clerks. But if the wit fail, saith Hippocrates, all other diligences are lost. But there is no man who hath better verified this than the good Marcus Cicero, who through grief of seeing his son such a do-nought,° with whom none of the means could prevail that he had procured to breed him wisdom, said in the end after this sort: 'What else is it, after the manner of the giants, to fight with the gods, than to resist against nature.' As if he should have said: 'What thing is there which better resembles the battle which the giants undertook against the gods than that a man who wanteth capacity should set himself to study?' For as the giants never overcame the gods, but were still vanquished by them, so whatsoever scholar will labour to overcome his own untoward nature shall rest vanquished by her. For which cause, the same Cicero counselleth us that we should not use force against our nature, nor endeavour to become orators if she assent not, for we shall undergo labour in vain.

CHAP. II.

That nature is that which makes a man of hability to learn.

It is an opinion very common and ordinary amongst the ancient philosophers to say that nature is she who makes a man of hability to learn, and that art with her precepts and rules gives a facility thereunto, but then use and experience, which he reaps of particular things, makes him mighty in working. Yet none of them ever showed in particular what thing this nature was, nor in what rank of causes it ought to be placed. Only they affirmed that this wanting in him who learned art, experience, teachers, books, and travail are of none avail. The ignorant vulgar seeing a man of great wit and readiness, straightways assign God to be the author thereof, and look no further, but hold every other imagination that goes beyond this for vanity. But natural philosophers despise this manner of talking, for put case that the same be godly, and contain therein religion and truth, yet it groweth from not knowing the order and disposition which God placed amongst natural things that day when they were created, and so cover their ignorance with a kind of warrantise, and in sort that none may reprehend or gainsay the same, they affirm that all befalls as God will, and that nothing succeeds which springs not from his divine pleasure. But though this be never so apparent a truth, yet are they worthy of reproof, because, as not every kind of demand, sayth Aristotle, is to be made after one fashion, so not every answer (though true) is to be given.[17]

Whilst a natural philosopher reasoned with a grammarian, there came to them an inquisitive gardener, and asked what the cause might be that, he cherishing the earth so charily, in delving, turning, dunging, and watering it, yet the same never well brought forth the herbage which he sowed therein; whereas the herbs which she bred of herself, she caused to increase with great facility. The grammarian answered this grew from the divine providence, and was so ordained through the good government of the world. At which answer, the natural philosopher laughed, seeing he reduced this to God, because he knew not the discourse of natural causes, nor in what sort they proceeded to their effects.

---

[17] 6. *Index*: 'It is an opinion very common [3] [...] is to be given' [censored].

The grammarian perceiving the other laugh asked whether he mocked him, or whereat else he laughed. The philosopher answered that he laughed not at him, but at the master who taught him so ill, for the knowledge and solution of things which spring from the divine providence (as are the works supernatural) appertain to the metaphysics (whom we now term divines) but this question propounded by the gardener is natural, and appertaineth to the jurisdiction of the natural philosophers, because there are certain ordered and manifest causes from which this effect may spring. And thus the natural philosopher answered saying that the earth is conditioned like a stepmother who very carefully brings up her own children which she breeds herself, but takes away the sustenance from those which appertain to her husband, and so we see that her own children are fat and fresh, and her stepchildren weak and ill-coloured. The herbs which the earth brings forth of herself are born of her proper bowels, and those which the gardener makes to grow by force are the daughters of another mother, wherethrough she takes from them the virtue and nourishment by which they ought to increase that she may give it to the herbs which are born of herself.

Hippocrates likewise reports that he going to visit the great philosopher Democritus, he told him the follies which the vulgar speak of physic, namely, that seeing themselves recovered from sickness, they would say, it was God who healed them, and that if His will were not, little had the good diligence of the physician availed.[18] This is so ancient a manner of talk, and the natural philosophers have so often refuted it, that the seeking to take the same away were superfluous, neither is it convenient. For the vulgar, who know not the particular causes of any effect, answereth better and with more truth as touching the universal cause, which is God, than to say some other unfitting thing. But I have often gone about to consider the reason and the cause whence it may grow that the vulgar sort is so great a friend to impute all things to God, and to reave them from nature, and do so abhor the natural means; and I know not whether I have been able to find it out. The vulgar (at least) gives hereby to understand that forasmuch as they know not what effects they ought to attribute to God immediately, and what to nature, they speak

---

[18] 7. *Index*: 'they would say [60] [...] the good diligence of the physician availed' [censored].

after this manner. Besides that, men are for the most part impatient, and desirous to accomplish speedily what they covet. But because the natural means are of such prolixity, and work with length of time, they possess not the patience to stand marking thereof, and knowing that God is omnipotent, and in a moment of time performeth whatsoever him pleaseth (whereof they find many examples) they would that He should give them health, as He did to the sick of the palsy; and wisdom, as to Solomon;[19] riches, as to Job,[20] and that He should deliver them from their enemy, as He did David.

The second cause is for that men are arrogant, and vain conceited, many of whom desire secretly in their hearts that God would bestow upon them some particular graces which should not befall after the common use (as is that the sun ariseth upon the good and bad, and that the rain falls upon all in general) for benefits are so much the more highly prized as they are the more rare. And for this cause we have seen many men to feign miracles in houses and places of devotion, for straightways, the people flocks unto them, and holds them in great reverence as persons of whom God makes a special account. And if they be poor, they favour them with large alms, and so some sin upon interest.

The third reason is that men have a liking to be well at their ease, whereas natural causes are disposed with such order and conceit that to obtain their effects it behoves to bestow labour. Wherefore they would have God demean Himself towards them after His omnipotence, and that (without sweating) they might come to the well-head of their desires. I leave aside the malice of those who require miracles at God's hand, thereby to tempt his almightiness, and to prove whether He be able to do it. And othersome, who to be revenged after their hearts' desire, call for fire from heaven, and such other cruel chastisements.

---

[19] Solomon] Salomon was David's second son by Bathsheba and third king of Israel *c.* 971–931 BC; author of *Proverbs*, *Ecclesiastes*, and *Song of Songs*. During his reign of forty years (he succeeded his father on the throne) the Hebrew monarchy reached its highest splendour. His reign was a period of great material prosperity as well as of intellectual activity, and he became reputed for his wisdom.

[20] Job] Job lived in great prosperity until he suddenly encountered a series of disasters against his health, family and wealth. Despite all his sufferings he maintained his integrity. God rewards his exemplary behaviour with greater prosperity.

The last cause is, for that many of the vulgar are religiously given, and hold dear, that God may be honored and magnified, which is much sooner brought about by way of miracles than by natural effects, but the common sort of men know not that works above nature and wonderful are done by God to show those who know it not that He is omnipotent, and that He serves Himself of them as an argument to prove His doctrine, and that this necessity once ceasing, He never doth it more. This may well be perceived considering that God doth no longer those unwanted things of the New Testament, and the reason is for that on His behalf He hath performed all necessary diligence that men might not pretend ignorance. And to think that He will begin anew to do the like miracles, and by them once again to prove His doctrine in raising the dead, restoring sight to the blind, and healing the lame and sick of the palsy is an error very great, for once God taught men what is behooful° and proved the same by miracles but returns not to do it anymore. God speaks once, saith Job, and turns not to a second replyal.

The token whereon I ground my judgement when I would discover whether a man have a wit appropriate to natural philosophy is to see whether he be addicted to reduce all matters to miracle without distinction; and contrariwise, such as hold not themselves contented until they know the particular cause of every effect leave no occasion to mistrust the goodness of their wit. These do well know that there are effects which must be reduced to God immediately (as miracles), and others to nature (and such are those, which have their ordinary causes from whence they accustom to spring) but speaking both of the one manner and the other, we always place God for author. For when Aristotle said that God and nature did nothing in vain, he meant not that nature was a universal cause endowed with a jurisdiction severed from God, but that she was a name of the order and concert which God hath bestowed in the frame of the world to the end that the necessary effects might follow for the preservation thereof. For in the same manner, it is usually said that the king and civil reason do no man wrong. In which kind of speech, no man conceiveth that this name 'reason' signifieth a prince which possesseth a several jurisdiction from that of the king; but a term which by his signification embraceth all the royal laws and constitutions ordained by the same king for the preservation of his commonwealth in peace. And as the king hath his special cases reserved to himself which cannot be decided by the law for that they are unusual and weighty, in like manner God left miraculous effects reserved for Himself, neither gave allowance unto natural causes that they might produce them. But here we must note that he who should know them for such, and difference them from natural works, behoves to be a great natural philosopher and to understand the ordinary causes that every effect may hold, and yet all this sufficeth not unless the Catholic church ratify them to be such. And as the doctors labour and study in reading this civil reason, preserving the whole in their memory that they may know and understand what the king's will was in the determination of such a case, so we natural philosophers (as doctors in this faculty) bestow all our study in knowing the discourse and order which God placed that day when He created the world so to contemplate and understand in what sort, and upon what cause, He would that things should succeed. And as it were a matter worthy laughter that a doctor should allege in his writings (though approved) that the king commands a case should be thus determined without showing the law and reason through which it was so decided, so natural philosophers laugh at such as say 'this is God's doing' without assigning the order and discourse of the particular causes whence they may spring. And as the king will give them no ear when they require him to break some just law, or to rule some case besides the order of justice which he hath commanded to be observed, so God will not hearken when any man demands of Him miracles and works besides natural order without cause why. For albeit the king every day abrogates and establisheth new laws and changeth judicial order (as well through the variation of times as for that it is the judgement of a frail man and cannot at one only time attain to perfect right and justice) notwithstanding the natural order of the universe, which we call nature, from that day wherein God created the world unto this hath had no need of adjoining or reaving any one jot, because He framed the same with such providence and wisdom, that to require this order might not be observed were to say that His works were imperfect.

To return then to that sentence so often used by natural philosophers that nature makes able, we must understand that there are wits and there are abilities which God bestoweth upon men besides natural order, as was the wisdom of the Apostles, who being simple and of base account, were miraculously enlightened and replenished with knowledge and learning. Of this sort of ability and wisdom it cannot be verified that nature makes able, for this is a work which is to be imputed immediately unto God, and not unto nature. The like is to be understood of the wisdom of the prophets, and of all those to whom God granted some grace infused. Another sort of hability is found in men which springs of their being begotten with that order and consent of causes which are established by God to this end, and of this sort it may be said with truth nature makes able. For, as we will prove in the last chapter of this work, there is to be found such an order and consent in natural things that if the fathers in time of procreation have regard to observe the same, all their children shall prove wise and none otherwise. But the whilst, this signification of nature is very universal and confused, and the understanding contents not itself, nor staieth until it conceive the particular discourse and the latest cause, and so it behoves to search out another signification of this name 'nature' which may be more agreeable to our purpose.

Aristotle and other natural philosophers descend into more particularities and call nature whatsoever substantial form which gives the being to anything and is the original of all the working thereof, in which signification our reasonable soul may reasonably be termed nature, for from her we receive our formal being, which we have of being men, and the self same is the beginning of whatsoever we do and work. But all souls being of equal perfection (as well that of the wiser, as that of the foolish) it cannot be affirmed that nature in this signification is that which makes a man able, for if this were true, all men should have a like measure of wit and wisdom. And therefore the same Aristotle found out another signification of nature, which is the cause that a man is able or unable, saying that the temperature of the four first qualities (hot, cold, moist, and dry) is to be called nature, for from this issue all the habilities of man, all his virtues and vices,[21] and this great variety of wits which we behold. And this is clearly proved by considering the age of a man when he is wisest, who in his childhood is no more than a brute beast, and useth none other powers than those of anger and concupiscence. But coming to youth, there begins to shoot out in him a marvellous wit, and we see that it lasteth till time certain, and no longer, for old age

---

[21] 8. *Index*: 'all his virtues and vices' [censored].

growing on, he goes every day losing his wit until it come to be quite decayed.

The variety of wits it is a matter certain that it springs not from the reasonable soul, for that is one self in all ages without having received in his forces and substance any alteration. But man hath in every age a diverse temperature and a contrary disposition by means whereof the soul doth other works in childhood, other in youth, and other in old age. Whence we draw an evident argument, that one self soul, doing contrary works in one self body, for that it partakes in every age a contrary temperature, when of young men, the one is able, and the other inapt, this grows for that the one of them enjoys a diverse temperature from the other. And this (for that it is the beginning of all the works of the reasonable soul) was by the physicians and the philosophers termed nature, of which signification this sentence is properly verified that nature makes able.

For confirmation of this doctrine, Galen wrote a book[22] wherein he proveth that the manners of the soul follow the temperature of the body in which it keeps residence, and that by reason of the heat, the coldness, the moisture, and the drought, of the territory where men inhabit, of the meats which they feed on, of the waters which they drink, and of the air which they breath, some are blockish, and some wise; some of worth, and some base; some cruel, and some merciful; many straight-brested, and many large; part liars, and part true speakers; sundry traitors, and sundry faithful; somewhere unquiet, and somewhere stayed; there double, here single; one pinching, another liberal; this man shamefast, that shameless; such hard, and such light of belief. And to prove this, he cites many places of Hippocrates, Plato, and Aristotle, who affirm that the difference of nations, as well in composition of the body, as in conditions of the soul, springeth from the variety of this temperature. And experience itself evidently showeth this, how far are different Greeks from Tartarians, Frenchmen from Spaniards, Indians from Dutch,° and Ethiopians from English. And this may be seen, not only in countries so far distant, but if we consider the provinces that environ all Spain, we may depart the virtues and vices which we have recounted amongst the inhabitants, giving each one his peculiar vice and virtue. And if

---

[22] a book] *Quod animi mores corporis temperaturam insequantur* [That Character Follows Bodily Temperament].

we consider the wit and manners of the Catalonians, Valencians, Murcians, Granadians, Andalusians, Extremadurans, Portuguese, Galicians, Asturians, Montagneses,[23] Biscayan, Navarrese, Aragonese, and of the kingdom of Castile, who sees not and knows not how far these are different amongst themselves, not only in shape of countenance and in feature of body, but even in the virtues and vices of the soul. Which all grows for that every of these provinces hath his particular and different temperature. And this variety of manners is known not only in countries so far off, but in places also that are not more than a little league in distance, it cannot be credited what odds there is found in the wits of the inhabitants. Finally, all that which Galen writeth in this his book is the ground-plot of this my treatise, albeit he declares not in particular the differences of the habilities which are in men, neither as touching the sciences which every one requires in particular. Notwithstanding, he understood that it was necessary to depart the sciences amongst young men, and to give each one that which to his natural hability was requisite, inasmuch as he said that well-ordered commonwealths ought to have men of great wisdom and knowledge who might in their tender age discover each one's wit and natural sharpness, to the end they might be set to learn that art which was agreeable and not leave it to their own election.

CHAP. III.

What part of the body ought to be well-tempered that a young man may have hability.

Man's body hath so many varieties of parts and powers (applied each to his end) that it shall not stray from our purpose but rather grows a matter of necessity to know first what member was ordained by nature for the principal instrument to the end man might become wise and advised. For it is a thing apparent that we discourse not with our foot, nor walk on our head, nor see with our nostrils, nor hear with our eyes, but that every of these

---

[23] Montagneses] in Spanish *montañeses*, i.e. the people of Cantabria, a region in the north of Spain between Asturias and the Basque Country.

parts hath his use and particular disposition for the work which it is to accomplish.

Before Hippocrates and Plato came into the world, it held for a general conceit amongst the natural philosophers that the heart was the principal part where the reasonable faculty made his residence, and the instrument wherewith the soul wrought the works of wisdom, of diligence, of memory, and of understanding. For which cause, the divine Scripture (applying itself to the ordinary speech of those times) in many places calls the heart the sovereign part of a man. But these two grave philosophers coming into the world gave evidence that this opinion was false, and proved by many reasons and experiments that the brain is the principal seat of the reasonable soul, and so they all gave hands° to this opinion, save only Aristotle, who (with a purpose of crossing Plato in all points) turned to revive the former opinion, and with topical° places to make it probable (with which of these opinions the truth swaieth, time serveth not now to discuss). For there is none of these philosophers that doubteth but that the brain is the instrument ordained by nature to the end that man might become wise and skilful, it sufficeth only to declare with what conditions this part ought to be endowed so as we may affirm that it is duly instrumentalized, and that a young man in this behalf may possess a good wit and hability.

Four conditions the brain ought to enjoy to the end the reasonable soul may therewith commodiously perform the works which appertain to understanding and wisdom. The first, good composition; the second, that his parts be well united; the third, that the heat exceed not the cold, nor the moist the dry; the fourth, that his substance be made of parts subtile and very delicate.

In the good composition are contained other four things: the first is good figure; the second, quantity sufficient; the third, that in the brain the four ventricles be distinct and severed, each duly bestowed in his seat and place; the fourth, that the capableness of these be neither greater nor less than is convenient for their workings.

Galen collects the good figure of the brain by an outward consideration, namely, the form and disposition of the head, which he sayth ought to be such as it should be if taking a perfect round ball of wax, and pressing it together somewhat on the sides, there will remain (after that manner) the forehead and the nape with a little bunchiness.° Hence it follows that the man who hath his forehead very plain, and his niddick[24] flat, hath not his brain so figured as is requisite for wit and hability. The quantity of the brain which the soul needeth to discourse and consider is a matter that breeds fear, for amongst all the brute beasts there is none found to have so much brain as a man, in sort, as if we join those of two the greatest oxen together, they will not equal that of one only man, be he never so little. And that whereto behoves more consideration is that amongst brute beasts, those who approach nearest to man's wisdom and discretion (as the ape, the fox, and the dog) have a greater quantity of brain than the other, though bigger bodied than they. For which cause, Galen said that a little head in any man is ever faulty because that it wanteth brain. Notwithstanding, I avouch that if his having a great head proceedeth from abundance of matter, and ill tempered, at such time as the same was shaped by nature, it is an evil token, for the same consists all of bones and flesh and contains a small quantity of brain, as it befalls in very big oranges, which opened, are found scarce of juice and hard of rind. Nothing offends the reasonable soul so much as to make his abode in a body surcharged with bones, fat, and flesh. For which cause Plato said that wise men's heads are ordinarily weak, and upon any occasion are easily annoyed, and the reason is for that nature made them of an empty skull with intention not to offend the wit by compassing it with much matter. And this doctrine of Plato is so true that albeit the stomach abides so far distant from the brain, yet the same works it offence when it is replenished with fat and flesh. For confirmation hereof, Galen allegeth a proverb which sayth 'a gross belly makes a gross understanding', and that this proceeds from nothing else than that the brain and the stomach are united and chained together with certain sinews by way of which they interchangeably communicate their damages. And contrariwise, when the stomach is dry and shrunk, it affords great aid to the wit, as we see in the hunger-starved,° and such as are driven to their shifts, on which doctrine (it may be) Persius founded himself when he said that the belly is that which quickens up the wit. But the thing most pertinent to be noted for this purpose is that if the other parts of the body be fat and fleshy, and therethrough a man grows over gross, Aristotle says it makes him to leese° his wit. For

---

[24] niddick] the nape of the neck.

which cause I am of opinion that if a man have a great head (albeit the same proceed for that he is endued with a very able nature, and that he is furnished with a quantity of well-tempered matter), yet he shall not be owner of so good a wit as if the same held a meaner size.

Aristotle is of a contrary opinion whilst he inquires for what cause a man is the wisest of all living creatures, to which doubt he answers that you shall find no creature which hath so little a head as man respecting withal the greatness of his body. But herein he swarved from reason, for if he had opened some man's head, and viewed the quantity of his brain, he should have found that two horses together had not so much brain as that one man. That which I have gathered by experience is that in little men it is best that the head incline somewhat to greatness, and in those who are big bodied, it proves best that they be little. And the reason is for that after this sort there is found a measurable quantity with which the reasonable soul may well perform his working.

Besides this, there are needful the four ventricles in the brain to the end the reasonable soul may discourse and philosophize: one must be placed on the right side of the brain, the second on the left, the third in the middle of these, and the fourth in the part behind the brain. Whereunto these ventricles serve and their large or narrow capableness for the reasonable soul, all shall be told by us a little hereafter when we shall entreat of the diversities of men's wits.

But it sufficeth not that the brain possess good figure, sufficient quantity, and the number of ventricles by us aforementioned with their capableness, great or little, but it behoves also that his parts holds a certain kind of continuedness[25] and that they be not divided. For which cause, we have seen in hurts of the head that some men have lost their memory, some their understanding, and others their imagination. And put case that after they have recovered their health, the brain reunited itself again, yet this notwithstanding, the natural union was not made which the brain before possessed.

The third condition of the fourth principal was that the brain should be tempered with measurable heat and without excess of

---

[25] continuedness] continuity.

the other qualities, which disposition we said heretofore that it is called good nature, for it is that which principally makes a man able and the contrary unable.

But the fourth (namely, that the brain have his substance or composition of subtle and delicate parts) Galen sayth is the most important of all the rest. For when he would give a token of the good disposition of the brain, he affirmeth that a subtle wit showeth that the brain is framed of subtle and very delicate parts, and if the understanding be dull, it gives evidence of a gross substance, but he makes no mention of the temperature. These conditions the brain ought to be endowed withal to the end the reasonable soul may therethrough shape his reasons and syllogisms. But here encounters us a difficulty very great, and this is that if we open the head of any beast, we shall find his brain composed with the same form and manner as a man's, without that any of the fore-reported conditions will be failing. Whence we gather that the brute beasts have also the use of prudence and reason by means of the composition of their brain, or else that our reasonable soul serves not itself of this member for the use of his operations, which may not be avouched. To this doubt, Galen answereth in this manner: 'Amongst the kinds of beasts it is doubted whether that which is termed unreasonable be altogether void of reason or not. For albeit the same want that which consists in voice (which is named "speech") yet that which is conceived in the soul and termed "discourse" of this it may be that all sorts of beasts are partakers, albeit the same is bestowed more sparingly upon some, and more largely on other some. But verily, how far man in the way of reason outgoeth all the rest, there is none who maketh question.' By these words Galen gives us to understand (albeit with some fearfulness) that brute beasts do partake reason, one more, and another less, and in their mind do frame some syllogisms and discourses though they cannot utter them by way of speech. And then the difference between them and man consisteth in being more reasonable and in using prudence with greater perfection.

The same Galen proves also by many reasons and experiments that asses (being of all brute beasts the bluntest) do arrive with their wit to the most curious and nice points which were devised by Plato and Aristotle, and thereon he collects[26] saying: 'I am

---

[26] collects] sums up

therefore so far from praysing the ancient philosophers in that they have found out some ample matter and of rare invention as when they say we must hold that there is self and diverse: one, and not one, not only in number, but also in kind. As I dare boldly affirm that even the very asses, who notwithstanding seem most blockish of all beasts, have this from nature.'

This self same meant Aristotle when he inquired the cause why man amongst all living creatures is wisest. And in another place he turns to doubt for what cause man is the most unjust of all living creatures, in which he gives us to understand the self same which Galen said, that the difference which is found between man and brute beast is the self same which is found between a fool and a wise man, which is nought else than in respect of the more and the less.[27] This truly is not to be doubted that brute beasts enjoy memory and imagination, and another power which resembles understanding, as the ape is very like a man and that his soul takes use of the composition of the brain it is a matter apparent, which being good, and such as is behooful,° performs his works very well and with much prudence, and if the brain be ill instrumentalized, it executes the same untowardly. For which cause we see that there be asses which in their knowledge are properly such, and others again are found so quick conceited and malicious that they pass the property of their kind. And amongst horses are found many jadishnesses° and good qualities, and some there are more trainable than the rest, all which grows from having their brain well or ill instrumentalized. The reason and solution of this doubt shall be placed in the chapter which followeth, for there we return to reason anew of this matter.

There are in the body some other parts from whose temperature as well the wit as the brain depend, of which we will reason in the last chapter of this work. But besides these and the brain, there is found in the body another substance whose service the reasonable soul useth in his operations and so requireth the three last qualities which we have assigned to the brain, that is, quantity sufficient, delicate substance, and good temperature. These are the vital spirits and arterial blood which go wandering through the whole body and remain evermore united to the imagination following his contemplation. The office of this spiritual substance

---

[27] 9. *Index*: 'Whence we gather that the brute beasts [143] [...] than in respect of the more and the less' [censored].

is to stir up the powers of man and to give them force and vigour that they may be able to work. This shall evidently be known to be their manner if we take consideration of the motions, of the imaginations, and of that which after succeeds in working. For if a man begin to imagine upon any injury that hath been profered him, the blood of the arteries runs suddenly to the heart and stirs up the wrathful part and gives the same heat and forces for revenge.

If a man stand contemplating any fair woman, or stay in giving and receiving by that imagination touching the venerious act, these vital spirits run forthwith to the genital members and raise them to the performance. The like befalls when we remember any delicate and savoury meat, which one called to mind, they straight abandon the rest of the body and fly to the stomach and replenish the mouth with water. And this their motion is so swift that if a woman with child long for any meat whatsoever and still retain the same in her imagination, we see by experience that she loseth her burthen if speedily it be not yielded unto her. The natural reason of this is because these vital spirits, before the woman conceived this longing, made abode in the belly, helping her there to retain the creature, and through this new imagination of eating, they hie to the stomach to raise the appetite, and in this space, if the belly have no strong retentive, it cannot sustain the same, and so by this means she leeseth her burthen.

Galen understanding this condition of the vital spirits, counsaileth physicians that they give not sick folk to eat when their humours are raw and upon digestion, for when they first feel the meat in the stomach, they straightways abandon the work about which before they were occupied and come thereunto to help it. The like benefit and aid the brain receives of these vital spirits when the reasonable soul is about to contemplate, understand, imagine, or perform actions of memory without which it cannot work. And like as the gross substance of the brain and his evil temperature brings the wit to confusion, so the vital spirits and the arterial blood (not being delicate and of good temperature) hinder in a man his discourse and use of reason. Wherefore Plato said that the suppleness and good temperature of the heart makes the wit sharp and quick-sighted (having proved before that the brain and not the heart is the principal seat of the reasonable soul). And the reason is because these vital spirits are engendered in the heart and partake of that substance and that temperature which rested in that which formed them. Of this arterial blood Aristotle meant when he said that those men are well compounded who have their blood hot, delicate, and pure, for they are also of good bodily forces and of a wit well disposed. These vital spirits are by the physicians termed 'nature', for they are the principal instrument with which the reasonable soul performeth his works, and of these also may that sentence be verified, nature makes able.

CHAP. IV.

It is proved that the soul vegetative, sensitive, and reasonable have knowledge without that anything be taught them if so be that they possess that convenient temperature which is requisite for their operation.

The temperature of the four first qualities, which we heretofore termed nature, hath so great force to cause that (of plants, brute beasts, and man) each one set himself to perform those works which are proper to his kind, that they arrive to that utmost bound of perfection which may be attained suddenly and without any others teaching them. The plants know how to form roots underground, and by way of them to draw nourishment to retain it, to digest it, and to drive forth the excrements. And the brute beasts likewise so soon as they are born know that which is agreeable to their nature, and fly the things which are naughty and noisome. And that which makes them most to marvel who are not seen in natural philosophy is that a man having his brain well tempered and of that disposition which is requisite for this or that science, suddenly and without having ever learned it of any, he speaketh and uttereth such exquisite matters as could hardly win credit. Vulgar philosophers, seeing the marvellous works which brute beasts perform, affirm it holds no cause of marvel because they do it by natural instinct, inasmuch as nature showeth and teacheth each in his kind what he is to do. And in this they say very well, for we have already alleged and proved that nature is nothing else than this temperature of the four first qualities, and that this is the schoolmaster who teacheth the souls in what sort they are to work. But they term instinct of nature a certain mass of things which rise from the niddick upward, neither could they ever expound or give us to understand what it is.[28] The grave philosophers (as Hippocrates, Plato and Aristotle) attribute all these marvellous works to heat, cold, moisture, and drought, and this they affirm of the first principle and pass no farther. And if you ask who hath taught the brute beasts to do these works which breed us such marvel and men to discourse with reason, Hippocrates answereth 'it is the natures of them all without any teacher'; as if he should say: 'The faculties or the temperature of which they consist are all given them without being taught by any other.' Which is clearly discerned if they pass on to consider the works of the soul vegetative, and of all the rest which govern man, who if it have a quantity of man's seed, well digested and seasoned with good temperature, makes a body so seemly and duly instrumentalized that all the carvers in the world cannot shape the like.

For which cause Galen wondering to see a frame so marvellous, the number of his several parts, the seating, the figure, and the use of each one by itself, grew to conclude it was not possible that the vegetative soul, nor the temperature, could fashion a workmanship so singular but that the author thereof was God, or some other most wise understanding. But this manner of speech is already by us heretofore refuted:[29] for it beseems not natural philosophers to reduce the effects immediately to God, and so to slip over the assigning of the second reasons, and especially in this case, where we see by experience that if man's seed consist of an evil substance, and enjoy not a temperature convenient, the vegetative soul runs into a thousand disorders. For if the same be cold and moist more than is requisite, Hippocrates sayth that the men prove eunuchs, or hermaphrodites; and if it be very hot and dry, Aristotle sayth that it makes them curl-pated,° crook-legged, and flat-nosed as are the Ethiopians, and if it be moist, the same Galen sayth that they grow long and lithy; and if it be dry, low of stature. All this is a great defect in mankind, and for such works we find little cause to give nature any commendation, or to hold her for advised, and if God were the author hereof, none of these qualities could divert him. Only the first men which the world

---

[28] 10. *Index*: 'and that this is the schoolmaster [27] [...] to understand what it is' [censored].

[29] 11. *Index*: 'But this manner of speech is already by us heretofore refuted' [censored].

possessed Plato affirms were made by God, but the rest were born answerable to the discourse of the second causes, which if they be well-ordered, the vegetative soul doth well perform his operations, and if they concur not in sort convenient, it produceth a thousand damageable effects.[30]

What the good order of nature for this effect must be is that the vegetative soul have an endowment of a good temperature, or else, let Galen and all the philosophers in the world answer me: what the cause is that the vegetative soul possesseth such skill and power in the first age of man to shape his body, and to increase and nourish the same, and when old age groweth on can yield the same no longer? For if an old man leese° but a tooth, he is past remedy of recovering another, but if a child cast them all, we see that natures return to renew them again. Is it then possible that a soul which hath done nought else in all the course of life than to receive food retain the same, digest it, and expel the excrements, new begetting the parts which fail should towards the end of life forget this, and want ability to do the same any longer? Galen for certain will answer that this skill and hability of the vegetative soul in youth springs from his possessing much natural heat and moisture, and that in age the same wants skill and power to perform it by means of the coldness and dryness to which a body of those years is subject. The knowledge of the sensitive soul takes his dependence also from the temperature of the brain, for if the same be such as his operations require that it should be, it can perform with due perfection. Otherwise, the same must also err no less than the soul vegetative. The manner which Galen held to behold and discern by eyesight the wisdom of the sensitive soul was to take a young kid, but newly kidded, which set on the ground begins to go (as if it had been told and taught that his legs were made to that purpose) and after that, he shakes from his back the superfluous moisture which he brought with him from his mother's belly, and lifting up the one foot, scrapes behind his ear, and setting before him sundry platters with wine, water, vinegar, oil, and milk, after he hath smelt them all, he fed only on that of milk. Which being beheld by diverse philosophers there present, they all with one voice cried out that Hippocrates had great reason to say that souls were skilful without the instruction

---

[30] 12. *Index*: 'and if God were the author hereof [64] [...] it produceth a thousand damageable effects' [censored].

of any teacher.[31] But Galen held not himself contented with this one proof, for two months after he caused the same kid, being very hungry, to be brought into the field, where smelling at many herbs, he did eat only those whereon goats accustomably feed.

But if Galen, as he set himself to contemplate the demeanour of this kid, had done the like with three or four together, he should have seen some gone better than other some, shrug themselves better, scratch better, and perform better all the other actions which we have recounted. And if Galen had reared two colts, bred of one horse and mare, he should have seen the one to pace with more grace than the other, and to gallop and stop better, and show more fidelity. And if he had taken an eyrie of falcons, and manned them, he should have found the first good of wing, the second good of prey, and the third ravening and ill-conditioned. The like shall we find in hounds, who being whelps of the same litter, the one for perfection of hunting will seem to want but speech, and the other have no more inclination thereunto than if he had been engendered by a herdman's bandog.

All this cannot be reduced to those vain instincts of nature which the philosophers feign. For if you ask for what cause one dog hath more instinct than another, both coming of one kind, and whelps of one sire, I cannot conjecture what they may answer save to fly back to their old leaning post, saying that God hath taught the one better than the other and given him a more natural instinct. And if we demand the reason, why this good hound, being yet but a whelp, is a perfect hunter, and growing in age hath no such sufficiency, and contrariwise, another being young cannot hunt at all and waxing old is wily and ready, I know not what they can yield in reply. Myself at least would say that the towardly hunting of one dog more than another grows from the better temperature of his brain, and again, that his well hunting whilst he is young and his decay in age is occasioned by means that in one age he partakes the temperature which is requisite to the qualities of hunting and in the other not. Whence we infer that sithence the temperature of the four first qualities is the reason

---

[31] 13. *Index*: 'Hippocrates had great reason to say that souls were skilful without the instruction of any teacher' [censored]. The censor suggested as an alternative to the supression of the previous sentence adding in the margins this note: 'La sentencia de estos filósofos es falsa' ('the judgment of these philosophers is false').

and cause for which one brute beast better performs the works of his kind than another, that this temperature is the schoolmaster which teacheth the sensitive soul what it is to do.

And if Galen had considered the demeanour and voyages of the ant, and noted his prudence, his mercy, his justice,[32] and his government, he would have taken astonishment to see a beast so little endowed with so great sageness without the help of any master or teacher to instruct him. But the temperature which the ant hath in his brain, being known, and how aptly it is appropriated to wisdom (as hereafter shall be shown) this wonderment will cease and we shall conceive that brute beasts with the temperature of their brain and the phantasms[33] which enter thereinto by the five senses, make such discourses[34] and partake those abilities which we do so note in them. And amongst beasts of one kind he which is most schoolable° and skilful is such because he hath his brain better tempered, and if through any occasion or infirmity the temperature of his brain incur alteration, he will suddenly leese his skill[35] and ability, as men also do.

But now we are to treat of a difficulty touching the reasonable soul, which is in what sort he hath this natural instinct for the operations of his kind (namely, sapience and prudence) and how on the sudden, by means of his good temperature, a man can be skilled in the sciences without the instruction of any other: seeing experience telleth us that if they be not gotten by learning, no man is at his birth endowed with them.

Between Plato and Aristotle, there is a weighty question as touching the verifying the reason or cause from whence the wisdom of man may spring. One sayth that the reasonable soul is more ancient than the body, for that before such time as nature endowed the same with these instruments, it made abode in heaven, in the company of God, whence it issued full of science and sapience. But when it entered to form this matter through the evil temperature which it found therein, it forewent the whole, until by process of time this ill temperature grew to amendment and there succeeded another instead thereof, with which (as more appliable to the sciences it had lost) it grew by little and little to

---

[32] 14. *Index*: 'his prudence, his mercy, his justice' [censored].
[33] phantasms] mental images, the immediate objects of sense perception.
[34] 15. *Index*: 'make such discourses' [censored].
[35] 16. *Index*: 'skill' [censored].

call that to remembrance which before it had forgotten. This opinion is false, and I much marvel that Plato being so great a philosopher could not render the reason of man's wisdom, considering that brute beasts have their prudencies[36] and natural habilities without that their soul departs from their body, or sties up[37] to heaven to learn them. In which regard he cannot go blameless, especially having read in Genesis (whereto he gave so great credit) that God instrumentalized the body of Adam before he created his soul. The self-same befalls also now, save that it is nature who begets the body, and in the last disposing thereof, God createth the soul in the same body without that it be sundered therefrom any time or moment.

Aristotle took another course affirming that every doctrine and every discipline comes from a foregoing knowledge, as if he would say: 'All that which men know and learn springs from that they have heard the same, seen it, smelt it, tasted it, or felt it. For there can grow no notice in the understanding which hath not first taken passage by some of the five senses.' For which cause he said that these powers issue out of the hands of nature as a plain table in which is no manner of painting, which opinion is also false as well as that of Plato. But that we may the better prove and make the same apparent, it behoves first to agree with the vulgar[38] philosophers that in man's body there rests but one soul, and that the same is reasonable, which is the original of whatsoever we do or effect. Albeit there are opinions, and there want not, who against this defend that in company of the reasonable soul there are associated some two or three more.[39]

This then standing thus in the works which the reasonable soul performs, as it is vegetative, we have already proved that the same knows how to shape man and to give him the figure which he is to keep, and knows likewise how to receive nourishment, to retain it, to digest it, and to expel the excrements, and if any part of the body do fail, she knows how to supply the same anew, and yield it that composition agreeable to the use which it is to hold. And

---

[36] 17. *Index*: 'have their prudencies' [censored].
[37] sties up] ascends
[38] 18. *Index*: 'vulgar' [censored].
[39] 19. *Index*: 'Albeit there are opinions, and there want not, who against this defend that in company of the reasonable soul there are associated some two or three more' [censored].

in the works of the sensitive and motive, the child so soon as it is born knows to suck and fashion his lips to draw forth the milk, and this so readily as not the wisest man can do the like. And herewithal, it assures the qualities which are incident to the preservation of his nature, shuns that which is noisome and damageable thereunto, knows to weep and laugh without being taught by any. And if this be not so, let the vulgar philosophers tell me awhile who hath taught the children to do these things, or by what sense they have learned it. Well I know they will answer that God hath given them this natural instinct as to the brute beasts, wherein they say not ill if the natural instinct be the self same with the temperature.

The proper operations of the reasonable soul, namely, to understand, to imagine and to perform actions of memory, a man cannot do them forthwith so soon as he is born, for the temperature of infancy serveth very unfitly therefore, and is merely appropriate to the vegetative and sensitive, as that of old age is appropriate to the reasonable soul and contrary to the vegetative and sensitive. And if as the temperature which serves for prudence is gotten in the brain by little and little, so the same could all be joined together at one instant, man should on the sudden have better skill to discourse and play the philosopher than if he had attained the same in the schools.

But because nature cannot perform this save by process of time, a man grows to gather wisdom by little and little, and that this is the reason and cause thereof is manifestly proved if we consider that a man after he hath been very wise grows by little and little into folly, for that he daily goes (till his decrepit age) accruing a contrary temperature. I for mine own part am of opinion that if nature, as she hath made man of seed hot and moist (and this is the temperature which directs the vegetative, and the sensitive, what they are to effectuate) so she had made him of seed cold and dry, even after his birth, he should straightways have been able to discourse and reason, and not have attended to suck inasmuch as this is the temperature agreeable to these operations. But for that we find by experience that if the brain have the temperature requisite for natural sciences, he hath no need of a master to teach him,[40] it falls out necessary that we mark one thing, which is that if a man fall into any disease by which his brain upon a sudden

---

[40] 20. *Index*: 'he hath no need of a master to teach him' [censored].

changeth his temperature (as are madness, melancholy, and frenzy) it happens that at one instant he leeseth, if he were wise, all his knowledge, and utters a thousand follies; and if he were a fool, he accrues more wit and ability than he had before.

I can speak of a rude country fellow who becoming frantic made a very eloquent discourse in my presence recommending his well doing to the bystanders, and that they should take care of his wife and children (if it pleased God to call him away in that sickness) with so many flowers of rhetoric and such apt choice of words as if Cicero had spoken in the presence of the Senate. Whereat the beholders marvelling asked me whence so great eloquence and wisdom might grow in a man who in his health time could scantly speak. And I remember I made answer that the art of oratory was a science which springs from a certain point or degree of heat, and that this country fellow, before sound, had by means of this infirmity attained thereunto.

I can also speak of another frantic person who for the space of more than eight days never uttered [a] word which I found not to carry his just quantity, and mostly he made couplements° of verses very well composed, whereat the bystanders wondering to hear a man speak in verse who in his health had never so much skill, I said it seldom fell out that he who was a poet in his health time should be so also in his sickness. For the temperature of the brain, by which when a man is whole, he becometh a poet, in sickness altereth and brings forth contrary operations. I remember that the wife of this frantic fellow and a sister of his named Margaret reproved him because he spoke ill of the saints, whereat the patient growing impatient said to his wife these words: 'I renounce God for the love of you, and St Mary for the love of Margaret, and St Peter for the love of John of Olmedo', and so he ran through a bead-roll of many saints whose names had consonance° with the other bystanders there present.[41]

---

[41] 'I renounce God for the love of you, and St Mary for the love of Margaret, and St Peter for the love of John of Olmedo'] This sentence rhymes in Spanish: 'Pues reniego de Dios, por amor de vos, y de sancta Maria, por amor de marigarcia, y de sāt Pedro, por amor de Juā d olmedo' (1575, fol. 52ʳ). Hence the reference later on in the sentence to the 'consonance' (OED, 'agreement of sounds; pleasing combination of sounds') of the names, which is lost in the translation into English because Carew — as Camilli before him (1586, C8ʳ) — decided to render the quotation word for word at the expense of the rhyme instead of privileging the rhyme and making up a new bead-roll of saint names.

But this is nothing and a matter of small importance in respect of the notable speeches uttered by a page of one of the great ones of this realm whilst he was mad, who in his health was reputed a youth of slender capacity, but falling into this infirmity he delivered such rare conceits, resemblances, and answers to such as asked him, and devised so excellent manners of governing a kingdom (of which he imagined himself to be sovereign) that for great wonder people flocked to see him and hear him, and his very master scarcely ever departed from his bed-head, praying God that he might never be cured. Which afterwards plainly appeared, for being recovered, his physician (who had healed him) came to take leave of his lord with a mind to receive some good reward, if of nothing else, yet at least in good words. But he encountered this greeting: 'I promise you master doctor that I was never more aggrieved at any ill success than to see this my page recovered, for it was not behooful° that he should change so wise folly for an understanding so simple as is this which in his health he enjoyed. Methinks that of one who tofore was wise and well-advised, you have made him a fool again, which is the greatest misery that may light upon any man.' The poor physician seeing how little thankfully his cure was accepted went to take leave of the page, who, amongst many other words that passed between them, told him this: 'Master doctor, I kiss your hands for so great a benefit bestowed on me in restoring mine understanding, but I assure you on my faith that in some sort it displeaseth me to have been cured. For whilst I rested in my folly, I led my life in the deepest discourses of the world, and imagined myself so great a lord as there reigned no king on the earth who was not my vassal, and were this a jest or a lie, what imported that whilst I conceived thereof so great a contentment as if it had been true? I rest now in far worse case, finding myself in truth to be but a poor page, and tomorrow I must begin again to serve one who whilst I was in mine infirmity I would have disdained for my footman.'

It skills not much whether the philosophers admit all this and believe that it may be so or not, but what if I should prove by very true stories that ignorant men stricken with this infirmity have spoken Latin, which they never learned in their health, and that a frantic woman told all persons who came to visit her their virtues and vices, and sometimes reported matters (with that assurance which they use to give who speak by conjectures and tokens), and for this cause, none almost durst come in to visit her, fearing to hear of those true tales which she would deliver? And (which is more to be marvelled at) when a barber came to let her blood: 'Friend, quoth she, have regard what you do, for you have but few days to live, and your wife shall marry such a man.' And this, though spoken by chance, fell out so true as it took effect before half a year came to an end.

Methinks I hear them who fly natural philosophy to say that this is a soul leasing and that (put case it were true) the devil as he is wise and crafty by God's sufferance entered into this woman's body and into the rest of those frantic persons whom I have mentioned, and caused them to utter those strange matters, and yet even to confess this, they are very loath, for the devil foreknoweth not what is to come because he hath no prophetical spirit. They hold it a very sufficient argument to avouch: 'This is false because I cannot conceive how it may be so.' As if difficult and quaint matters were subject to blunt wits, and came within the reach of their capacities. I pretend not hereby to take those to task who have defect of understanding, for that were a bootless labour, but to make Aristotle himself confess that men endowed with the temperature requisite for such operations may conceive many things without having received thereof any particular perseverance or learned the same at the hands of any other. 'Sundry also because this heat is a neighbour to the seat of the mind are wrapped in the infirmity of sottishness, or are heated by some furious instinct whence grew the sibyls and bacchants, and all those who men think are egged on by some divine inspiration, whereas this takes his original not from any disease but from a natural distemperature. Marcus, a citizen of Syracuse, was excellentest poet after he lost his understanding, and those in whom this abated heat approacheth least to mediocrity are verily altogether melancholic, but thereby much the wiser.' In these words Aristotle clearly confesseth that when the brain is excessively heated, many thereby attain the knowledge of things to come (as were the sibyls) which Aristotle sayth grows not by reason of any disease but through the inequality of the natural heat, and that this is the very reason and cause thereof, he proves apparently by an example, alleging that Mark, a citizen of Syracuse, was a poet in most excellency at such time as through excessive heat of the brain he fell besides himself, and when he returned to a more moderate temperature, he lost his versifying, but yet remained more wise and advised. Insomuch that Aristotle not only admits the temperature of the brain for the principal occasion of these extravagant successes, but also reproves them who hold the same for a divine revelation and no natural cause.

The first who termed these marvellous matters by the name of divinesse° was Hippocrates: 'And that if any such point of divinesse be found in the disease that it manifesteth also a providence.' Upon which sentence he chargeth physicians that if the diseased utter any such divine matters, they may thereby know in what case she rests and prognosticate what will become of him. But that which in this behalf drives me to most wonder is that demanding of Plato how it may come pass that of two sons begotten by one father, one hath the skill of versifying without any other teaching, and the other, toiling in the art of poetry can never beget so much as one verse, he answereth that he who was born a poet is possessed and the other not. In which behalf, Aristotle had good cause to find fault with him, for that he might have reduced this to the temperature as elsewhere he did.

The frantic person's speaking of Latin, without that he ever learned the same in his health time, shows the consonance which the Latin tongue holds with the reasonable soul, and (as we will prove hereafter) there is to be found a particular wit appliable to the invention of languages and Latin words, and the phrases of speech in that tongue are so fitting with the ear, that the reasonable soul possessing the necessary temperature for the invention of some delicate language suddenly encounters with this. And that two devisers of languages may shape the like words (having the like wit and hability) it is very manifest, presupposing that when God created Adam and set all things before him, to the end he might bestow on each his several name whereby it should be called, he had likewise at that instant molded another man with the same perfection and supernatural grace. Now I demand: if God had placed the same things before this other man that he might also set them names whereby they should be called, of what manner those names should have been? For mine own part, I make no doubt but he would have given these things, those very names which Adam did. And the reason is very apparent, for both carried one self eye to the nature of each thing which of itself was no more but one. After this manner might the frantic person light upon the Latin tongue and speak the same without ever having learned it in his health, for the natural temperature of his brain, conceiving alteration, through the infirmity it might (for a space) become like his, who first invented the Latin tongue and feign the like words (but yet not with that concert and continued finesse, for this would give token that the devil moved that tongue, as the church teacheth her exorcists). This self, sayth Aristotle, befell some children who at their birth-time spoke some words very plainly and afterward kept silence, and he finds fault with the vulgar philosophers of his time, who, for that they knew not the natural cause of this effect, imputed it to the devil.

The cause why children speak so soon as they are born and after forthwith turn to hold their peace Aristotle could never find out though he went much about it, but yet it could never sink into his brain that it was a device of the devil's, nor an effect above nature, as the vulgar philosophers held opinion, who seeing themselves hedged in with the curious and nice points of natural philosophy, make them believe who know little that God or the devil are authors of the prodigious and strange effects of whose natural cause they have no knowledge and understanding.

Children which are engendered of seed cold and dry (as are those begotten in old age) some few days and months after their birth begin to discourse and philosophize, for the temperature cold and dry (as we will hereafter prove) is most appropriate to the operations of the reasonable soul, and that which process of time and many days and months should bring about is supplied by the present temperature of the brain, which for many causes anticipateth what it was to effect. Other children there are, sayth Aristotle, who as soon as they are born begin to speak, and afterwards hold their peace until they attain the ordinary and convenient age of speaking. Which effect floweth from the same original and cause that we recounted of the page, and of those furious and frantic persons, and of him who spoke Latin on a sudden without having learned it in his health. And that children whilst they make abode in their mother's belly, and so soon as they are born, may undergo these infirmities is a matter past denial. But whence that divining of the frantic woman proceeded, I can better make Cicero to conceive than these natural philosophers, for he describing the nature of man, said in this manner: 'The creature foresightful, searchful,° apt for many matters, sharp conceited, mindful, replenished with reason and counsel, whom we call by the name of man.' And in particular he affirmeth that there is found a certain nature in some men which in foreknowing things to come exceedeth other men, and his words are these: 'For there is found a certain force and nature which foretells things to come, the force and nature of which is not by reason to be unfolded.' The error of the natural philosophers consisteth in not considering (as Plato did) that man was made to the likeness of God, and that he is a partaker of his divine providence, and that [he has] the power of discerning all the three differences of time: memory for the past, conceiving for the present, and imagination and understanding for those that are to come. And as there are men superior to others in remembering things past, and others in knowing the present, so there are also many who partake a more natural hability for imagining of what shall come to pass. One of the greatest arguments which forced Cicero to think that the reasonable soul is uncorruptible was to see the certainty with which the diseased tell things to come, and especially when they are near their end. But the difference which rests between a prophetical spirit and this natural wit is that that which God speaks by the mouth of his prophets is infallible, for it is the express word of God, but that which man prognosticateth by the power of his imagination holds no such certainty.

Those who say that the discovering of their virtues and vices by the frantic woman to the persons who came to visit her was a trick of the devil's playing, let them know that God bestows on men a certain supernatural grace to attain and conceive which are the works of God and which of the devil. The which Saint Paul placed amongst the divine gifts, and calls it the imparting of spirits. Whereby we may discern whether it be the devil or some good angel that intermeddleth with us. For many times the devil sets to beguile us under the cloak of a good angel, and we have need of this grace and supernatural gift to know him and difference him from the good. From this gift they are farthest sundered who have not a wit capable of natural philosophy, for this science and that supernatural infused by God fall under one self ability, to weet,° the understanding at least, if it be true, that God in bestowing his graces do apply himself to the natural good of everyone, as I have afore rehearsed.[42]

Jacob lying at the point of death (at which time the reasonable soul is most at liberty to see what is to come) all his twelve children entered to visit him, and he to each of them in particular

---

[42] 21. *Index*: 'From this gift [473] [...] as I have afore rehearsed' [censored].

recited their virtues and vices and prophesied what should befall as touching them and their posterity.[43] Certain it is that he did all this inspired by God, but if the divine Scripture and our faith had not ascertained us hereof, how would these natural philosophers have known this to be the work of God, and that the virtues and vices which the frantic woman told to such as came to visit her were discovered by the power of the devil whilst this case in part resembles that of Jacob?[44]

They reckon that the nature of the reasonable soul is far different from that of the devil and that the powers thereof (understanding, imagination, and memory) are of another very diverse kind, and herein they be deceived. For if a reasonable soul inform° a well instrumentalized body (as was that of Adam) his knowledge comes little behind that of the subtillest devil, and without the body he partakes as perfect qualities as the other. And if the devils foresee things to come, conjecturing and discoursing by certain tokens, the same also may a reasonable man do when he is about to be freed from his body, or when he is endowed with that difference of temperature which makes a man capable of this providence. For it is a matter as difficult for the understanding to conceive how the devil can know these hidden things, as to impute the same to the reasonable soul. It will not fall in these men's heads that in natural things there may be found out certain signs by means of which they may attain to the knowledge of matters to come. And I affirm there are certain tokens to be found which bring us to the notice of things past and present, and to forecast what is to follow, yea, and to conjecture some secrets of the heaven: 'Therefore we see that his things invisible are understood by the creatures of the world by means of the things which have been created.' Whosoever shall have power to accomplish this shall attain thereunto, and the other shall be such as Homer spoke of: 'The ignorant understandeth the things passed but not the things to come.' But the wise and discreet is the ape[45] of God, for he imitates Him in many matters, and albeit he cannot accomplish them with so great perfection, yet he carries some resemblance unto Him by following Him.

---

[43] Jacob] Jacob was the third patriarch of the Hebrew people. Jacob's wives had twelve sons; the offspring of his sons became the tribes of Israel after the Exodus.
[44] 22. *Index*: 'whilst this case in part resembles that of Jacob?' [censored].
[45] ape] imitator.

CHAP. V.

It is proved that from the three qualities, hot, moist, and dry, proceed all the differences of men's wits.

The reasonable soul making abode in the body, it is impossible that the same can perform contrary and different operations if for each of them it use not a particular instrument. This is plainly seen in the power of the soul, which performeth diverse operations in the outward senses, for every one hath his particular composition: the eyes have one, the ears another, the smelling another, and the feeling another. And if it were not so, there should be no more but one sort of operations, and that should all be seeing, tasting, or feeling, for the instrument determines and rules the power for one action and for no more.

By this so plain and manifest a matter, which passeth through the outward senses, we may gather what that is in the inward. With this self power of the soul, we understand, imagine, and remember. But if it be true that every work requires a particular instrument, it behoveth of necessity that within the brain there be one instrument for the understanding, one for the imagination, and another different from them for the memory. For if all the brain were instrumentalized after one self manner, either the whole should be memory, or the whole understanding, or the whole imagination. But we see that these are very different operations, and therefore it is of force that there be also a variety in the instruments. But if we open by skill and make an anatomy of the brain, we shall find the whole compounded after one manner, of one kind of substance, and alike, without parts of other kinds, or a different sort. Only there appear four little hollownesses who (if we well mark them) have all one self composition and figure without anything coming between which may breed a difference.

What the use and profit of these may be, and whereto they serve in the head, is not easily decidable, for Galen and the anatomists, as well new as ancient, have laboured to find out the truth, but none of them hath precisely nor in particular expressed whereto the right ventricle serveth, nor the left, nor that which is placed in the middest of these two, nor the fourth, whose seat in the brain keeps the hinder part of the head. They affirm only (though with some doubt) that these four concavities are the shops where the vital spirits are digested and converted into animals, so to give sense and motion to all the parts of the body. In which operation, Galen said once that the middle ventricle was the principal, and in another place he unsays it again, affirming that the hindermost is of greatest efficacy and valure.

But this doctrine is not true, nor founded on good natural philosophy, for in all man's body there are not two so contrary operations, nor that so much hinder one another, as are discoursing and digestion of nourishment. And the reason is because contemplation requireth quiet, rest, and a clearness in the animal spirits; and digestion is performed with great stirring and travail, and from this action rise up many vapours which trouble and darken the animal spirits, so as by means of them the reasonable soul cannot discern the figures. And nature was not so unadvised as in one self place to conjoin two actions which are performed with so great repugnancy. But Plato highly commends the wisdom and knowledge of him who shaped us, for that he severed the liver from the brain by so great a distance, to the end that by the rumbling there made, whilst the nourishments are mingled, and by the obscureness and darkness occasioned through the vapours in the animal spirits, the reasonable soul might not be troubled in his discourses and considerations. But though Plato had not touched this point of philosophy, we see hourly by experience that because the liver and the stomach are so far from the brain, presently upon meat, and some space thereafter, there is no man that can give himself to study.

The truth of this matter is that the fourth ventricle hath the office of digesting and altering the vital spirits and to convert them into animal, for that end which we have before remembered. And therefore nature hath severed the same by so great a distance from the other three and made that brain sundered apart, and so far off (as appeareth) to the end that by his operation he hinder not the contemplation of the rest. The three ventricles placed in the forepart I doubt not but that nature made them to none other end than to discourse and philosophize. Which is apparently proved for that in great studyings and contemplations always that part of the head finds itself aggrieved which answereth these three concavities. The force of this argument is to be known by consideration that when the other powers are weary of performing their works, the instruments are always aggrieved whose service they used, as in our much looking the eyes are pained, and with much going, the soles of the feet wax sore.

Now the difficulty consists to know in which of these ventricles the understanding is placed, in which the memory, and in which the imagination, for they are so united and near neighboured that neither by the last argument, nor by any other notice, they can be distinguished or discerned. Then considering that the understanding cannot work without the memory be present (representing unto the same the figures and fantasies agreeable thereunto: 'It behoveth that the understanding part busy itself in beholding the phantasms'), and that the memory cannot do it if the imagination do not accompany the same (as we have already heretofore declared) we shall easily understand that all the powers are united in every several ventricle and that the understanding is not solely in the one, nor the memory solely in the other, nor the imagination in the third, as the vulgar philosophers have imagined, but that this union of powers is accustomably made in man's body, inasmuch as the one cannot work without the aid of the other, as appeareth in the four natural abilities (digestive, retentive, attractive, and expulsive) where, because each one stands in need of all the residue, nature disposed to unite them in one self place and made them not divided or sundered.

But if this be true, then to what end made nature those three ventricles and joined together the three reasonable powers in every of them, seeing that one alone sufficed to understand and to perform the actions of the memory?[46] To this may be answered that there riseth a like difficulty in scanning whence it commeth that nature made two eyes, and two ears, sithence in each of them is placed the whole power of sight and hearing, and we can see having but one eye. Whereto may be said that the powers ordained for the perfection of a creature, how much the greater number they carry, so much the better assured is that their perfection, for upon some occasion, one or two may fail, and therefore it serves well to the purpose that there remain some others of the same kind which may be applied to use.

In an infirmity which the physicians term resolution,[47] or palsy of the middle side, the operation is ordinarily lost of that ventricle which is stricken on that side, and if the other two remained not sound, and without endamageance,° a man should thereby

---

[46] 'To imagine' is missing in the enumeration in English, in Italian and in all Spanish editions.
[47] resolution] paralysis or paresis of a muscle.

become witless and void of reason. And yet for all this, by wanting that only ventricle, there is a great abatement discerned in his operations, as well in those of the understanding as of the imaginative and memory, as they shall also find in the loss of one sight who were wont to behold with two, whereby we clearly comprise that in every ventricle are all the three powers, sithence by the annoyance of any one, all the three are weakened. Seeing then all the three ventricles are of one self composition, and that there rests not amongst them any variety of parts, we may not leave to take the first qualities for an instrument and to make so many general differences of wits as they are in number. For to think that the reasonable soul (being in the body) can work without some bodily instrument to assist her is against all natural philosophy.[48] But of the four qualities (heat, cold, moisture, and drought), all physicians leave out cold as unprofitable to any operation of the reasonable soul wherethrough it is seen by experience in the other habilities that if the same mount above heat, all the powers of man do badly perform their operations, neither can the stomach digest his meat, nor the cods° yield fruitful seed, nor the muscles move the body, nor the brain discourse. For which cause, Galen said: 'Coldness is apparently noisome to all the offices of the soul.' As if he should say: 'Cold is the ruin of all the operations of the soul', only it serves in the body to temper the natural heat and to procure that it burn not overmuch. And yet Aristotle is of a contrary opinion where he affirmeth: 'It is a matter certain that that blood carrieth most forcible efficacy which is thickest and hottest, but the coldest and thinnest hath a more accomplished force to perceive and understand.' As if he would say: 'The thick and hot blood makes great bodily forces, but the pure and cold is cause that man possesseth great understanding.' Whereby we plainly see that from coldness springeth the greatest difference of wit that is in any man, namely, in the understanding.

Aristotle moreover moves a doubt, and that is why men who inhabit very hot countries (as Egypt) are more witty and advised than those who are born in cold regions. Which doubt he resolves in this manner: that the excessive heat of the country fretteth and consumeth the natural heat of the brain, and so leaves it cold,

---

[48] 23. *Index*: 'it behoveth of necessity that within the brain [18] [...] against all natural philosophy' [censored].

whereby man grows to be full of reasonableness. And that contrariwise, the much cold of the air fortifieth the much natural heat of the brain, and yields it not place to resolve. For which cause, sayth he, such as are very hot brained cannot discourse nor philosophize but are giddy-headed, and not settled in any one opinion. To which opinion it seems that Galen leaneth saying that the cause why a man is unstable and changeth opinion at every moment is for that he hath a hot brain. And contrariwise, his being stable and firm springs from the coldness of his brain. But the truth is that from this heat there groweth not any difference of wit, neither did Aristotle mean that the cold blood, by his predominance, did better the understanding but that which is less hot. True it is that man's variableness springs from his partaking of much heat, which lifts up the figures that are in the brain and makes them to boil, by which operation there are represented to the soul many images of things which invite him to their contemplation, and the soul to possess them all leaves one and takes another. Contrariwise, it befalls in coldness, which for that it imprints inwardly these figures and suffers them not to rise, makes a man firm in one opinion, and it proves so because none other presents itself to call the same away. Coldness hath this quality that it not only hindereth the motions of bodily things but also makes that the figures and shapes which the philosophers call spiritual be unmovable in the brain. And this firmness seemeth rather a negligence than a difference of hability. Alike true it is that there is found another diversity of firmness which proceeds from possessing an understanding well compacted together and not from the coldness of the brain. So there remain drought, moisture, and heat for the service of the reasonable faculty. But no philosopher as yet wist° to give to every difference of wit determinately that which was his. Heraclitus said: 'A dry brightness makes a most wise mind', by which sentence he gives us to understand that dryness is the cause why a man becomes very wise, but he declares not in what kind of knowledge.

The self same meant Plato when he said that the soul descended into the body endowed with great wisdom, and through the much moisture which it there found grew to become dull and untoward. But this wearing away in the course of age, and purchasing dryness, the soul grew to discover the knowledge which he tofore enjoyed. Amongst brute beasts, sayth Aristotle, those are wisest whose temperature is most inclined to cold and dry, as are the ants and bees, who for wisdom concur with those men that partake most of reason. Moreover, no brute beast is found of more moisture or less wit than a hog, wherethrough the poet Pindar,[49] to gibe at the people of Boeotia and to handle them as fools, said thus:

> Th'untoward folk which now is nam'd
> Boeotia were once called Hogs.

Moreover, blood through his much moisture, saith Galen, makes men simple. And for such, the same Galen recounts that the comics jested at Hippocrates' children saying of them that they had much natural heat, which is a substance moist and very vaporous. This is ordinarily incident to the children of wise men, and hereafter I will make report of the cause whence it groweth. Amongst the four humours which we enjoy, there is none so cold and dry as that of melancholy, and whatsoever notable men for learning have lived in the world, sayth Aristotle, they were all melancholic. Finally, all agree in this point that dryness makes a man very wise, but they express not to which of the reasonable powers it affordeth greatest help. Only Esay[50] the prophet calls it by his right name, where he sayth that 'travail gives understanding', for sadness and affliction not only diminisheth and consumeth the moisture of the brain, but also drieth up the bones, with which quality the understanding groweth more sharp and sightful.° Whereof we may gather an example very manifest by taking into consideration many men who, cast into poverty and affliction, have therethrough uttered and written sentences worth the marvelling at, and afterwards rising to better fortune, to eat and drink well, would never once open their mouths. For a delicious life, contentment, and good success, and to see that all things fall out after our liking, looseneth and maketh the brain moist. And this is it which Hippocrates said 'mirth looseneth the heart', as if he would have said that the same enlargeth and giveth it heat and grossness.

---

[49] Pindar] Pindar (*c.* 522–443 BC) was a Greek lyric poet from Thebes. Thebes was the largest city of the region of Boeotia (in Central Greece). The Boeotian people were portrayed as customarily dull by the Athenians, even if Boeotia was the home of Pindar, Hesiod and Plutarch, among others.

[50] Esay] Esay is the King James Version for Isaiah. Isaiah was a prophet from the eighth century BC.

230 And the same may easily be proved another way, for if sadness and affliction dry up and consume the flesh, and for that reason man gaineth more understanding, it falls out a matter certain that his contrary, namely mirth, will make the brain moist and diminish the understanding. Such as have purchased this manner of wit are suddenly inclined to pastimes, to music, and to pleasant conversations, and fly the contrary, which at other times gave them a relish and contentment. Now by this the vulgar sort may conceive whence it grows that a wise and virtuous man attaining to some great dignity (whereas at first he was but poor
240 and base) suddenly changeth his manners and his fashion of speech. And the reason is because he hath gotten a new temperature, moist and full of vapours, whence it follows that the figures are cancelled which tofore he had in his brain and his understanding dulled.

From moisture, it is hard to know what difference of wit may spring sithence it is so far contrary to the reasonable faculty. At least (after Galen's opinion) all the humours of our body, which hold overmuch moisture, make a man blockish and foolish, for which cause he said: 'The readiness of mind and wisdom grows
250 from the humour of choler; the humour of melancholy is author of firmness and constancy; blood, of simplicity and dullness; the phlegmatic complexion availeth nothing to the polishing of men.' Insomuch that blood with his moistures and the phlegm cause an impairing of the reasonable faculty.

But this is understood of the faculties or reasonable wits which are discoursive[51] and active, and not of the passive, as is the memory, which depends as well on the moist as the understanding doth on the dry. And we call memory a reasonable power because without it the understanding and the imaginative are of no valure.
260 It ministereth matter and figures to them all, whereupon they may syllogize conformably to that which Aristotle sayth: 'It behoves that the understander go beholding the phantasms' and the office of the memory is to preserve these phantasms to the end that the understanding may contemplate them. And if this be lost, it is impossible that the powers can work and that the office of memory is none other than to preserve the figures of things without that it appertains thereto to devise them. Galen

---

[51] discoursive] rational.

expresseth in these words: 'Memory verily lays up and preserveth in itself the things known by the sense and by the mind, and is therin as it were their storehouse and receiving place and not their inventor.' And if this be the use thereof, it falls out apparent that the same dependeth on moisture, for this makes the brain pliant and the figure is imprinted by way of straining. To prove this, we have an evident argument in boyage, in which any one shall better con° by heart than in any other time of life, and then doth the brain partake greatest moisture. Whence Aristotle moveth this doubt: 'Why in old age we have better wit and in young age we learn more readily?' As if he should say: 'What is the cause that when we are old we have much understanding and when we are young we learn with more towardliness?' Whereto he answereth that the memory of old men is full of so many figures of things which they have seen and heard in the long course of their life that when they would bestow more therein it is not capable thereof, for it hath no void place where to receive it. But the memory of young folk, when they are newly born, is full of plaits, and for this cause they receive readily whatsoever is told or taught them. And he makes this plainer by comparing the memory of the morning with that of the evening, saying that in the morning we learn best because at that time our memory is empty, and at the evening illy because then it is full of those things which we encountered during the day. To this problem Aristotle wist not how to answer, and the reason is very plain, for if the spices and figures which are in the memory had a body and quantity to occupy the place, it would seem that this were a fitting answer, but being undivided and spiritual, they cannot fill nor empty any place where they abide. Yea, we see by experience that by how much more the memory is exercised every day receiving new figures, so much the more capable it becometh. The answer of this problem is very evident after my doctrine, and the same importeth, that old men partake much understanding because they have great dryness, and fail of memory for that they have little moisture, and by this means the substance of the brain hardeneth and so cannot receive the impression of the figures, as hard wax with difficulty admitteth the figure of the seal, and the soft with easiness. The contrary befalls in children, who through the much moisture wherewith the brain is endowed, fail in understanding, and through the great suppleness of their brain abound in memory. Wherein by reason of the moisture, the shapes and figures that come from without make a great, easy, deep and well formed impression.

That the memory is better the morning than the evening cannot be denied, but this springeth not from the occasion alleged by Aristotle, but the sleep of the night passed hath made the brain moist, and fortified the same, and by the waking of the whole day, it is dried and hardened. For which cause, Hippocrates affirmeth: 'Those who have great thirst at night shall do well to drink', for sleep makes the flesh moist and fortifieth all the powers which govern man. And that sleep so doth, Aristotle himself confesseth.

By this doctrine is perfectly seen that the understanding and memory are powers opposite and contrary in sort, that the man who hath a great memory shall find a defect in his understanding, and he who hath a great understanding cannot enjoy a good memory, for it is impossible that the brain should of his own nature be at one self time dry and moist.[52] On this maxim Aristotle grounded himself to prove that memory is a power different from remembrance, and he frames his argument in this manner: 'Those who have much remembrance are men of great understanding, and those who possess a great memory find want of understanding; so then memory and remembrance are contrary powers.' The former proposition after my doctrine is false, for those who have much remembrance are of little understanding and have great imaginations, as soon hereafter I will prove. But the second proposition is very true, albeit Aristotle knew not the cause whereon was founded the enmity which the understanding hath with the memory.

From heat, which is the third quality, groweth the imaginative, for there is no other reasonable power in the brain, nor any other quality to which it may be assigned besides that the sciences which appertain to the imaginative are those which such utter as dote in their sickness, and not of those which appertain to the understanding or to the memory. And frenzy, peevishness, and melancholy, being hot passions of the brain, it yields a great argument to prove that imagination consists in heat. One thing breeds me a difficulty herein, and that is that the imagination carrieth a contrariety to the understanding, as also to the memory, and the reason hereof is not to be gotten by experience, for in the

---

[52] 24. *Index*: 'By this doctrine is perfectly seen [319] […] at one self time dry and moist' [censored].

brain may very well be united much heat and much dryness, and so likewise much heat and much moisture to a large quantity. And for this cause, a man may have a great understanding and a great imagination, and much memory with much imagination. And verily, it is a miracle to find a man of great imagination who hath a good understanding and a sound memory. And the cause thereof behoves to be for that the understanding requires that the brain be made of parts very subtile and delicate, as we have proved heretofore out of Galen, and much heat frets and consumes what is delicate, and leaves behind the parts gross and earthly. For the like reason, a good imagination cannot be united with much memory, for excessive heat resolveth the moisture of the brain and leaveth it hard and dry, by means whereof it cannot easily receive the figures. In sort that in man there are no more but three general differences of wits, for there are no more but three qualities whence they may grow. But under these three universal differences, there are contained many other particulars by means of degrees of access which heat, moisture and dryness may have.

Notwithstanding there springs a difference in wits from every degree of these three qualities, for the dry, the hot, and the moist may exceed in so high a degree that it may altogether disturb the animal power, conformable to that sentence of Galen: 'every excessive distemperature resolves the forces', and so it is. For albeit dryness give help to the understanding, yet it may be that the same shall consume his operations. Which Galen and the ancient philosophers would not admit, but affirm that if old men's brains grew not cold, they should never decay though they became dry in the fourth degree. But they have no reason for this, as we will prove in the imaginative, for albeit his operations be performed with heat, yet if it pass the third degree forthwith the same begins to resolve, and the like doth the memory through overmuch moisture.

How many differences of wits grow by means of the superabounding of each of these three qualities cannot for this present be particularly recited, except tofore we recount all the operations and actions of the understanding, the imagination, and the memory. But the whilst we are to know that the principal works of the understanding are three: the first, to discourse; the second, to distinguish; and the third, to choose. Hence comes it that they place also three differences in the understanding. Into three other is the memory divided: one receives with ease, and suddenly forgetteth; another is slow to receive, but a long time retaineth; and the last receiveth with ease and is very slow to forget.

The imagination containeth many more differences, for he hath three, no less than the understanding and memory, and from each degree ariseth three other. Of these we will more distinctly discourse hereafter when we shall assign to each the science which answereth it in particular.

But he that will consider three other differences of wit shall find that there are habilities in those who study, some which have a disposition for the clear and easy contemplations of the art which they learn, but if you set them about matters obscure and very difficult, it will prove a lost labour for the teacher to shape them a figure thereof by fit examples, or that they frame themselves the like by their own imagination, for they want the capacity.

In this degree are all the bad scholars of whatsoever faculty, who being demanded touching the easy points of their art answer to the purpose, but coming to matters of more curiousness they will tell you a hundred follies. Other wits advance themselves one degree higher, for they are pliant and easy in learning things and they can imprint in themselves all the rules and considerations of art, plain, obscure, easy, and difficult. But as for doctrine, argument, doubting, answering, and distinguishing, they are all matters wherewith they may in no wise be cumbered. These need to learn sciences at the hands of good teachers, well skilled in knowledge, and to have plenty of books, and to study them hard, for so much the less shall their knowledge be as they forbear to read and take pains. Of these may be verified that so famous sentence of Aristotle: 'Our understanding is like a plain table wherein nothing is portrayed.' For whatsoever they are to know and attain, it behoves that first they hear the same of some other and are barren of all invention themselves. In the third degree, nature maketh some wits so perfect that they stand not in need of teachers to instruct them, nor to direct in what sort they are to philosophize, for out of one consideration indicted to them by their schoolmaster, they will gather a hundred, and without that ought be bestowed unto them, they fill their wit with science and knowledge. Those wits beguiled Plato and made him to say that our knowledge is a certain spice° of remembrance when he heard them speak and say that which never fell into consideration with other men.

To such, it is allowable that they write books, and to others not: for the order and concert which is to be held to the end that sciences may daily receive increase and greatest perfection is to join the new invention of ourselves, who live now, with that which the ancients left written in their books. For dealing after this manner, each in his time, shall add an increase to the arts, and men who are yet unborn shall enjoy the invention and travail of such as lived before. As for such who want invention, the commonwealth should not consent that they make books, nor suffer them to be printed, because they do nought else save heap up matters already delivered and sentences of grave authors returning to repeat the self things, stealing one from hence, and taking another from thence. And there is no man but after such a fashion may make a book.

Wits full of invention are by the Tuscans called goatish, for the likeness which they have with a goat in their demeanour and proceeding. These never take pleasure in the plains, but ever delight to walk alone through dangerous and high places, and to approach near steep downfalls, for they will not follow any beaten path, nor go in company. A property like this is found in the reasonable soul when it possesseth a brain well instrumentalized and tempered, for it never resteth settled in any contemplation but fareth forthwith unquiet, seeking to know and understand new matters. Of such a soul is verified the saying of Hippocrates, 'the going of the soul is the thought of men.' For there are some who never pass out of one contemplation and think not that the whole world can discover another such. These have the property of a beast who never forsakes the beaten path, nor careth to walk through desert and unhaunted places but only in the high market way and with a guide before him. Both these diversities of wits are ordinary amongst professors of learning. Some others there are of high searching capacities, and estranged from the common course of opinions, they judge and entreat of matters with a particular fashion, they are frank in delivering their opinion and tie not themselves to that of any other. Some sorts are close, moist, and very quiet, distrusting themselves and relying upon the judgement of some grave man whom they follow, whose sayings and sentences they repute as sciences and demonstrations, and all things contrarying the same they reckon vanity and leasings.

470   These two differences of wits are very profitable if they be united, for as amongst a great drove of cattle, the herdsmen accustom to mingle some dozen of goats to lead them and make them trot apace to enjoy new pastures that they may not suffer scarcity; so also it behoveth that in human learning there be some goat-like wits who may discover to the cattle like understandings through secrets of nature, and deliver unto them contemplations not heard of wherein they may exercise themselves, for after this manner, arts take increase and men daily know more and more.

CHAP. VI.

Certain doubts and arguments are propounded against the doctrine of the last chapter, and their answer.

One of the causes for which the wisdom of Socrates hath been so famous till this day is for that after he was adjudged by the oracle of Apollo to be the wisest man of the world, he said thus: 'I know this only, that I know nothing at all', which sentence, all those that have seen and read, passed it over as spoken by Socrates, for that he was a man of great humbleness, a despiser of worldly
10 things, and one to whom, in respect of divine matters, all else seemed of no valure. But they verily are beguiled for none of the ancient philosophers possessed the virtue of humility, nor knew what thing it was, until God came into the world and taught the same.

The meaning of Socrates was to give to understand how little certainty is contained in human sciences and how unsettled and fearful the understanding of a philosopher is in that which he knoweth, seeing by experience that all is full of doubts and arguments, and that we can yield assent to nothing without
20 fearing that it may be contrary. For it was said: 'The thoughts of men are doubtful, and our foreseeings uncertain.' And he who will attain the true knowledge of things it behoves that he rest settled and quiet without fear or doubt of being deceived, and the philosopher who is not thus wise-grounded may with much truth affirm that he knoweth nothing.

This same consideration had Galen when he said: 'Science is a convenient and firm notice which never departeth from reason; therefore, thou shalt not find it amongst the philosophers, especially when they consider the nature of things. But verily much less in matters of physic, nay rather (to speak all in one word) it never makes his full arrival where men are.'

Hereby it seemeth that the true notice of things fails to come this way, and to man ariveth only a certain opinion which makes him to walk uncertain and with fear whether the matter which he affirmeth be so or no.[53] But that which Galen noteth more particularly touching this is that philosophy and physic are the most uncertain of all those wherewith men are to deal. And if this be true, what shall we say touching the philosophy whereof we now entreat where with the understanding we make an anatomy of a matter so obscure and difficult as are the powers and faculties of the reasonable soul, in which point are offered so many doubts and arguments that there remains no clear doctrine upon which we may rely?

One of which, and the principal, is that we have made the understanding an instrumental power as the imagination and the memory, and have given dryness to the brain as an instrument with which it may work. A thing far repugnant to the doctrine of Aristotle and all his followers, who placing the understanding severed from the bodily instrument, prove easily the immortality of the reasonable soul, and that the same issuing out of the body endureth forever. Now the contrary opinion being disputable, the way hereby is stopped up, so that this cannot be proved.[54] Moreover, the reasons on which Aristotle groundeth himself to prove that the understanding is not an instrumental power carry such efficacy as other than that cannot be concluded. For to this power appertaineth the knowing and understanding the nature and being of whatsoever material things in the world, and if the same should be conjoined with any bodily thing, that self would hinder the knowledge of the residue. As we see in the outward senses that if the taste be bitter, all the things which the tongue toucheth partake the same savour, and if the crystalline humour be green or yellow, all that the eye seeth it judgeth to be of the same colour. The reason of this is for that the thing within breeds an impediment to that without.

Aristotle sayth moreover that if the understanding were

---

[53] 25. *Index*: 'Hereby it seemeth [32] [...] he affirmeth be so or no' [censored].
[54] 26. *Index*: 'In which point are offered [41] [...] so that this cannot be proved' [censored].

mingled with any bodily instrument, it would retain some quality, for whatsoever uniteth itself with heat or cold it is of force that it partake of the same quality. But to say that the understanding is hot, cold, moist, or dry, is to utter a matter abominable to the ears of all natural philosophers.

The second principal doubt is that Aristotle and all the peripatetics bring in two other powers besides the understanding, the imagination, and the memory; namely, remembrance, and common sense, grounding upon that rule that the powers are known by way of the actions. They said that besides the operations of the understanding, the imagination and the memory, there are also two other different. So then the wit of man taketh his original from five powers and not from three only, as we did prove.

We said also in the last chapter, after the opinion of Galen, that the memory doth none other work in the brain save only to preserve the shapes and figures of things in such sort as a chest preserveth and keepeth apparel and what so else is put thereinto. And if by such a comparison we are to understand the office of this power, it is requisite also to prove another reasonable faculty which may fetch out the figures from the memory and represent them to the understanding, even as it is necessary that there be one to open the chest and to take out what hath been laid up therein.

Besides this, we said that the understanding and the memory are contrary powers and that the one chaseth away the other,[55] for the one loveth great dryness and the other much moisture and a suppleness of the brain. And if this be true, wherefore said Aristotle and Plato that men who have their flesh tender enjoy great understanding, seeing this suppleness is an effect of moisture?

We said also that for effecting that a memory may be good it was necessary the brain should be endowed with moisture, for the figures ought to be printed therein by way of compression, and the same being hard, they cannot so easily make a sign therein. True it is that to receive figures with readiness it requireth that the brain be pliant, but to preserve the shapes some long time, all affirm that it is necessary the same be hard and dry, as it

---

[55] 27. *Index*: 'we said that the understanding and the memory are contrary powers and that the one chaseth away the other' [censored].

appeareth in outward things where the figure printed in a pliant substance is easily cancelled, but in the dry and hard, it never perisheth. Wherethrough we see many men who con° by heart with great readiness but forget again very speedily. Of which Galen rendering a reason sayth that such through much moisture have the substance of their brain tender and not settled, for the figure is soon cancelled, as if it were sealed in water. And contrariwise, other learn by heart with difficulty, but what they have once learned, they never forget again. Wherethrough it seemeth a matter impossible that there should be that difference of memory which we speak of, which should learn with ease and preserve a long time.

It is also hard to understand how it is possible that so many figures being sealed together in the brain, the one should not cancel the other, for if in a piece of softened wax there be printed many seals of diverse figures, it falls out certain that some cancel othersome by the intermingling of these figures.

And that which breedeth no less difficulty is to know whence it proceedeth that the memory by exercising itself becometh the more easy to receive figures, it being certain that not only bodily exercise but spiritual much more drieth and soaketh the flesh.

It is also hard to conceive in what sort the imagination is contrary to the understanding (if there be none other more urgent cause than to say that excessive heat resolveth the subtile parts of the brain, leaving an earthly and gross remnant), seeing that melancholy is one of the grossest and earthliest humours of our body. And Aristotle sayth that the understanding useth the service of none so much as of that. And this difficulty is increased considering that melancholy is a gross humour, cold and dry, and choler is of a delicate substance and of temperature hot and dry, and yet for all this melancholy is more appropriate to the understanding than choler. Which seemeth repugnant to reason, for this humour aideth the understanding with two qualities and gainsetteth itself only with one, which is heat. But melancholy aideth it with his dryness and with none other, and opposeth itself by his cold and by his gross substance, which is a thing that the understanding most abhorreth. For which cause, Galen assigneth more wit and prudence to choler than to melancholy, saying thus: 'Readiness and prudence spring from the humour of choler, and the melancholic humour is author of integrity and constancy.'

Lastly, the cause may be demanded whence it may grow that

toiling and continual contemplation of study maketh many wise in whom at the beginning the good nature of these qualities, which we speak of, was wanting. And so, by giving and receiving with the imagination, they come to make themselves capable of many verities which tofore they knew not, nor had the temperature which thereto was requisite. For if they had possessed the same, so much labour should not have been needful.

All these difficulties and many other besides are contrary to the doctrine of the last chapter. For natural philosophy hath not so certain principles as the mathematical sciences, wherein the physician and the philosopher (if he be also a mathematician) may always make demonstration. But coming afterwards to the cure which is conformable to the art of physic, he shall commit therein many errors and yet not always through his own fault (sithence in the mathematics he always followed a certainty) but through the little assurance of the art, for which cause Aristotle said: 'The physician though he always cure not, is not therefore a bad one, provided that he forslow not to perform any of those points which appertain to the art.' But if he should commit any error in the mathematics, he would be void of excuse. For performing in this science all the diligences which it requireth, it is impossible that the truth should not appear. In sort that albeit we yield not a manifest demonstration of this doctrine, yet the whole fault is not to be laid on our want of capacity, neither may it straightways be recounted as false that we deliver.

To the first principal doubt, we answer that if the understanding were severed from the body and had nought to do with heat, cold, moist, and dry, nor with the other bodily qualities, it would follow that all men should partake equal understanding and that all should equally discourse. But we see by experience that one man understandeth and discourseth better than another, then this groweth for that the understanding is an instrumental power and better disposed in one than in another, and not from any other occasion. For all reasonable souls and their understandings (sundered from the body) are of equal perfection and knowledge.[56] Those who follow Aristotle's doctrine, seeing by experience that some discourse better than othersome, have found an excuse in appearance saying that the discoursing of one better

---

[56] 28. *Index*: 'To the first principal doubt [170] [...] of equal perfection and knowledge' [censored].

than another is not caused for that the understanding is an instrumental power, and that the brain is better disposed in some than in othersome, but for that the understanding (whilst the reasonable soul remaineth in the body) standeth in need of the phantasms and figures which are in the imagination and in the memory. Through default whereof, the understanding falls to discourse illy and not through his own fault, nor for that it is joined with a matter badly instrumentalized. But this answer is contrary to the doctrine of Aristotle himself, who proveth that by how much the memory is the worse, by so much the understanding is the better, and by how much the memory is bettered, by so much the understanding is impaired (and the same we have heretofore proved as touching the imagination). In confirmation of that which Aristotle demandeth, what the cause is that the waxing old have so bad a memory and so good an understanding, and when we are young it falls out contrary that we possess a great memory and small understanding. Hereof, in one thing we see the experience, and Galen noteth it, that when in a disease the temperature and good disposition of the brain is impaired, many times the operations of the understanding are thereby lost, and yet those of the memory and the imagination remain sound, which could not come to pass if the understanding enjoyed not a particular instrument for itself besides this which the other powers do partake.

To this I know not what may be yielded in answer unless it be by some metaphysical relation compounded of action and power which neither themselves know what it meaneth, nor is there any other man that understands it.[57] Nothing more endamageth man's knowledge than to confound the sciences, and what belongs to the metaphysics, to entreat thereof in natural philosophy, and matters of natural philosophy in the metaphysics.

The reasons whereupon Aristotle grounded himself are of small moment, for the consequence followeth not to say that the understanding, because it must know material things, should not therefore enjoy a bodily instrument. For the bodily qualities which serve for the composition of the instrument make no alteration of the power, nor from them do the phantasms arise[58]

---

[57] 29. *Index*: 'unless it be by some metaphysical relation [206] [...] any other man that understands it' [censored].

[58] 30. *Index*: 'The reasons whereupon Aristotle grounded himself [213] [...] arise' [censored].

even as the sensible, placed above the sense, causeth not the self sense. This is plainly seen in touching, for notwithstanding that the same is compounded of four material qualities, and that the same hath in it quantity and hardness or softness, for all this, the hand discerneth whether a thing be hot or cold, hard or soft, great or little. And if you ask in what sort the natural heat which is in the hand hindereth not the touching that it may discern the heat which is in the stove, we answer that the qualities which serve for the composition of the instrument do not alter the instrument itself, neither from them do there issue any shapes whereby to know them. Even as it appertaineth to the eye to know all figures and qualities of things, and yet we see that the eye itself hath his proper figure and quantity, and of the humours and skins which go to his composition, some have colours, and some are diaphane and transparent, all which hindereth not but that we with our sight may discern the figures and quantities of all the things which shall appear before us. And the reason is for that the humours, the skins, the figure, and the quantity serve for the composition of the eye and such things cannot alter the sightful° power and therefore trouble not nor hinder the knowledge of the outward figures. The like we affirm of the understanding, that his proper instrument (though the same be material and joined with it) cannot enlarge it, for from it issue no understandable shapes which have force to alter it. And the reason is for that the understandable placed above the understanding causeth not the understanding, and so it remaineth at liberty to understand all the outward material things without that it encounter ought to hinder the same. The second reason wherein Aristotle grounded himself is of less importance than the former, for neither the understanding, nor any other accident, can be qualitylike, for of themselves they cannot be the subject of any quality. For which cause, it little skilleth that the understanding possess the brain for an instrument together with the temperature of the four first qualities, that therefore it may be called qualitylike, inasmuch as the brain and not the understanding is the subject of the heat, the cold, the moist, and the dry.

To the third difficulty which the peripatetics allege saying that by making the understanding an instrumental power we reave one of those principles which serve to prove the immortality of the reasonable soul, we answer that there are other arguments of more soundness whereby to prove the same, whereof we will treat in the chapter following.

260   To the second argument, we answer that not every difference of operations argueth a diversity of powers, for (as we will prove hereafter) the imaginative performeth matter so strange that if this maxim were true in sort as the vulgar philosophers had it, or admitting the interpretation which they give it, there should be in the brain ten or twelve powers more. But because all these operations are to be marshalled under one general reason, they argue no more than one imaginative, which is afterwards divided into many particular differences by the means of the sundry operations which it performeth: the composing of the shapes, in
270 the presence or the absence of the objects, not only argueth not a diversity of the general powers (as are the common sense and the imaginative) but even not of the very particulars.

    To the third argument we answer that the memory is nothing else but a tenderness of the brain disposed with a certain kind of moisture to receive and preserve that which the imaginative apprehendeth. With the like proportion that white or blue paper[59] holds with him who writeth,[60] for as the writer writeth in the paper the things which he would not forget, and after he hath written them returns to read them, even so we ought to conceive that the
280 imagination writeth in the memory the figures of the things known by the five senses and by the understanding, as also some others of his own framing. And when it will remember ought, saith Aristotle, it returneth to behold and contemplate them. With this manner of comparison Plato served himself when he said that fearing the weak memory of old age he hastened to make another of paper (namely, books) to the end his travails ought not to be lost, but that he might have that which might represent them unto him when he list to read them. This self doth the imaginative of writing in the memory, and returning to read it when it would
290 remember the same. The first who uttered this point was Aristotle, and the second Galen who said thus: 'Forasmuch as that part of the soul which imagineth whatsoever the same be, seemeth to be the self that also remembereth.' And so verily it seemeth to be, for the things which we imagine with long thinking are well fixed in the memory, and that which we handle with light consideration

---

[59] blue paper] sensitized paper used for copying maps and plans, made by saturating the paper with potassium ferrocyanide.
[60] 31. *Index*: 'The like we affirm of the understanding [238] [...] with him who writeth' [censored].

also soon we forget the same again. And as the writer, when he writeth fair, the better assureth it to be read, so it befalls to the imaginative that if it seal with force, the figure remaineth well imprinted in the brain. Otherwise it can scarcely be discerned. The like also chanceth in old deeds which being found in part, and in part perished by time, cannot well be read unless we gather much by reason and conjecture. So doth the imaginative when in the memory some figures remain and some are perished, where Aristotle's error had his original, who for this cause conceived that remembrance was a different power from the memory. Moreover, he affirmed that those who have great remembrance are likewise of great understanding, which is also false, for the imaginative, which is that that makes the remembrance, is contrary to the understanding; in sort, that to gather memory of things and to remember them after they are known is a work of the imagination, as to write and return to read it is a work of the scrivener and not of the paper. Whereby it falleth out that the memory remaineth a power passive and not active, even as the blue and the white of the paper is none other than a commodity whereby to write.

To the fourth doubt may be answered that it maketh little to the purpose as touching the wit, whether the flesh be hard or tender, if the brain partake not also the same quality, the which we see many times hath a distinct temperature from all the other parts of the body. But when they concur in one self tenderness, it is an evil token for the understanding and no less for the imagination. And if we consider the flesh of women and children, we shall find that in tenderness it exceedeth that of men, and this notwithstanding commonly men have a better wit than women, and the natural reason hereof is for that the humours which make the flesh tender are phlegm and blood, because they are both moist (as we have above specified), and of them Galen said that they make men simple and dullards. And contrariwise, the humours which harden the flesh are choler and melancholy, and hence grow the prudence and sapience which are found in man. In sort that it is rather an ill token to have the flesh tender than dry and hard. And so in men who have an equal temperature throughout their whole body, it is an easy matter to gather the quality of their wit by the tenderness or hardness of their flesh. For if it be hard and rough, it giveth token either of a good understanding or a good imagination, and if smooth and supple of the contrary, namely, of good memory and small understanding and less imagination. And to understand whether the brain have correspondence, it behoveth to consider the hair, which being big, black, rough, and thick, yieldeth token of a good imagination or a good understanding. And if soft and smooth, they are a sign of much memory and nothing else. But who so will distinguish and know whether the same be understanding or imagination (when the hair is of this sort) it must be considered of what form the child is in the act of laughter, for this passion discovereth much of what quality he is in the imagination.

What the reason and cause of laughter should be, many philosophers have laboured to conceive and none of them hath delivered ought that may well be understood, but all agree that the blood is a humour which provoketh a man to laugh, albeit none express with what quality this humour is endowed more than the rest, why it should make a man addicted to laughter. The follies which are committed with laughing are less dangerous, but those which are done with labour are more perilous, as if he[61] should say: 'When the diseased become giddy and doting do laugh, they rest in more safety than if they were in toil and anguish', for the former cometh of blood, which is a most mild humour, and the second of melancholy. But we grounding upon the doctrine, whereof we entreat, shall easily understand all that which in this case may be desired to be known. The cause of laughter (in my judgement) is nought else but an approving, which is made by the imagination, seeing or hearing somewhat done or said which accordeth very well. And this power remaineth in the brain when any of these things give it contentment, suddenly it moveth the same, and after it all the muscles of the body, and so, many times we do allow of witty sayings by bowing down of the head. When then the imagination is very good, it contents not itself with every speech, but only with those which please very well, and if they have some little correspondence and nothing else, the same receiveth thereby rather pain than gladness. Hence it groweth that men of great imagination laugh very seldom, and the point most worthy of noting is that jesters, and natural counterfeiters, never laugh at their own merriments nor at that which they hear others to utter: for they have an imagination so

---

[61] he] Hippocrates, as it explicitly appears in the original in Spanish.

delicate that not even their own pleasantries can yield that correspondence which they require.

Hereto may be added that merriments (besides that they must have a good proportion, and be uttered to the purpose) must be new and not tofore heard or seen. And this is the property not only of the imagination but also of all the other powers which govern man, for which cause we see that the stomach when it hath twice fed upon one kind of meat, straightways loatheth the same. So doth the sight one self shape and colour; the hearing one concordance, how good soever; and the understanding one self contemplation. Hence also it proceedeth that the pleasant conceited man laugheth not at the jests which himself uttereth, for before he send them forth from his lips, he knew what he would speak. Whence I conclude that those who laugh much are all defective in their imagination, wherethrough whatsoever merriment and pleasantry (how cold soever) with them carrieth a very good correspondence. And because the blood partaketh much moisture (whereof we said before that it breedeth damage to the imagination) those who are very sanguine are also great laughers. Moisture holdeth this property that because the same is tender and gentle, it abateth the force of heat and makes that it burn not overmuch. For which cause, it partakes better agreement with dryness because it sharpeneth his operations. Besides this, where there is much moisture it is a sign that the heat is remiss, seeing it cannot resolve nor consume the same, and the imagination cannot perform his operations with a heat so weak. Hence we gather also that men of great understanding are much given to laughter, for that they have defect of imagination, as we read of that great philosopher Democritus and many others whom myself have seen and noted. Then by means of this laughter we shall know if that which men or boys have of flesh hard and tough, and of hair black, thick, hard, and rough, betoken either the imagination or the understanding. In sort, that Aristotle in this doctrine was somewhat out of the way.

To the fifth argument we answer that there are two kinds of moisture in the brain, one which groweth of the air (when this element predominateth in the mixture) and another of the water, with which the other elements are amassed. If the brain be tender by the first moisture, the memory shall be very good: easy to receive, and mighty to retain the figures for a long time. For the moisture of the air is very supple and full of fatness, on which the shapes are tacked with sure holdfast, as we see in pictures which are limned in oil, who being set against the sun and the water, receive thereby no damage at all. And if we cast oil upon any writing, it will never be wiped out but marreth the same, and that which cannot be read, with oil is made legible by yielding thereto a brightness and transparence. But if the difference of the brain spring from the second kind of moisture, the argument frameth very well, for if it receive with facility with the same readiness, it turneth again to cancel the figure, because the moisture of the water hath no fatness wherein the figures may fasten themselves. These two moistures are known by the hair. For that which springs from the air maketh them to prove unctious and full of oil and fat, and the water maketh them moist and very supple.

To the sixth argument may be answered that the figures of things are not printed in the brain as the figure of the seal is in wax, but they pierce thereinto to remain there affixed in sort as the sparrows are attached to birdlime or the flies stick in honey. For these figures are bodiless and cannot be mingled nor corrupt one the other.

To the seventh difficulty we answer that the figures amass and mollify the substance of the brain in such sort as wax groweth soft by plying the same between our fingers. Besides that the vital spirits have virtue to make tender and supple the hard and dry members as the outward heat doth the iron. And that the vital spirits ascend to the brain when anything is learned by heart, we have proved heretofore, and every bodily and spiritual exercise doth not dry. Yea, the physicians affirm that the moderate fatteneth.

To the eighth argument we answer that there are two species of melancholy: one natural, which is the dross of the blood, whose temperature is cold and dry, accompanied with a substance very gross, this serves not of any value for the wit but maketh men blockish, sluggards, and grinners, because they want imagination. There is another sort which is called choler adust[62] or atrabile,[63] of which Aristotle said that it made man exceeding wise, whose temperature is diverse as that of vinegar. Sometimes it performeth

---

[62] choler adust] black choler, black bile; thick black and acrid fluid secreted by the renal glands; the cause of melancholy.

[63] atrabile] black bile supposed to be secreted by the renal or atrabiliary glands, or by the spleen, and to be the cause of melancholy.

the effects of heat, lightning the earth, and sometimes it cooleth, but always it is dry and of a very delicate substance. Cicero confesseth that he was slow-witted because he was not melancholic adust, and he said true, for if he had been such, he should not have possessed so rare a gift of eloquence. For the melancholic adust want memory, to which appertaineth the speaking with great preparation. It hath another quality which much aideth the understanding, namely, that it is clear like the agate stone, with which clearness it giveth light within to the brain and maketh the same to discern well the figures. And of this opinion was Heraclitus when he said: 'A dry clearness maketh a most wise mind', with which clearness natural melancholy is not endowed, but his black is deadly, and that the reasonable soul there within the brain standeth in need of light to discern the figures and the shapes, we will prove hereafter.

To the ninth argument we answer that the prudence and readiness of the mind which Galen speaketh of appertaineth to the imagination, whereby we know that which is to come, whence Cicero said: 'Memory is of things passed, and prudence of those to come.' The readiness of the mind is that which commonly they call a sharpness in imagining, and by other names, craftiness, subtility, cavilling, wiliness, wherefore Cicero said: 'Prudence is a subtilty which with a certain reason can make choice of good things and of evil.' This sort of prudence and readiness men of great understanding do want because they lack imagination. For which reason we see by experience in great scholars in this sort of learning which appertaineth to the understanding that taking them from their books they are not worth a rush to yield or receive in traffic of worldly affairs. This spice° of prudence Galen said very well that it came of choler, for Hippocrates recounting to Damagetus his friend in what case he found Democritus when he went to visit him for curing him, writeth that he lay in the field under a plane tree, bare-legged and without breeches, leaning against a stone with a book in his hand and compassed about with brute beasts, dead and dismembered.[64] Whereat Hippocrates marvelling asked him whereto those beasts of that fashion served, and he then answered that he was about to search what humour it was which made man to be headlong, crafty, ready, double, and

---

[64] Hippocrates recounting to Damagetus] This is a reference to the 'Epistle to Damagetus', a spurious work of Hippocrates.

cavillous, and had found (by making an anatomy of those wild beasts) that choler was the cause of so discommendable a property, and that to revenge himself of crafty persons he would handle them as he had done the fox, the serpent and the ape. This manner of prudence is not only odious to men, but also St Paul sayth of it: 'The wisdom of the flesh is enemy to God.' The cause is assigned by Plato, who affirmeth that knowledge which is removed from justice ought rather to be termed subtility than prudence, as if he should have said: 'It is no reason that a knowledge which is severed from justice should be called wisdom, but rather craft, or maliciousness.' Of this, the devil evermore serveth himself to do men damage, and St James said that this wisdom came not from heaven but is earthly, beastly, and devilish.

There is found another spice of wisdom conjoined with reason and simplicity, and by this, men know the good and shun the evil, the which Galen affirmeth doth appertain to the understanding, for this power is not capable of maliciousness, doubleness, nor subtilty, nor hath the skill how to do naught, but is wholly upright, just, gentle, and plain. A man endowed with this sort of wit is called upright and simple, wherethrough when Demosthenes went about to creep into the good liking of the judges in an oration which he made against Eschines, he termed them upright and simple in respect of the simplicity of their duty, concerning which, Cicero sayth: 'Duty is simple and the only cause of all good things.' For this sort of wisdom, the cold and dry of melancholy is a serving instrument, but it behoveth that the same be composed of parts very subtile and delicate.

To the last doubt may be answered that, when a man setteth himself to contemplate some truth, which he would fain know and cannot by and by find it out, the same groweth for that the brain wanteth his convenient temperature; but when a man standeth ravished in a contemplation, the natural heat that is in the vital spirits and the arterial blood run forthwith to the head, and the temperature of the brain enhanceth itself until the same arrive to the term behooful.° True it is that much musing, to some doth good, and to some harm, for if the brain want but a little to arrive to that point of convenient heat, it is requisite that he make but small stay in the contemplation, and if it pass that point straightways, the understanding is driven into a garboil by the over plentiful presence of the vital spirits, and so he cannot attain to the notice of the truth. For which cause, we see many men who upon the sudden speak very well but with advisement are nothing worth. Others have their understanding so base (either through too much coldness, or too much drought), that it is requisite the natural heat abide a long time in the head, to the end the temperature may lift itself up to the degrees which are wanting, wherethrough they speak better upon deliberation than on the sudden.

CHAP. VII.[65]

It is showed that though the reasonable soul have need of the temperature of the four first qualities, as well for his abiding in the body, as also to discourse and syllogize, yet for all this, it followeth not that the same is corruptible and mortal.

Plato held it for a matter very certain that the reasonable soul is a substance bodiless and spiritual not subject to corruption or mortality as that of brute beasts, the which departing from the body, possesseth another better and more quiet life. But this is to be understood, saith Plato, if a man have led his life conformable to reason, for otherwise it were better that the soul had remained still in the body, there to suffer the torments with which God chastiseth the wicked. This conclusion is so notable and catholic that if he attained the knowledge thereof by the happiness of his wit with a just title he came to be called the divine Plato. But albeit the same is such as we see, yet for all this Galen could never bring within his conceit that it was true, but held it always doubtful, seeing a wise man through the heat of his brain to dote, and by applying cold medicines unto him, he cometh to his wits again. In respect whereof, he said he could wish that Plato were now living to the end he might ask him how it was possible that the reasonable soul should be immortal, seeing it altered so easily with heat, with cold, with moisture, and with drought. And principally, considering that the same departs from the body through overmuch heat, or when a man giveth over himself excessively to lasciviousness, or is forced to drink poison, and such other bodily

---

[65] 32. *Index*: Chapter VII [censored].

alterations, which accustomably bereave the life. For if it were bodiless and spiritual (as Plato affirmeth) heat, being a material quality, could not make the same to leese° his powers, nor set his operations in a garboil.

These reasons brought Galen into a confusion and made him wish that some Platonist would resolve him these doubts, and I believe that in his lifetime he met not with any, but after his death experience showed him that which his understanding could not conceive. For it is a thing certain that the infallible certainty of our immortal soul is not gathered from human reasons or from arguments which prove that it is corruptible, for to the one and the other an answer may easily be shaped: it is only our faith which maketh us certain and assured that the same endureth forever. But Galen had small reason to intricate himself in arguments of so slight consequence, for the works which seem to be performed by means of some instrument, it cannot well be gathered in natural philosophy that it proceedeth from a defect in the principal agent if they take not perfection. That painter who portrayeth well when he hath a pencil requisite for his art falleth not in blame if with a bad pencil he draw ill favoured shapes and of bad delineation.° And it is no good argument to say that the writer had an imperfection in his hand when through default of a well-made pen he is forced to write with a stick. Galen, considering the marvellous works which are in the universe, and the wisdom and providence by which they were made and ordained, concluded thereof that in the world there was a God, though we behold him not with our corporal eyes, of whom he uttered these words: 'God was not made at any time, inasmuch as he is everlastingly unbegotten.' And in another place he sayth that the frame and composition of man's body was not made by the reasonable soul, nor by the natural heat, but by God or by some very wise understanding.

Out of which there may be framed an argument against Galen, and his false consequence be overthrown, and it is thus: thou hast suspected that the reasonable soul is corruptible because if the brain be well-tempered it fitteth well to discourse and philosophize, and if the same grow hot or cold beyond due,[66] it doteth and uttereth a thousand follies. The same may be inferred

---

[66] due] due quality or character.

considering the works which thou speakest of as touching God, for if He make a man in places temperate (where the heat exceedeth not the cold, nor the moist the dry) He produceth him very witty and discreet, and if the country be untemperate, He breedeth them all fools and doltish. For the same Galen affirmeth that it is a miracle to find a wise man in Scythia, and in Athens they are all born philosophers. To suspect then that God is corruptible because with one quality he performeth these works well, and with the contrary they prove ill, Galen himself would not confess, for as much as he said before, that God was everlasting.

Plato held another way of more certainty, saying that albeit God be everlasting, almighty, and of infinite wisdom, yet he proceedeth in his works as a natural agent, and makes himself subject to the disposition of the four first qualities. In sort, that to beget a man very wise, and like to himself, it behoveth that he provide a place the most temperate of the whole world where the heat of the air may not exceed the cold, nor the moist the dry, and therefore he said: 'But God as desirous of war, and of wisdom, having chosen a place which should produce men like unto Himself, would that the same should be first inhabited.' And though God would shape a man of great wisdom in Scythia, or in any other intemperate country, and did not herein employ his omnipotence, he should of necessity yet prove a fool through the contrariety of the first qualities. But Plato would not have inferred (as Galen did) that God was alterable[67] and corruptible, for that the heat and coldness would have brought an impediment to his work. The same may be collected when a reasonable soul, for that it is seated in a brain inflamed, cannot use his discretion and wisdom, and not to think that in respect thereof the same is subject to mortality and corruption. The departure out of the body, and the not being able to support the great heat, nor the other alterations which are wont to kill men, showeth plainly that the same is an act and substantial form of man's body, and that to abide therein, it requireth certain material dispositions fitteth to the being, which it hath of the soul, and that the instruments with which it must work be well composed and well united, and of that temperature which is requisite for his operations, all which

---

[67] alterable] able to produce a change in something.

failing, it behoveth of force that it err in them, and depart from the body.

The error of Galen consisteth in that he would verify by the principles of natural philosophy whether the reasonable soul, issuing out of the body, do forthwith die or not, this being a question which appertaineth to another superior science and of more certain principles, in which we will prove that it is no good argument nor concludeth well that the soul of man is corruptible because the same dwelleth quietly in a body endowed with these qualities and departeth when they do fail. Neither is this difficult to be proved, for other spiritual substances of greater perfection than the reasonable soul do make choice of place, altered with material qualities in which it seemeth they take abode with their content. And if there succeed any contrary dispositions, forthwith they depart because they cannot endure it: for it is a thing certain that there are to be found some dispositions in a man's body which the devil coveteth with so great eagerness as to enjoy them, he entereth into the man where they rested, wherethrough he becometh possessed. But the same being corrupted and changed by contrary medicines, and an alteration being wrought in these black, filthy, and stincking humours, he naturally comes to depart. This is plainly discerned by experience, for if there be in a house, great, dark, foul, putrefied, melancholic, and void of dwellers to make abode therein, the devils soon take it up for their lodging. But if the same be cleansed, the windows opened, and the sunbeams admitted to enter, by and by they get them packing, and especially if it be inhabited by much company, and that there be meetings and pastimes, and playing on musical instruments.

How greatly harmony and good proportion offendeth the devil is apparently seen by the authority of the divine Scripture, where we find recounted that David, taking a harp and playing thereupon, straightways made the devils run away and depart out of Saul[68] his body. And albeit this matter have his spiritual understanding, yet I conceive thereby that music naturally molesteth

---

[68] Saul] Saul was the first king of a united Kingdom of Israel. He was anointed by the prophet Samuel. Eventually, he was accused of disobedience and of rejecting the word of God. As a result, he was punished by an evil spirit from God that tormented him. When the evil spirit troubled Saul, David was sent to play on an harp before him to relieve his suffering. Thus he was introduced to the court of Saul.

the devil, wherethrough he cannot in any sort endure it. The people of Israel knew before by experience that the devil was enemy to music, and because they had notice hereof, Saul's servants spoke these words: 'Behold, the evil spirit of the Lord tormenteth thee, let my Lord the king therefore command that thy servants, who wait in thy presence, search out a man who can play on the cithern to the end that when the evil spirit of the Lord taketh thee, he may play with his hand and thou thereby mayst receive ease.' In the self manner as there are found out words and conjurations which make the devil to tremble and not to hear them, he abandoneth the place which he chose for his habitation. So Josephus[69] recounteth that Solomon left in writing certain manners of conjuration by which he not only chased away the devil for the present, but he never had the hardiness to return again to that body from whence he was once so expelled. The same Solomon showed also a root of so abominable savour in the devil's nose that if it were applied to the nostrils of the possessed, he would forthwith shake his cares and run away. The devil is so slovenly, so melancholic, and so much an enemy to things neat, cheerful, and clear, that when Christ entered into the region of Gerasans, St Matthew recounteth how certain devils met him in dead carcasses which they had caught out of their graves, crying, and saying: 'Jesus thou son of David, what hast thou to do with us that thou art come beforehand to torment us? We pray thee that if thou be to drive us out of this place where we are, thou wilt yet let us enter into that herd of swine which is yonder.' For which reason, the Holy Scripture termeth them unclean spirits, whence we plainly discern that not only the reasonable soul requireth such dispositions in the body that they may inform° it, and be the beginning of all his operations, but also hath need to sojourn therein as in a place befitting his nature.

The devils then (being a substance of more perfection) abhor some bodily qualities and in the contrary take pleasure and contentment. In sort, that this of Galen is no good argument (the reasonable soul through excess of heat departs from the body, ergo it is corruptible) inasmuch as the devil doth the like (as we have said) and yet for all this is not mortal.

But that which to this purpose deserveth most note is that the

---

[69] Josephus] Flavius Josephus (37–100 AD) was a Jewish-Roman historical writer.

devil not only coveteth places alterable with bodily qualities to sojourn there at his pleasure, but also when he will work anything which much importeth him, he serves himself with such bodily qualities as are aidable to that effect. For if I should demand now wherein the devil grounded himself when minding to beguile Eve, he entered rather into a venomous serpent than into a horse, a bear, a wolf, or any other beast which were not of so ghastly shape, I wot not what might be given in answer, well I know that Galen admitteth not the sentences of Moses, nor of Christ our redeemer because, sayth he, they both spoke without making demonstration. But I have always desired to learn from some Catholic the solution of this doubt and none hath yet satisfied me.

This is certain (as already we have proved) that burnt and inflamed choler is a humour which teacheth the reasonable soul in what sort to practise treasons and treacheries, and amongst brute beasts there is none which so much partaketh of this humour as the serpent wherethrough more than all the rest, sayth the Scripture, he is crafty and guileful. The reasonable soul although it be the meanest of all the intelligences, partakes yet the same nature with the devil and the angels. And in like manner, as there it takes the service of venomous choler to make a man wily and subtle, so the devil (being entered into the body of this cruel beast) made himself the more cunning and deceitful. This manner of philosophizing will not stick much in the natural philosophers' stomachs because the same carrieth some appearance that it may be so. But that which will breed them more astonishment is that when God would draw the world out of error, and easily teach them the truth (a work contrary to that which the devil went about) he came in the shape of a dove, and not of an eagle, nor a peacock, nor of any other birds of fairer figure. And the cause known is this, that the dove partaketh much of the humour which inclineth to uprightness, to plainness, to truth, and to simplicity, and wanted choler, the instrument of guile and maliciousness.

None of these things are admitted by Galen nor by natural philosophers, for they cannot conceive how the reasonable soul and the devil (being spiritual substances) can be altered by material qualities as are heat, coldness, moisture, and drought. For if fire bring in heat to the wood, it is because they both possess a body and a quantity whereof they are the subject, the which faileth in spiritual substances and admit (as a thing yet impossible) that bodily qualities might alter a spiritual substance, what eyes hath the devil, or the reasonable soul, wherewith to see the colours and shapes of things? Or what smelling to receive savours or what hearing for music? Or what feeling to rest offended with much heat, seeing that for all these bodily instruments are behooful?° And if the reasonable soul being severed from the body remain aggrieved and receive anguish and sadness, it is not possible that his nature should rest free from alteration or not come to corruption. These difficulties and arguments perplexed Galen and the other philosophers of our times, but with me they conclude nothing. For when Aristotle affirmed that the chiefest property which substance had was to be subject to accidents, he restrained the same, neither to bodily nor to spiritual, for the property of the general is equally partaked by the special, and so he said that the accidents of the body pass to the substance of the reasonable soul, and those of the soul, to the body. On which principle he grounded himself to write all that which he uttered as touching physiognomy, especially that the accidents by which the powers receive alteration are all spiritual, without body, and without quantity or matter, and so they grow to multiply in a moment through their mean, and pass through a glass window without breaking the same, and two contrary accidents may be extended in one self subject as much as possibly they can be. In respect of which self quality, Galen termeth them undividable, and the vulgar philosophers intentional, and the matter being in this sort they may be very well proportioned with the spiritual substance.

    I cannot forgo to think that the reasonable soul, severed from the body, as also the devil, hath a power sightful,° smelling, hearing, and feeling. The which (me seemeth) is easy to be proved. For if it be true that their powers be known by means of their actions, it is a thing certain that the devil had a smelling power when he smelled that root which Solomon commanded should be applied to the nostrils of the possessed. And likewise that he had a hearing power, seeing he heard the music which David made to Saul. To say then that the devil received these qualities by his understanding it is a matter not avouchable in the doctrine of the vulgar philosophers, for this power is spiritual and the objects of the five senses are material. And so it behoveth to seek out some other powers in the reasonable soul and in the devil to which they may carry proportion. And if not, put case that the soul of the rich glutton had obtained at the hands of Abraham that the soul of Lazarus should return to the world to preach to his brethren, and persuade them that they should become honest men, to the end they might not pass to that place of torments where himself abode. I demand now in what manner the soul of Lazarus should have known to go to the city, and to those men's houses, and if the same had met them by the way, in company with others, whether it could have known them by sight and been able to diversify them from those who came with them? And if those brethren of the rich glutton had inquired of the same who it was, and who had sent it, whether the same did partake any power to hear their words? The same may be demanded of the devil when he followed after Christ our redeemer, hearing him to preach, and seeing the miracles which he did, and in that disputation which they had together in the wilderness, with what ears the devil received the words and the answers which Christ gave unto him?

Verily, it betokens a want of understanding to think that the devil, or the reasonable soul (sundered from the body), cannot know the objects of the five senses albeit they want the bodily instruments. For by the same reason, I will prove unto them that the reasonable soul severed from the body cannot understand, imagine, nor perform the actions of memory. For if whilst the same abideth in the body, it cannot see being deprived of eyes, neither can it discourse or remember if the brain be inflamed. To say then, that the reasonable soul severed from the body cannot discourse because it hath no brain is a folly very great, the which is proved by the self history of Abraham: 'Son remember that thou hast enjoyed good things in thy lifetime, and Lazarus likewise evil, but now he is comforted and thou art tormented. And besides all this, there is placed betwixt you and us a great chaos in sort that those who would pass from hence to you cannot, nor from you to us.' And he said: 'I pray thee then, oh Father, that thou wilt send to my father's house, for I have five brothers that he may yield testimony unto them, so as they come not also to this place of torments.' Whence I conclude that as these two souls discoursed between themselves, and the rich glutton remembered that he had five brothers in his father's house and Abraham brought to his remembrance the delicious life which he had lived in the world, together with Lazarus's penance, and this without use of the brain, so also the souls can see without bodily eyes, hear without ears, taste without a tongue, smell without nostrils, and touch without sinews and without flesh, and that much better beyond comparison. The like may be understood of the devil, for he partaketh the same nature with the reasonable soul.

All these doubts the soul of the rich glutton will very well resolve, of whom St Luke recounteth that being in hell, he lifted up his eyes and beheld Lazarus, who was in Abraham's bosom, and with a loud voice said: 'Father Abraham have mercy on me, send Lazarus that he may dip the point of his finger in water and cool my tongue, for I am tormented in this flame.' Out of the past doctrine, and out of that which is there read, we gather that the fire of hell burneth the souls and is material as this of ours, and that the same annoyed the rich glutton and the other souls (by God's ordinance) with his heat, and that if Lazarus had carried to him a pitcher of fresh water, he should have taken great refreshment thereof. And the reason is very plain, for if that soul could not endure to abide in the body through excessive heat of the fever, and when the same drank fresh water, the soul felt refreshment, why may not we conceive the like when the soul is united with the flames of the fire infernal? The rich glutton's lifting up of his eyes, his thirsty tongue, and Lazarus's finger, are all names of the powers of the soul that so the Scriptures might express them. Those who walk not in this path, and ground not themselves on natural philosophy, utter a thousand follies, but yet hence it cannot be concluded that if the reasonable soul partake grief and sorrow (for that his nature is altered by contrary qualities) therefore the same is corruptible or mortal. For ashes, though they be compounded of the four elements, and of action and power, yet there is no natural agent in the world which can corrupt them, or take from them the qualities that are agreeable to their nature. The natural temperature of ashes we all know to be cold and dry, but though we cast them never so much into the fire, they will not leese their radical coldness which they enjoy. And albeit they remain 100,000 years in the water, it is impossible that (being taken thence) they hold any natural moisture of their own. And yet for all this, we cannot but grant that by fire they receive heat, and by water, moisture. But these two qualities are superficial in the ashes, and endure a small time in the subject, for taken from the fire forthwith they become cold, and from the water, they abide not moist an hour.

But there is offered a doubt in this discourse and reasoning of the rich glutton with Abraham, and that is how the soul of Abraham was endowed with better reason than that of the rich man, it being alleged before that all reasonable souls (issued out of the body) are of equal perfection and knowledge. Whereto we may answer in one of these two manners. The first is that the science and knowledge which the soul purchaseth, whilst it remaineth in the body, is not lost when a man dieth, but rather groweth more perfect, for he is freed from some errors. The soul of Abraham departed out of this life replenished with wisdom and with many revelations and secrets which God communicated unto him as his very friend. But that of the rich glutton it behoved that of necessity it should depart away ignorant. First, by reason of his sin, which createth ignorance in a man; and next, for that riches herein work a contrary effect unto poverty: this giveth a man wit, as hereafter we may well prove, and prosperity reaveth it away. There may also another answer be given after our doctrine, and it is this, that the matter of which these two souls disputed was school divinity,° for to know whether abiding in hell, there were place for mercy, and whether Lazarus might pass unto hell, and whether it were convenient to send a deceased person to the world who should give notice to the living of the torments which the damned there endured, are all school-points whose decision appertaineth to the understanding, as hereafter I will make proof, and amongst the first qualities, there is none which so much garboileth this power as excessive heat with which the rich glutton was so tormented. But the soul of Abraham made his abode in a place most temperate where it enjoyed great delight and refreshment, and therefore it bred no great wonder that the same was better able to dispute. I concluding then that the reasonable soul and the devil in their operations use the service of material qualities, and that by some they rest aggrieved, and by othersome they receive contentment. And for this reason, they covet to make abode in some places and fly from some other, and yet notwithstanding are not corruptible.

CHAP. VIII.

How there may be assigned to every difference of wit his science, which shall be correspondent to him in particular, and that which is repugnant and contrary, be abandoned.

All arts, saith Cicero, are placed under certain universal principles, which being learned with study and travail finally we so grow to attain unto them. But the art of poesy is in this so special as if God or nature make not a man a poet, little avails it to deliver him the precepts and rules of versifying. For which
10 cause he said thus: 'The studying and learning of other matters consisteth in precepts and in arts, but a poet taketh the course of nature itself and is stirred up by the forces of the mind, and as it were inflamed by a certain divine spirit.' But herein Cicero swerved from reason, for verily there is no science or art devised in the commonwealth which if a man wanting capacity for himself to apply he shall reap any profit thereof, albeit he toil all the days of his life in the precepts and rules of the same. But if he apply himself to that which is agreeable with his natural ability, we see that he will learn in two days. The like we say of poesy without
20 any difference, that if he who hath any answerable nature give himself to make verses, he performeth the same with great perfection, and if otherwise, he shall never be good poet.

This being so, it seemeth now high time to learn by way of art what difference of science is answerable in particular to what difference of wit, to the end that everyone may understand with distinction (after he is acquainted with his own nature) to what art he hath a natural disposition. The arts and sciences which are gotten by the memory are these following, Latin, grammar, or of whatsoever other language, the theory of the laws, divinity
30 positive, cosmography, and arithmetic.

Those which appertain to the understanding are school divinity,° the theory of physic, logic, natural and moral philosophy, and the practice of the laws, which we term pleading. From a good imagination spring all the arts and sciences, which consist in figure, correspondence, harmony, and proportion, such are poetry, eloquence, music, and the skill of preaching, the practice of physic, the mathematics, astrology, and the governing of a commonwealth, the art of warfare, painting, drawing, writing, reading, to be a man gracious, pleasant, neat, witty in managing, and all the engines and devices which artificers make. Besides a certain special gift, whereat the vulgar marvelleth, and that is to indite diverse matters unto four, who write together, and yet all to be penned in good sort. Of all this, we cannot make evident demonstration, nor prove every point by itself. For it were an infinite piece of work, notwithstanding by making proof thereof in three or four sciences, the same reason will afterwards prevail for the rest.

In the catalogue of sciences, which we said appertained to the memory, we placed the Latin tongue and such other as all the nations in the world do speak, the which no wise man will deny, for tongues were devised by men that they might communicate amongst themselves and express one to another their conceits, without that in them there lie hid any other mystery or natural principles, for that the first devisers agreed together and after their best liking (as Aristotle saith) framed the words and gave to every each his signification. From hence arose so great a number of words and so many manners of speech so far besides rule and reason, that if a man had not a good memory, it were impossible to learn them with any other power. How little the understanding and the imagination make for the purpose to learn languages and manners of speech is easily proved by childhood, which being the age wherein man most wanteth these two powers, yet, saith Aristotle, children learn any language more readily than elder men, though these are endowed with a better discourse of reason. And without farther speech, experience plainly proveth this, for so much as we see that if a Biscayan of 30 or 40 years age come to dwell in Castile, he will never learn this language, but if he be but a boy, within two or three years you would think him born in Toledo. The same befalls in the Latin tongue, and in those of all the rest of the world, for all languages hold one self consideration. Then if in the age when memory chiefly reigneth, and the understanding and the imagination least, languages are better learned than when there grows defect of memory and an increase of understanding, it falls out apparent that they are purchased by the memory and by none other power. Languages, saith Aristotle, cannot be gathered out by reason, nor consist in discourse or disputations, for which cause it is necessary to hear the word from another and the signification which it beareth, and to keep the same in mind, and so he proveth that if a man be born deaf, it follows of necessity that he be also dumb, for he cannot hear from another the articulation of the names, nor the signification which was given them by the first deviser.

That languages are apleasure and a conceit of men's brains and nought else is plainly proved, for in them all may the sciences be taught and in each is to be said and expressed that which by the other is inferred. Therefore none of the grave authors attended the learning of strange tongues thereby to deliver their conceits, but the Greeks wrote in Greek, the Romans in Latin, the Hebrews in the Hebrew language, and the Moors in Arabic, and so do I in my Spanish because I know this better than any other. The Romans as lords of the world, finding it was necessary to have one common language by which all nations might have commerce together and themselves be able to hear and understand such as came to demand justice and things appertaining to the government, commanded that in all places of their empire there should schools be kept where the Latin tongue might be taught, and so this usage hath endured even to our time.

School divinity it is a matter certain that it appertaineth to the understanding, presupposing that the operations of this power are to distinguish, conclude, discourse, judge, and make choice, for nothing is done in this faculty which is not to doubt for inconveniences, to answer with distinction, and against the answer to conclude that which is gathered in good consequence, and to return to replication, until the understanding find where to settle. But the greatest proof which in this case may be made is to give to understand with how great difficulty the Latin tongue is joined with school divinity, and how ordinarily it falleth not out that one self man is a good Latinist and a profound scholar, at which effect some curious heads who have lighted hereon much marvelling procured to search out the cause from whence the same might spring, and by their conceit found that school divinity, being written in an easy and common language, and the great Latinists, having accustomed their ear to the well-sounding and fine style of Cicero, they cannot apply themselves to this other. But well should it fall out for the Latinists if this were the cause, for forcing their hearing by use, they should meet with a remedy for this infirmity; but to speak truth, it is rather a headache than an ear-sore.° Such as are skilful in the Latin tongue, it is necessary that they have a great memory, for otherwise they can never become so perfect in a tongue which is not theirs. And because a great and happy memory is as it were contrary to a great and high raised understanding, in one subject where the one is placed, the other is chased away.

Hence remaineth it that he who hath not so deep and lofty an understanding (a power whereto appertaineth to distinguish, conclude, discourse, judge, and choose) cannot soon attain the skill of school divinity. Let him that will not allow this reason for current payment, read St Thomas, Scotus,[70] Durand,[71] and Cajetan,[72] who are the principal in this faculty, and in them he shall find many excellent points indited and written in a style very easy and common. And this proceeded from none other cause than that these grave authors had from their childhood a feeble memory for profiting in the Latin tongue. But coming to logic, metaphysic and school divinity, they reaped that great fruit which we see because they had great understanding.

I can speak of a school divine (and many other can verify the same that knew and conversed with him) who being a principal man in this faculty, not only spoke not finely, nor with well shaped sentences in imitation of Cicero, but whilst he read in a chair, his scholars noted in him that he had less than a mean knowledge in the Latin tongue. Therefore they counseled him (as men ignorant of this doctrine) that he should secretly steal some hour of the day from school divinity and employ the same in reading of Cicero. Who knowing this counsel to proceed from his good friends, not only procured to remedy it privily, but also publicly after he had read the matter of the Trinity[73] and how the divine Word might take flesh, he meant to hear a lecture of the Latin tongue. And it fell out a matter worthy consideration that

---

[70] Scotus] John Duns, Scotus or Duns Scotus (*c.* 1266–1308) is one of the major theologians of the High Middle Ages. Scotus is best known for his theories on the univocity of being (on the concept of existence). Scotus's works were widely published in early modern Europe, and Scotism grew particularly in Catholic countries. Its presence was notable at Alcalá.

[71] Durand] Durand of Saint-Pourçain (*c.* 1275–1332) was a French philosopher and theologian from the Dominican Order. In theology he advocated a separation of natural knowledge from knowledge obtained through faith and revelation.

[72] Cajetan] Thomas Cajetan (1469–1534), Italian philosopher, theologian and cardinal. He was the Pope's Legate in Wittenberg, from where he opposed Martin Luther and the Protestant Reformation. His commentary on the Thomas Aquinas's *Summa Theologica* (1540) is one of his most well-known works.

[73] the Trinity] *De Trinitate* [On the Trinity], by St Augustine, written *c.* 400 AD.

in the long time while he did so, he not only learned nothing of
new, but grew well-near to leese° that little Latin which he had
before, and so at last was driven to read in the vulgar. Pius the
fourth, inquiring what divines were of most special note at the
Council of Trent, he was told of a most singular Spanish divine
whose solutions, answers, arguments, and distinctions were
worthy of admiration. The Pope therefore desirous to see and
know so rare a man sent word unto him that he should come to
Rome and render him account of what was done in the Council.
He came and the Pope did him many favours, amongst the rest,
commanded him to be covered, and taking him by the hand, led
him walking to Castle St Angelo, and speaking very good Latin
showed him his devise touching certain fortifications which he
was then about to make the Castle stronger, asking his opinion
in some particulars. But he answered the Pope so intricately, for
that he could not speak Latin, that the Spanish ambassador, who
at that time was Don Lewes de Requesens,[5] great Commander of
Castile, was fain° to step forth to grace him with his Latin, and to
turn the Pope's discourse into another matter. Finally, the Pope
said to his chamberlains it was not possible that this man had so
much skill in divinity as they made report, seeing he had so little
knowledge in the Latin tongue. But if as he proved him in this
tongue, which is a work of memory, and in platforming and
building, which belong to the imagination, so he had tried him in
a matter appertaining to the understanding, he would have
uttered divine considerations.

In the catalogue of sciences which appertain to the imagination,
we placed poetry amongst the first, and that not by chance nor
for want of consideration, but thereby to give notice how far off
those who have a special gift in poetry are from understanding.
For we shall find that the self difficulty, which the Latin tongue
holdeth in uniting with school divinity, is also found (yea, and
beyond comparison far greater) between this faculty and the art

---

Don Lewes de Requesens] Don Luis de Requeséns y Zúñiga (1528–1576) was a Spanish diplomat in Rome from 1563 to 1569, the right hand of John of Austria during the suppression of the Morisco Revolt in the Alpujarras as well as during the Battle of Lepanto. In 1572 he was appointed Governor of the Duchy of Milan, and two years later succeeded Fernando Álvarez de Toledo, Third Duke of Alba, as governor of the Spanish Netherlands. After his death in Brussels in 1576, he was replaced in this position by John of Austria.

of versifying, and the same is so contrary to the understanding that by the self reason, for which man is likely to prove singular therein he may take his leave of all the other sciences which appertain to this power, and also to the Latin tongue through the contrariety which a good imagination beareth to a great memory.

For the first of these two, Aristotle found not the reason but yet confirmed mine opinion by experience saying: 'Marke, a citizen of Syracuse, was best poet when he lost his understanding', and the cause is for that the difference of the imagination to which poetry belongeth is that which requireth three degrees of heat, and this quality so extended (as we have before expressed) breeds an utter loss of the understanding, the which was observed by the same Aristotle. For he affirmeth that this Mark the Syracusane, growing to more temperate, enjoyed a better understanding, but yet he attained not to versify so well through default of heat, with which this difference of the imagination worketh. And this Cicero wanted when going about to describe in verse the heroical actions of his consulship and the happy birth of Rome in that she was governed by him. He said thus: *O fortunatam natam me consule Romam*.[74] For which cause, Juvenal not conceiving that to a man endowed with so rare a wit as Cicero poetry was a matter repugnant, did satirically nip him saying: 'If thou hadst rehearsed the Philippics against Mark Antony, answerable to the tune of so bad a verse, it should not have cost thy life.'[75]

But worse did Plato understand the same when he said that poetry was no human science but a divine revelation. For if the poets were not ravished besides themselves or full of God, they could not make nor utter anything worthy regard. And he proveth it by a reason, avouching that whilst a man abideth in his sound judgement, he cannot versify. But Aristotle reproveth him for affirming that the art of poetry is not an ability of man but a revelation of God. And he admitteth that a wise man and who is free possessed of his judgement cannot be a poet. And the reason is because where there resteth much understanding it behoveth of force that there befall want of the imagination, whereto appertaineth the art of versifying, which may the more apparently

---

[74] *O fortunatam natam me consule Romam*] O Rome most blest, established in my consulship!
[75] Juvenal] Juvenal was a Roman poet of the late first and early second century AD, author of the *Satires*. This quotation is taken from Satire X.

be proved, knowing that Socrates, after he had learned the art of poetry, for all his precepts and rules could not make so much as one verse. And yet, notwithstanding, he was by the oracle of Apollo adjudged the wisest man of the world.

I hold it then for certain that the boy who will prove of a notable vain for versifying and to whom, upon every slight consideration, consonances° offer themselves, shall ordinarily incur hazard not to learn well the Latin tongue, logic, philosophy, physic, school divinity, and the other arts and sciences which appertain to the understanding and to the memory. For which cause we see by experience that if we charge such a boy to form a nominative without book, he will not learn it in two or three days, but if there be a leaf of paper written in verse to be recited in any comedy, in two turns he fixeth them in his memory. These lose themselves by reading books of chivalry: Orlando,[76] Boccace,[77] Diana of Monte maggior,[78] and such other devices, for all these are works of the imagination. What shall we say then of the harmony of the organs, and of the singing men of the chapel, whose wits are most unprofitable for the Latin tongue and for all other sciences which appertain to the understanding and to the memory? The like reason serveth in playing on instruments and all sorts of music. By these three examples, which we have yielded of the Latin, of school divinity, and of poetry, we shall understand this doctrine to be true, and that we have duly made this partition albeit we make not the like mention in the other arts.

Writing also discovereth the imagination, and so we see that few men of good understanding do write a fair hand, and to this purpose I have noted many examples. And specially I have known a most learned school divine who, shaming at himself to see how bad a hand he wrote, durst not write a letter to any man, nor to answer those which were sent to him, so as he determined with himself to get a scrivener secretly to his house who should teach

---

[76] Orlando] This a reference to the Italian epic poem *Orlando furioso* (first published in full in 1532), by Ludovico Ariosto (1474–1533).

[77] Boccace] Boccaccio.

[78] Diana of Monte maggior] Diana of Montemayor. Jorge de Montemayor (1520?–1561) was the author of the famous pastoral romance *Los Siete Libros de la Diana* (The Seven Books of the Diana) (*c.* 1559), the first pastoral novel published in Spain. Despite the fact that he was Portuguese, Montemayor mostly wrote in Spanish. The *Diana* was a success in early modern Europe, and got translated into other languages, among them, English.

250 him to frame a reasonable letter that might pass, and having for many days taken pains herein, it proved lost labour, and he reaped no profit thereby. Wherefore, as tired out, he forsook the practice, and the teacher, who had taken him in hand, grew astonished to see a man so learned in his profession to be so untoward for writing. But myself, who rest well assured that writing is a work of the imagination, held the same for a natural effect. And if any man be desirous to see and note it, let him consider the scholars who get their livings in the universities by copying out of writings in good form, and he shall find that they
260 can little skill of grammar, logic, and philosophy, and if they study physic or divinity, they fish nothing near the bottom. The boy, then, who with his pen can trick a horse to the life, and a man in good shape, and can make a good pair of [79] it serves little to employ him in any sort of learning, but will do best to set him to some painter who by art may bring forward his nature.

To read well and with readiness discovereth also a certain spice of the imagination, and if the same be very effectual, it booteth little to spend much time at his book, but shall do better to set him to get his living by reading of processes. Here a thing
270 noteworthy offereth itself, and that is that the difference of the imagination which maketh men eloquent and pleasant is contrary to that which is behooful° for a man to read with facility, wherethrough none who is prompt-witted can learn to read without stumbling and putting too somewhat of his own head.

To play well at primero[80] and to face[81] and vie,[82] and to hold and give over when time serveth, and by conjectures to know his adversaries' game, and the skill of discarding, are all works of the imagination. The like we say of playing at cent, and at triumph, though not so far-forth as the primero of Almaine, and the same
280 not only maketh proof and demonstration of the difference of the wit, but also discovereth all the virtues and vices in a man. For at

---

[79]] Carew leaves a blank space in his translation. In the Italian, it corresponds to 'scarabotolli', and in the Spanish, to 'lazos y rasgos' ('laces and strokes'). In sixteenth-century Spain, 'lazos' meant ornaments of intertwined lines in letters, pieces of writing, paintings and illuminated manuscripts.

[80] primero] gambling game in which each player was dealt four cards from a forty-card pack, and players bet on the combinations of the cards they were dealt.

[81] face] to bluff in the card game of primero.

[82] vie] a sum ventured or staked on one's cards.

every moment, there are offered occasions in this play by which a man shall discover what he would do in matters of great importance if opportunity served.

Chess-play is one of the things which best discovereth the imagination, for he that makes ten or twelve fair draughts° one after another on the chess-board gives an evil token of profiting in the sciences which belong to the understanding and to the memory, unless it fall out that he make a union of two or three powers, as we have already noted. And if a very learned school-divine (of mine acquaintance) had been skilled in this doctrine, he should have got notice of a matter which made him very doubtful. He used to play often with a servant of his, and lighting mostly on the loss, told him, much moved: 'Sirrah, how comes it to pass that thou who canst skill neither of Latin, nor logic, nor divinity, though thou hast studied it, yet beatest me that am full of Scotus and St Thomas? Is it possible that thou shouldst have a better wit than I? Verily I cannot believe it, except the devil reveal unto thee what draughts thou shouldst make.' And the mystery was that he had great understanding with which he attained the delicacies of Scotus and Thomas, but wanted that difference of imagination which serveth for playing chess, whereas his servant had an ill understanding and a bad memory, but a good imagination. The scholars who have their books well righted and their chamber well dressed and clean kept, everything in his due place and order, have a certain difference of imagination very contrary to the understanding and to the memory.

Such a like wit have men who go neat and handsomely apparelled, who look all about their cape for a mote, and take dislike at any one wry plait of their garment, this assuredly springeth from their imagination. For if a man that had no skill in versifying, nor towardliness thereunto, chance to fall in love, suddenly, saith Plato, he becomes a poet and very trim and handsome, for love heateth and drieth his brain, and these are qualities which quicken the imagination. The like, as Juvenal noteth, anger doth effect, which passion heateth also the brain:

Anger makes verse, if nature but deny.

Gracious talkers and imitators, and such as can hold at bay, have a certain difference of imagination, very contrary to the understanding and to the memory. For which cause they never prove learned in grammar, logic, school divinity, physic, or the laws. If then they be witty in managing toward for every matter they take in hand, ready in speech, and answering to the purpose, these are fit to serve in courts of justice,[83] for solicitors, attorneys, merchants, and factors to buy and sell but not for learning. Herein the vulgar is much deceived seeing them so ready at all hands, and them seemeth that if such gave themselves to learning, they would prove notable fellows. But in substance there is no wit more repugnant to matters of learning than these. Children that are slow of speech have a moistness in their tongue and also in their brain, but that wearing away, in process of time they become very eloquent and great talkers, through the great memory which they get when that moisture is tempered.

This we know by the things tofore rehearsed befell that famous orator Demosthenes, of whom we said that Cicero marvelled how being so blunt of speech when he was a boy, growing greater he became so eloquent. Children also who have a good voice and warble in the throat are most untoward for all sciences, and the reason is for that they are cold and moist. The which two qualities, being united, we said before, that they breed a damage in the reasonable part. Scholars, who learn their lesson in such manner as their master delivereth it, and so recite the same, it shows a token of a good memory, but the understanding shall abye the bargain. There are offered in this doctrine some problems and doubts, the answer whereunto will perhaps yield more light to conceive that what we have propounded doth carry truth. The first is: whence it groweth that great Latinists are more arrogant and presumptuous on their knowledge than men very well skilled in that kind of learning which appertaineth to the understanding? In sort, that the proverb to let us know what manner of fellow a grammarian is sayth that a grammarian is arrogance itself. The second is: whence it cometh that the Latin tongue is so repugnant to the Spanish capacities and so natural to the French, Italian, Dutch,° English, and other northernly nations? As we see in their works, which by their good Latin phrase straightways prove the author to have been a stranger, and by the barbarousness and ill composition we know the same for

---

[83] 'courts of justice'] corresponds to the Italian 'palazzo' (1586, H6ᵛ), which is an accurate translation of the Spanish 'palacio' ('palace') (1575, fol. 124ʳ). Thus, in Spanish the reference is not to the Courts of Justice but rather to the Court — and therefore to a job in the administration and not in the judicial system.

a Spaniards. The third is: for what reason the things that are spoken and written in the Latin tongue sound better, carry a more loftiness, and have greater delicacy than any other language how good soever? We having avouched before that all languages are nought else but a conceit at pleasure of those who first devised them without holding any foundation in nature. The fourth doubt is: seeing all sciences which appertain to the understanding are written in Latin, how it can frame that such as want memory may read and study them in those books whilst the Latin is (by this reason) so repugnant unto them?

To the first problem we answer that to know whether a man have defect of understanding, there falls out no token more certain than to see him lofty, big looked, presumptuous, desirous of honour, standing on terms, and full of ceremonies. And the reason is for that all these be works of a difference of the imagination which requireth no more but one degree of heat, wherewith the much moisture (which is requisite for the memory) accordeth very fitly, for it wanteth force to resolve the same. Contrariwise, it is an infallible token that if a man be 'naturally lowly', despiser of himself and his own matters, and that not only he vaunteth not, nor praiseth himself, but feels displeasure at the commendations given him by others, and takes shame of places and ceremonies pertaining to honour, such a one may well be pointed at for a man of great understanding, but of small imagination and memory. I said naturally lowly, for if he be so by cunning, this is no certain sign. Hence it cometh that as the grammarians are men of great memory, and make a union with this difference of the imagination, so it is of force that they fail in understanding and be such as the proverb[84] paints them forth.

To the second problem may be answered that Galen inquiring out the wit of men by way of the temperature of the region where they inhabit, saith that those who make abode under the north have all of them want of understanding, but those who are seated between the north and the burned zone[85] are of great wisdom, which situation answereth directly to our region. And verily so it

---

[84] the proverb] Implicitly, the aforementioned proverb 'a grammarian is arrogance itself'; *grammaticus ipsa arrogantia est*, in the Spanish text.
[85] zone] each of the five encircling regions, distinguished by climate, into which the surface of the earth is divided by the tropics and the polar circles.

is, for Spain is not so cold as the places subjected to the Pole, nor so hot as the burned zone. The same sentence doth Aristotle produce, demanding for what cause such as inhabit very cold regions partake less understanding than those who are born in the hotter, and in the answer he very homely handles the Flemish, Dutch, English, and French, saying that their wits are like those of drunkards, for which cause they cannot search out, nor understand the nature of things, and this is occasioned by the much moisture wherewith their brain is replenished and the other parts of the body, the which is known by the whiteness of the face, and the golden colour of the hair, and by that it is a miracle to find a Dutchman bald, and above this they are generally great and of tall stature, through the much moisture which breedeth increase of flesh. But in the Spaniards, we discern the quite contrary: they are somewhat brown, they have black hair, of mean stature, and for the most part we see them bald, which disposition, saith Galen, groweth for that the brain is hot and dry. And if this be true, it behoveth of force that they be endowed with a bad memory and a good understanding, but the Dutchmen possess a great memory and small understanding. For which cause, the one can [acquire] no skill of Latin, and the other easily learn the same. The reason which Aristotle alleged to prove the slender understanding of those who dwell under the north is that the much cold of the country calleth back° the natural heat inward, by counterposition,° and suffereth not the same to spread abroad. For which cause, it partaketh much moisture and much heat, and these unite a great memory for the languages and a good imagination, with which they make clocks, bring the water to Toledo, devise engines and works of rare skill which the Spaniards through defect of imagination cannot frame themselves unto. But set them to logic, to philosophy, to school divinity, to physic, or to the laws, and beyond comparison a Spanish wit with his barbarous terms will deliver more rare points than a stranger. For if you take from them this finesse and quaint phrase of writing, there is nothing in them of rare invention or exquisite choice.

For confirmation of this doctrine, Galen said that in Scythia one only man became a philosopher, but in Athens there were many such, as if he should say that in Scythia, which is a province under the north, it grew a miracle to see a philosopher, but in Athens, they were all born wise and skilful. But albeit philosophy and the other sciences rehearsed by us be repugnant to the northern people, yet they profit well in the mathematicals and in astrology because they have a good imagination.

The answer of the third problem dependeth upon a question much hammered° between Plato and Aristotle: the one saith that there are proper names which by their nature carry signification of things, and that much wit is requisite to devise them. And this opinion is favoured by the divine Scripture, which affirmeth that Adam gave every of those things which God set before him the proper name that best was fitting for them. But Aristotle will not grant that in any tongue there can be found any name, or manner of speech, which can signify ought of its own nature, for that all names are devised and shaped after the conceit of men. Whence we see by experience that wine hath above sixty names, and bread as many, in every language his, and of none we can avouch that the same is natural and agreeable thereunto, for then all in the world would use but that. But for all this, the sentence of Plato is truer, for put case that the first devisers feigned the words at their pleasure and will, yet was the same by a reasonable instinct communicated with the ear with the nature of the thing and with the good grace and well sounding of the pronunciation, not making the words over short or long, nor enforcing an unseemly framing of the mouth in time of utterance, settling the accent in his convenient place, and observing the other conditions which a tongue should possess to be fine and not barbarous.

Of this self opinion with Plato was a Spanish gentleman who made it his pastime to write books of chivalry, because he had a certain kind of imagination which enticeth men to feigning and leasings. Of him it is reported that being to bring into his works a furious giant, he went many days devising a name which might in all points be answerable to his fierceness. Neither could he light upon any until playing one day at cards in his friend's house, he heard the owner of the house say: 'Ho, sirrah, boy, *tra qui tantos.*' The gentleman so soon as he heard this name *traquitantos*, suddenly he took the same for a word of full sound in the ear, and without any longer looking arose saying: 'Gentlemen, I will play no more, for many days are past sithence I have gone seeking out a name which might fit well with a furious giant whom I bring into those volumes which I now am making, and I could not find the same until I came to this house, where ever I receive all courtesy.' The curiosity of this gentleman in calling the giant *Traquitantos* had also those first men who devised the Latin tongue, in that they found out a language of so good sound to the ear. Therefore we need not marvel that the things which are spoken and written in Latin do sound so well, and in other tongues so ill, for their first inventers were barbarous.

480 The last doubt I have been forced to allege for satisfying of diverse who have stumbled thereon, though the solution be very easy, for those who have great understanding are not utterly deprived of memory inasmuch as if they wanted the same, it would fall out impossible that the understanding could discourse or frame reasons, for this power is that which keepeth in hand the matter and the fantasies, whereon it behoveth to use speculation. But for that the same is weak of three degrees of perfection, whereto men may attain in the Latin tongue (namely, to understand, to write, and to speak the same perfectly) it can
490 hardly pass the first without fault and stumbling.

CHAP. IX.

How it may be proved that the eloquence and finesse of speech cannot find place in men of great understanding.

One of the graces by which the vulgar is best persuaded, and thinketh that a man hath much knowledge and wisdom, is to hear him speak with great eloquence: to have a smooth tongue, plenty of sweet and pleasant words, and to allege many examples fit for the purpose that is in hand. But this verily springeth from a union which the memory maketh with the imagination, in a degree and
10 measure of heat that cannot resolve the moisture of the brain and serveth to lift up the figures and cause them to boil, wherethrough are discovered many conceits and points to be uttered. In this union it is impossible that discourse may be found, for we have already said and proved heretofore that this power greatly abhorred heat, and moisture cannot support it. Which doctrine, if the Athenians had known, they would not so much have marvelled to see so wise a man as Socrates not to have the gift of utterance, of whom, those who understood how great his knowledge was, said that his words and his sentences were like a
20 wooden chest knobby and nothing trimmed on the outside, but that in opening the same, within held lineaments and portraitures of rare admiration. In the same ignorance rest they who,

attempting to render a reason of Aristotle's bad style and obscureness, said that of set purpose, and because he would that his works should carry authority, he wrote under riddles and with so slender ornament of words and simple manner of deliverance. And if we consider also the so harsh proceeding of Plato, and the briefness with which he writeth, the obscurity of his reasons, and the ill placing of the parts of his tale, we shall find that nought else save this occasioned the same. For such also we find the works of Hippocrates, the thefts which he committeth of nouns and verbs, the ill disposition of his sentences, and the weak foundation of his reasons to stuff out the empty places of his doctrine. What will you more? Unless that when he would yield a very particular reckoning to his friend Damagetus, how Artaxerxes, King of Persia, had sent for him, promising him as much gold and silver as he lift himself, and to make him one of the great ones of his kingdom, having plenty of answers to so many demands, he writ only thus: 'The king of Persia hath sent for me not knowing that with me the respect of wisdom is greater than that of gold, farewell.' Which matter, if it had passed through the hands of any other man of good imagination and memory, a whole leaf of paper would not have sufficed to set it forth.

But who would have been so hardy to allege for the purpose of this doctrine the example of St Paul, and to affirm that he was a man of great understanding and little memory, and that with these his forces, he could not skill of tongues, nor deliver his mind in them polishedly and with gracefulness if himself had not so said: 'I reckon not myself to have done less than the greatest Apostles, for though I be ignorant of speech, yet am I not so in understanding.' As if he should say: 'I confess that I have not the gift of utterance, but for science and knowledge, none of the greatest Apostles goeth beyond me.' Which difference of wit was so appropriate to the preaching of the Gospel that choice could not be made of a better, for that a preacher should be eloquent, and have great furniture of quaint terms, is not a matter convenient, for the force of the orators of those days appeared in making the hearers' repute things false for true, and what the vulgar held for good and behooful,° they, using the precepts of their art, persuaded the contrary, and maintained that it was better to be poor than rich, sick than whole, fond than wise, and other points manifestly repugnant to the opinion of the vulgar.

For which cause the Hebrews termed them *geragnin*, that is to say, deceivers. Of the same opinion was Cato the more,[86] and held the abode of these in Rome for very dangerous, inasmuch as the forces of the Roman Empire were grounded on arms, and they began then to persuade that the Roman youth should abandon those, and give themselves to this kind of wisdom. Therefore, in brief, he procured them to be banished out of Rome, forbidding them ever to return again.

If God then had sought out an eloquent preacher who should have used ornament of speech, and that he had entered into Athens or Rome avouching that in Jerusalem the Jews had crucified a man who was very God, and that he died of his own accord to redeem sinners, and rose again the third day, and ascended into heaven where he now sitteth, what would the hearers have thought, save that these things were some of those follies and vanities which the orators were wont to persuade by the force of their art? For which cause, St Paul said: 'For Christ sent me not to baptize but to preach the gospel, and that not in wisdom of words, least the cross of Christ might prove in vain.' The wit of St Paul was appropriate to this service, for he had a large discourse to prove in the synagogues and amongst the gentiles that Jesus Christ was the Messiah promised in the law, and that it was bootless to look for any other. And herewithal he was of slender memory, and therefore he could not skill to speak with ornament and sweet and well-relished° terms, and this was that which was behooful for preaching of the Gospel. I will not maintain for all this that St Paul had not the gift of tongues, but that he could speak all languages as he did his own, neither am I of opinion that to defend the name of Christ, the forces of his great understanding sufficed, if there had not been joined therewithal the mean of grace and a special aid which God to that purpose bestowed upon him. It sufficeth me only to say that supernatural gifts work better when they light upon an apt disposition than if a man were of himself untoward and blockish. Hereto alludeth that doctrine of St Jerome which is found in his proem upon Esay[87] and Jeremiah,[88] where asking what the cause

---

[86] Cato the more] Cato the Elder, or Cato the Censor (234 BC–149 BC) was a conservative Roman statesman who firmly opposed Hellenization.
[87] Esay] Isaiah. See note 50.
[88] Jeremiah] Jeremiah was one of the greater prophets of the Old Testament. He

100 is that it being one self Holy Ghost which spoke by the mouth of Jeremiah and of Esay, one of them propounded the matters which he wrote with so great elegancy, and Jeremiah scarcely wist° how to speak. To which doubt he answereth that the Holy Ghost applieth itself to the natural manner of proceeding of each prophet, without that his grace varieth their nature or teacheth them the language wherein they are to publish their prophecy. Therefore we must understand that Esay was a noble gentleman brought up in court, and in the city of Jerusalem, and for this cause had ornament and polishedness° of speech. But Jeremiah
110 was born and reared in a village of Jerusalem called Anathochites, blunt and rude in behaviour as a country person, and of such a style the Holy Ghost used the service in the prophecy which he commanded unto him. The same may be said of St Paul's Epistles, that the Holy Ghost dwelled in him when he wrote them to the end he might not err, but the language and manner of speech was St Paul's natural, applied to the doctrine which he wrote, for the truth of school divinity° abhorreth many words. But the practice of languages and the ornament and polishment° of speech may very well be joined with positive° divinity, for this faculty
120 appertaineth to the memory and is nought else save a mass of words and Catholic sentences taken out of the holy doctors and the divine Scripture, and preserved in this power, as the grammarian doth with the flowers of the poets Virgil, Horace, Terence,[89] and other Latin authors whom he readeth, who meeting occasion to rehearse them, he comes out straightways with a shred of Cicero or Quintilian whereby he makes his hearers know what he is able to do.

   Those that are endowed with this union of the imagination and of the memory, and travail in gathering the fruit of whatsoever
130 hath been said or written in their profession, and serve themselves therewith at convenient occasions with great ornament of words and gracious fashions of speech for that so many things are

---

assisted King Josiah of Judah in his work of reformation and repentance after the idolatrous practices of his father and grandfather. Jeremiah undertook as his main project to reveal the sins of the people and warn them about the impending disaster: the destruction and captivity brought about by the Babylonian army.

[89] Terence] Publius Terentius Afer (*c.* 190–159 BC) was a Roman playwright, author of six plays.

already found out in all the sciences, it seemeth to them who know not this doctrine that they are of great profoundness, whereas in truth they hold much of the ass, for if you grow to try them in the foundations of that which they allege and affirm, they then discover their wants. And the reason is because so great a flowing of speech cannot be united with the understanding, whereto appertaineth to search out the bottom of the truth. Of these the divine Scripture said: 'Where there is plenty of words, there reigneth great scarcity', as if He had said that a man of many words ordinarily wanteth understanding and wisdom.

Those who are endowed with this union of the imagination and memory enter with great courage to interpret the divine Scripture, it seeming to them that because they understand well the Hebrew, Greek and Latin tongues, they have the way made smooth to gather out the very spirit of the letter. But verily they ruinate themselves; first, because the words of the divine text and his manners of speech have many other significations besides those which Cicero understood in Latin. And then because their understanding is defective, which power verifieth whether a sense be catholic or depraved, and this is it which may make choice by the grace supernatural of two or three senses that are gathered out of the letter, which is most true and Catholic.

Beguilings, saith Plato, never befall in things unlike and very different, but when many things meet which carry near resemblance, for if we set before a sharp sight, a little salt, sugar, meal, and lime, all well pounded and beaten to powder, and each one severally by itself, what should he do who wanted taste, if with his eyes he should be set to discern every of these powders from other without erring, saying, 'this is salt', 'this sugar', 'this meal', and 'this lime'? For my part I believe he would be deceived through the great resemblance which these things have between themselves. But if there were a heap of salt, one of sugar, one of corn, one of earth, and one of stones, it is certain he would not be deceived in giving each of these heaps his name, though his sight were dim, for each is of a diverse figure. The same we see befalleth every day in the senses and spirits which the divines give to the Holy Scripture: of which two or three being looked on, at first sight they all carry a show to be Catholic, and to agree well with the letter, but yet in truth are not so, neither the Holy Ghost so meant. To choose the best of these senses, and to refuse the bad, it is a thing assured that the divine employeth not his memory not his imagination, but his understanding. Wherefore I avouch that the positive° divine ought to confer with the schoolman and to inquire at his hands that of these senses he may choose that which shall appear to be soundest, unless he will be sent to the Holy House.[90] For this cause do heretics so much abhor school divinity and learn to banish it out of the world, for by distinguishing, inferring, framing of reasons, and judging, we attain to understand the truth and to discover falsehood.

CHAP. X.

How it is proved that the theoric of divinity appertaineth to the understanding and preaching (which is his practice) to the imagination.

It is a problem often demanded, not only by folk learned and wise, but also the vulgar will put in their oar, and every day bring in question, for what cause a divine being a great man in the schools, sharp in disputing, ready in answering, and in writing and lecturing of rare learning, yet getting up into the pulpit, cannot skill of preaching. And contrariwise, if one prove a gallant[91] preacher, eloquent, gracious, and that draws the people after him, it seems a miracle if he be deeply seen in school divinity.° Wherefore they admit not for a sound consequence: 'such a one is a great school divine, therefore he will prove a good preacher'; and contrariwise, they will not grant he is a good preacher, therefore he hath skill in school divinity. For to reverse the one and other of these consequences, there may be alleged for each more instances than are hairs on our head.

No man hitherto hath been able to answer this demand, save after the ordinary guise, viz., to attribute the whole to God and to the distribution of his graces, and to my liking they do very well, inasmuch as they know not any more particular occasion thereof. The answer of this doubt (in some sort) is given by us in the foregoing chapter, but not so particularly as is requisite, and it was that school divinity appertaineth to the understanding. But now we affirm and will prove that preaching, and his practice, is

---

[90] Holy House] The Inquisition.
[91] gallant] full of showy expressions.

a work of the imagination. And as it falls out a difficult matter to join in one self brain a good understanding and much imagination, so likewise it will hardly fall that one self man be a great school divine and a famous preacher. And that school divinity is a work of the understanding hath tofore been proved when we proved the repugnancy which it carried to the Latin tongue. For which cause it shall not now be necessary to prove the same anew, only it shall suffice to give to understand that the grace and delightfulness which good preachers have, whereby they draw their audience unto them and hold them well pleased, is altogether a work of the imagination, and part thereof of a good memory. And to the end I may better expound myself, and cause it as it were to be felt with the hand, it behoveth first to presuppose that man is a living creature capable of reason, of company, and of civility, and to the end that his nature might be the more abled by art, the ancient philosophers devised logic to teach him how he might frame his reasons with those precepts and rules, how he should define the nature of things, distinguish, divide, conclude, argue, judge, and choose, without which works it grows impossible that the artist[92] can go forward. And that he might be companiable and civil, it behoved him to speak and to give other men to weet° the conceits which he framed in his mind. And for that he should not deliver them without disposition and without order, they devised another art which they termed rhetoric, which by his precepts and rules might beautify the speech with polished words, with fine phrases, and with stirring affections and gracious colours. But as logic teacheth not a man to discourse and to argue in one science alone, but without difference in all alike, so also rhetoric instructeth how to speak in divinity, in physic, in skill of the laws and in all other sciences and conversations which men intermeddled withal. In sort, that if we will feign a perfect logician, or an accomplished orator, he cannot fall into due consideration unless he be seen in all the sciences, for they all appertain to his jurisdiction, and in which soever of them he may exercise his rules without distinction, not as physic (which hath his matter limited whereof it must entreat), and so likewise natural philosophy and moral, metaphysic, astrology, and the rest. And therefore Cicero said: 'The orator wheresoever he abideth

---

[92] artist] a person skilled or proficient at a particular task or occupation; an expert.

dwelleth in his own.' And in another place he affirmeth: 'In a perfect orator is found all the knowledge of the philosophers,' and therefore the same Cicero avouched that there is no art more difficult than that of a perfect orator, and with more reason he might so have said if he had known with how great hardness all
70 the sciences are united in one particular subject.

Anciently the doctors of the law were adorned with the name of orator, for the perfection of pleading required the notice and furniture of all the arts in the world, for the laws do judge them all. Now to know the defence reserved for every art by itself, it was necessary to have a particular knowledge of them all, for which cause Cicero said: 'No man ought to be reputed in the number of orators who is not well seen in all the arts.' But seeing it was impossible to learn all sciences, first, through the shortness of life, and then because man's wit is so bounded, they let them pass, and
80 of necessity held themselves contented to give credit to the skilful in that art whereof they made profession and no farther.

After this manner of defending causes, straightways succeeded the evangelical doctrine, which might have been persuaded by the art of oratory better than all the sciences of the world besides, for that the same is the most certain and truest. But Christ our redeemer charged St Paul that he should not preach it with wisdom of words to the end the gentiles should not think it was a well couched leasing, as are those which the orators use to persuade by the force of their art. But when the faith had been
90 received, many years after it was allowed to preach with places of rhetoric and to use the service of eloquent speech, for that then the inconvenience fell not in consideration which was extant when St Paul preached. Yea, we see that the preacher reapeth more fruit who hath the conditions of a perfect orator, and is more haunted than he that wanteth them. And the reason is very plain, for if the ancient orators gave the people to understand things false for true (using those their precepts and rules) more easily shall the Christian auditory be drawn when by art they are persuaded to that which already they understand and believe. Besides that the
100 Holy Scripture (after a sort) is all things, and to yield the same a true interpretation, it behoveth to have all the sciences conformable to that so oft said saw: 'he sent his damsels to call to the castle.' This fitteth not to be remembered to the preachers of our time, nor to advise them that now they may do it, for their particular study (besides the fruit which they pretend to bring with

their doctrine) is to seek out a good text to whose purpose they may apply many fine sentences taken out of the divine Scripture, the holy doctors, poets, historians, physicians, and lawyers, without forbearing any science, and speaking copiously with quietness and pleasant words, and with all these things they go amplifying and stuffing their matter an hour or two if need be. Of this saith Cicero the orators of his time made profession. The force of an orator, saith he, and the self art of well speaking, seemeth that it undertaketh and promiseth to speak with copiousness and ornament of whatsoever matter that shall be propounded. Then if we shall prove that the graces and conditions which a perfect orator ought to have do all appertain to the imagination and to the memory, we shall also know that the divine who is endowed with them will be an excellent preacher, but being set to the doctrine of St Thomas and Scotus can little skill thereof, for that the same is a science belonging to the understanding, in which power of necessity it holdeth little force.

What the things be which appertain to the imagination, and by what signs they are to be known, we have heretofore made mention. Now we will return to a replication of them that they may the better be refreshed to the memory. All that which may be termed good figure, good purpose and provision, comes from the grace of the imagination, as are merry jests, resemblances, quips, and comparisons.

The first thing which a perfect orator is to go about (having matter under hand) is to seek out arguments and convenient sentences whereby he may dilate and prove, and that not with all sorts of words, but with such as give a good consonance° to the ear. And therefore Cicero said: 'I take him for an orator who can use in his discourses words well turning with the ear, and sentences convenient for proof.' And this for certain appertaineth to the imagination, sithence therein is a consonance of well pleasing words, and a good direction in the sentences.

The second grace which may not be wanting in a perfect orator is to possess much invention, or much reading, for if he rest bound to dilate and confirm any matter whatsoever, with many speeches and sentences applied to the purpose, it behoveth that he have a very swift imagination and that the same supply (as it were) the place of a brach,[93] to hunt and bring the game to his hand, and

---

[93] brach] a hound which hunts by scent.

when he wants what to say, to devise somewhat as if it were material. For this cause we said before that heat was an instrument with which the imagination worketh, for this quality lifteth up the figures and maketh them to boil. Here is discovered all that which in them may be seen, and if there fell out nought else to be considered, this imagination hath force not only to compound a figure possible with another, but doth join also (after the order of nature) those which are impossible, and of them grows to shape mountains of gold, and calves that fly. In lieu of their own invention, orators may supply the same with much reading, forasmuch as their imagination faileth them. But in conclusion, whatsoever books teach is bounded and limited, and the proper invention is a good fountain which always yieldeth forth new and fresh water. For retaining the things which have been read, it is requisite to possess much memory, and to recite them in the presence of the audience with readiness cannot be done without the same power. For which cause Cicero said: 'He shall (in mine opinion) be an orator worthy of so important a name, who with wisdom, with copiousness, and with ornament, can readily deliver every matter that is worth the hearing.' Heretofore we have said and proved that wisdom appertaineth to the imagination, copiousness of words and sentences to the memory, ornament and polishment to the imagination. To recite so many things without faltering or stopping, for certain is achieved by the goodness of the memory.

To this purpose, Cicero avouched that the good orator ought to rehearse by heart and not by book. It falleth not besides the matter to let you understand that M. Antony of Lebrissa[94] through old age grew to such a decay of memory, that he read his rhetoric lecture to his scholars out of a paper, and for that he was so excellent in his profession, and with good proofs confirmed his points propounded, it passed for current. But that which might no way be tolerated was that where he died suddenly of an apoplexy, the university of Alcalá recommended the making of

---

[94] M. Antony of Lebrissa] the 'M.' stands for 'Maestro', both in Spanish (1575, fol. 144r) as well as in Italian (1586, K1r). Antonio de Lebrija or Antonio de Nebrija (1441–1522) was a Spanish humanist — grammarian, historian, and poet — who taught grammar and rhetoric first at the University of Salamanca and subsequently at Alcalá. His most renowned work is his *Gramática castellana* (1492), the first grammar of the Spanish language.

his funeral oration to a famous preacher, who invented and disposed what he had to say the best he could. But time so pressed him as it grew impossible for him to con° the same without book. Wherefore getting up into the pulpit with his paper in his hand, he began to speak in this sort: 'That which this notable man used to do whilst he read to his scholars, I am now also resolved to do in his imitation, for his death was so sudden, and the commandment to me of making his funeral sermon so late, as I had neither place nor time to study what I might say, nor to con° it by heart. Whatsoever I have been able to gather with the travail of this night, I bring here written in this paper, and beseech your masterships that you will hear the same with patience, and pardon my slender memory.'

This fashion of rehearsing with paper in the hand so highly displeased the audience as they did nought else than smile and murmur. Therefore, very well said Cicero that it behoved to rehearse by heart and not by book. This preacher verily was not endowed with any invention of his own, but was driven to fetch the same out of his books; and to perform this, great study and much memory were requisite. But those who borrow their conceits out of their own brain stand not in need of study, time, or memory, for they find all ready at their fingers' ends. Such will preach to one self audience all their life long without repeating any point touched in twenty years before, whereas those that want invention, in two Lents cull° the flowers out of all the books in a whole world, and ransack to the bottom all the writings that can be gotten, and at the third Lent must go and get themselves a new auditory except they will hear cast in their teeth: 'This is the same which you preached unto us in the year before.'

The third property that a good orator ought to have is that he know how to dispose his matter, placing every word and sentence in his fit room, in sort that the whole may carry an answerable proportion and one thing bring in another. And to this purpose, Cicero said: 'Disposition is an order and distribution of things which showeth what ought in what places to be bestowed.' Which grace when it is not natural, accustomably breedeth much cumber to the preachers, for after they have found in their books many things to deliver, all of them cannot skill to apply this provision readily to every point. This property of ordering and distributing is, for certain, a work of the imagination, since in effect it is nought else but figure and correspondence.

The fourth property wherewith good orators should be endowed, and the most important of all, is action, wherewith they give a being and life to the things which they speak, and with the same do move the hearers, and supple them to believe how that is true which they go about to persuade. For which cause Cicero said: 'Action is that which ought to be governed by the motion of the body, by the gesture, by the countenance, and by the confirmation and variety of the voice.' As if he should say: 'Action ought to be directed in making the motions and gestures which are requisite for the things that are spoken, lifting up and falling with the voice, growing passionate, and suddenly turning to appeasement; one while speaking fast, another while leisurely reproving and cherishing, moving the body, sometimes to the one side, sometimes to the other, plucking in the arms, and stretching them out, laughing and weeping, and upon some occasions beating the hands together.' This grace is so important in preachers that by the same alone (wanting both invention and disposition) of matters of small value and ordinary, they make a sermon which filleth the audience with astonishment, for that they have this action which otherwise is termed spirit or pronunciation. Herein falleth a thing worth the marking, whereby is discovered how much this grace can prevail, and it is that the sermons which through the much action and much spirit do please much, when they be set down in writing are nothing worth, nor will any wellnear vouchsafe their reading. And this groweth because with the pen it is impossible to portray those motions and those gestures which in the pulpit so far wan men's likings. Other sermons show very well in paper, but at their preaching no man listeth to give ear because that action is not given them which is requisite at every close. And therefore Plato said that the style wherewith we speak is far different from that which we write well, wherethrough we see many men who can speak very well do yet indite but meanly; and others contrariwise, indite very well and discourse but harshly. All which is to be reduced to action, and action for certain is a work of the imagination, for all that which we have uttered thereof, maketh figure, correspondence, and good consonance.

The fifth grace is to know how to assemble and allege good examples and comparisons which better contenteth the hearers' humour than anything else. For by a fit example they easily understand the doctrine, and without the same it soon slippeth out of their mind. Whereon Aristotle propounded this question: 'Whence it riseth that men (in making speeches) are better pleased with examples and fables than with conceits?' As if he should say: 'For what occasion do such as come to hear orators make more reckoning of the examples and fables which they allege to prove the things that they strive to persuade than of the arguments and reasons which they frame?' And to those he answereth that by examples and fables men learn best because it is a proof which appertaineth to the sense, but arguments and reasons hold not the like reason, for that they are a work whereto is requisite much understanding. And for this cause Christ our redeemer in his sermons used so many parables and comparisons, because by them he gave to understand many divine secrets. This point of devising fables and comparisons it is a things certain that the same is performed by the imagination, for it is figure, and denoteth good correspondence and similitude.

The sixth property of a good orator is to have a ready tongue of his own, and not affected choice words, and many gracious sorts of utterance, of which graces we have entreated oftentimes heretofore, proving that the one part of them appertaineth to the imagination, and the other to a good memory.

The seventh property of a good orator is that which Cicero speaketh of: 'Furnished with voice, with action, and with comeliness', the voice full and ringing, pleasing to the hearers, not harsh, not hoarse, nor sharp. And although it be true that this springeth from the temperature of the breast and the throat, and not from the imagination, yet sure it is that from the same temperature from which a good imagination groweth (namely, heat), a good voice also fetcheth his original. And to know this, importeth much for our purpose, for the school divines, in that they are of a cold and dry complexion, cannot have their voice a good instrument, and this is a great defect in a pulpit.

This same Aristotle also proveth, alleging the example of old men, by reason of their coldness and dryness. To have a full and clear voice, much heat is requisite to enlarge the passages, and measurable moisture which may supple and soften them. And also Aristotle demandeth: 'Why all who by nature are hot, are also big-voiced?' For which cause we see the contrary in women and eunuchs, who through the much coldness of their complexion, sayth Galen, have their throat and voice very delicate, in sort that when we hear a good voice, we can straightways say it comes of much heat and moisture in the breast, which two qualities, if they pass so far as the brain, make the understanding to decay, and the memory and imagination to increase, which are the two powers whereof the good preacher serveth himself to content his auditory.

The eighth property of a good orator, sayth Cicero, is to have tongue at will, ready, and well exercised, which grace cannot befall men of great understanding, for that it may be ready, it behoveth the same to partake much heat and mean drought. And this cannot light in the melancholic, either natural or by adustion.[95] Aristotle proveth it by asking this question: 'Whence commeth it that such as have an impediment in their speech are reputed to be of complexion melancholic?' To which problem he answereth very untowardly saying that the melancholic have a great imagination and that the tongue cannot haste to utter so fast as the imagination conceiveth, wherethrough they stammer and stumble. Which yet proceedeth from nought else save that the melancholic have ever their mouth full of froth and spittle, through which disposition their tongue is moist and slipper, which thing may evidently be discerned considering the often spitting of such. This self reason did Aristotle render when he demanded: 'Whence it groweth that some are so slow tongued?' And he answered that such have their tongue very cold and moist, which two qualities breed an impediment therein, and make it subject to the palsy. And so you see his conceit of the imagination cannot follow, for this he yieldeth a profitable remedy, viz., to drink a little wine, or at first to hallow somewhat loud before they speak in the presence of their audience, for thereby the tongue getteth heat and drieth.

But Aristotle sayth further, that not to speak plain may grow from having the tongue very hot and very dry, and voucheth the example of choleric persons who, growing in choler, cannot speak, and when they are void of passion and choler, they are very eloquent. The contrary betideth to the phlegmatic, who, being quiet, cannot talk, and when they are angered utter speeches of great eloquence. The reason of this is very manifest, for although it is true that heat aideth the imagination and the tongue also, yet the same may also breed them damage. First, for that they want

---

[95] adustion] state of the humours of being adust; process of becoming adust. See note 99.

supply of replies and witty sentences, as also because the tongue cannot pronounce plainly through overmuch dryness, wherethrough we see that after a man hath drunk a little water, he speaketh better.

The choleric, being quiet, deliver very well, for they then retain that point of heat which is requisite for the tongue and the good imagination. But in anger, the heat groweth beyond due° and turneth the imagination topsy-turvy. The phlegmatic unincensed° have their brain very cold and moist, and therefore are set a ground what to say, and their tongue is over slipper through too much moisture. But when they are set on fire and in choler, the heat forthwith getteth up and so lifteth up the imagination, by which means there comes to their mind much what to deliver, and the tongue giveth no hindrance for that it is heated. These have no great vain in versifying, for that they are cold of brain, who yet (once angered) do then make verses best and with most facility, against such as have stirred them, and to this purpose Juvenal said:

Anger makes verse, if nature but deny.

Through the defect of tongue, men of great understanding cannot be good orators or preachers, and specially for that action requireth a speech sometimes high, and sometimes low, and those who are slow tongued cannot pronounce but with loud voice, and in a manner crying out, and this is one of the things which soonest cloyeth the hearers. Whereon Aristotle moveth this doubt: 'Whence it springeth that men of slow tongue cannot speak soft?' To which problem he answereth very well, saying that fastened to the tongue which is the roof of the mouth, by reason of much moisture, is better loosened with a force than if you put thereto but little might, as if one would lift up a lance taking the same by the point, he shall sooner raise it at one push and with a force, than taking it up by little and little.

Me seemeth I have sufficiently proved that the good natural qualities which a perfect orator ought to have spring for the most part from a good imagination, and some from the memory. And if it be true that the good preachers of our time content their audience because they have these gifts, it followeth very well that whosoever is a great preacher can small skill of school divinity, and a great scholar will hardly away with preaching through the contrariety which the understanding carrieth to the imagination and to the memory. Well knew Aristotle by experience that although the orator learned natural and moral philosophy, physic, metaphysic, the laws, the mathematics, astrology, and all the arts and sciences, notwithstanding he was seen of all these but in the flowers and choice sentences, without piercing to the root of the reason and occasion of any of them. But he thought that this not knowing the divinity, nor the cause of things which is termed *propter quid*, grew for that they bent not themselves thereunto, and therefore propounded this demand: 'Why do we imagine that a philosopher is different from an orator?' To which problem he answereth that the philosopher placeth all his study in knowing the reason and cause of every effect, and the orator in knowing the effect and no farther. And verily it proceedeth from nought else than for that natural philosophy appertaineth to the understanding, which power the orators do want, and therefore in philosophy they can pierce no farther than into the upper skin of things. This self difference there is between the school divine and the positive,° that the one knoweth the cause of whatsoever importeth his faculty, and the other the propositions which are verified and no more. The case then standing thus, it falleth out a dangerous matter that the preacher enjoyeth an office and authority to instruct Christian people in the truth, and that their auditory is bound to believe them, and yet they want that power through which the truth is digged up from the root. We may say of them (without lying) those words of Christ our redeemer: 'Let them go, they are blind, and do guide the blind; and if the blind guide the blind, both fall into the ditch.' It is a thing intolerable to behold with how great audacity such set themselves to preach who cannot one jot of school divinity, nor have any natural ability to learn the same.

Of such St Paul greatly complaineth, saying: 'But the end of the commandement is charity from a pure heart and good conscience, and faith unfeigned. From which verily some straying have turned aside to vain babbling, who would be doctors in the law and yet understand not the things which they speak, nor which they avouch.'[96]

Besides this, we have proved tofore that those who have much

---

[96] 33. *Index*: 'for that natural philosohy appertaineth to the understanding [394] [...] por no tener entendimiento para alcanzar la verdad ('for having no understanding to reach truth')', within the fragment cut by Carew [censored].

imagination are choleric, subtle, malignant, and cavillers, and always inclined to evil, which they can compass with much readiness and craft. Touching the orators of his time, Aristotle propoundeth this demand: 'Why we use to call an orator crafty, and give not this name to a musician, nor to a comical poet?' And more would this difficulty have grown if Aristotle had understood that music and the stage appertain to the imagination. To which problem he answereth that musicians and stage players shoot at none other butt than to delight the hearers, but the orator goes about to purchase somewhat for himself, and therefore it behoveth him to use rules and readiness to the end the hearers may not smell out his fetch and bent.

Such properties as these be had those false preachers of whom St Paul spoke writing to the Corinthians: 'But I fear that as the serpent beguiled Eve with his subtlety, so their senses are led astray, for these false Apostles are guileful workmen who transform themselves into the Apostles of Christ, and this is no wonder, for Satan transformed himself into an angel of light, and therefore it is no great matter for his ministers to transform themselves as ministers of justice whose end shall be their work.' As if he should say: 'I have great fear, my brethren, that as the serpent beguiled Eve with his subtlety and malice, so they also intricate their judgement and perseverance, for these false Apostles are like pottage made of a fox. Preachers who speak under wiles represent very perfectly a kind of holiness, seem the Apostles of Jesus Christ, and yet are disciples of the devil, who can skill so well to represent an angel of light that there needeth not a supernatural gift to discover what he is. And since the master can play his part so well, it is not strange that they also who have learned his doctrine practise the semblable, whose end shall be none other than their works.' All these properties are well known to appertain to the imagination, and that Aristotle said very well that orators are subtle and ready because they are ever in hand to get somewhat for themselves.

Such as possess a forcible imagination we said before that they are of complexion very hot, and from this quality spring three principal vices in a man: pride, gluttony, and lechery, for which cause the Apostle said: 'Such served not our Lord Jesus Christ but their belly.'

And that these three evil inclinations spring from heat, and the contrary virtues from cold, Aristotle proveth, saying thus: 'And therefore it holdeth the same force to shape conditions, for heat and cold (more than anything else which is in the body) do season manners, and therefore printeth and worketh in us the qualities of manners', as if he should say: 'From heat and cold spring all the conditions of man, for these two qualities do more alter our nature than any other.' For which cause men of great imagination are ordinarily bad and vicious, for they abandon themselves to be guided by their natural inclination, and have wit and ability to do lewdly. For which cause the same Aristotle asketh whence it groweth that a man being so much instructed is yet the most unjust of all living creatures. To which problem he maketh answer that man hath much wit and a great imagination, and for this he findeth many ways to do ill, and (as by his nature he coveteth delights, and to be superior to all and of great happiness) it is of force that he offend, for these things cannot be achieved but by doing wrong to many. But Aristotle wist° not how to frame this problem, nor to yield a fitting answer.

Better might he have inquired for what cause the worst people are commonly of great wit, and amongst those, such as are best furnished with ability commit the lewdest pranks, whereas of due,° a good wit and sufficiency should rather incline a man to virtue and goodliness than to vices and misdoing. The answer hereto is for that those who partake much heat are men of great imagination, and the same quality which maketh them witty traineth them to be naughty and vicious. But when the understanding overruleth, it ordinarily inclineth a man to virtue because this power is founded on cold and dry, from which two qualities bud many virtues, as are continence, humility, temperance, and from heat the contrary. And if Aristotle had known this point of philosophy, he should have been able to answer this problem which saith: 'Whence may it proceed that that sort of men whom we call craftsmen of Bacchus or stage players are for the most part ill conditioned?' As if he should say. 'For what cause are such as gain their living on the stage, innkeepers and butchers, and those whose service is used about feasts and banquets to order the cates,° ordinarily naught and vicious?' To which problem he answereth saying that such by being occupied in these belly-cheer offices leave themselves no leisure to study, and therefore pass over their life in incontinence. And hereto is poverty also aiding, which accustomably bringeth with it many evils. But verily this is not the reason, but playing on the stage and ordering of feasts springeth from the difference of the imagination which inviteth a man to this manner of life. And because this difference of imagination consisteth in heat, all of them have very good stomachs and great appetite to eat and drink. These, although they gave themselves to learning, should thereby reap little fruit, and had they been never so wealthy, yet would they howsoever have cast their affection to these services, were they even baser than they are, for the wit and ability draweth every one to that art which answereth it in proportion.

For this cause Aristotle demanded what the reason was why there are men who more willingly addict themselves to the profession of which they have made choice (though somewhiles unworthy) than to the more honourable. As for example, to be rather a juggler, a stage player, or a trumpeter, than an astrologer or an orator. To which problem he answereth very well saying that a man soon discerneth to what art he is disposed and inclined of his own nature because he hath somewhat within that teacheth him, and nature can do so much with her pricks that albeit the art and office be unseemly for the calling of the learner, yet he cleaveth unto that and not to others of greater estimation. But sithence we have put by this manner of wits from the function of preaching, and that we are bound to give and bestow upon every difference of ability that sort of learning which is answerable thereto in particular, we must likewise determine what sort of wit he ought to be endowed withal unto whose charge the function of preaching is to be committed, which is the thing that most importeth the Christian commonwealth, for we must conceive that albeit we have proved heretofore that it is a matter repugnant in nature to find a great wit accompanied with much imagination and memory, notwithstanding this rule holdeth not so universally in all arts, but that it admitteth his exceptions and sometimes commeth short.

In the last chapter of this work save one, we will prove at full that if nature be possessed of her due force, and have no impediment cast athwart° to stop her, she maketh so perfect a difference of wit as the same uniteth in one self subject a great understanding with much imagination and memory, as if they were not contrary, nor held any natural opposition.

This should be a fitting ability and convenient for the function of preaching if there could be found many subjects to be endowed therewith, but (as we will show in the place alleged) they are so few, that of 100,000 whom I have measured, I can meet but with one of the size. Therefore it behoveth to seek out another more familiar difference of wit, though not so far stepped° in perfection as the former. We must then weet that between the physicians and philosophers riseth a great diversity in opinions, for resolving the temperature and the quality of vinegar, of choler adust, and of ashes, inasmuch as these things sometimes work the effect of heat, and sometimes of cold, and thereon they divided themselves into diverse sects. But the truth is that all these things which suffer adustion, and are consumed and burned by the fire, have a variable temperature. The greater part of the subject is cold and dry, but there are also other parts intermingled, so subtle and delicate, and of such fervency and heat, that albeit they contain little in quantity, yet they carry more efficacy in working than all the rest of the subject.

So we see that vinegar and melancholy through adustion open and leaven the earth by means of the heat, and close it not though the more part of these humours be cold. Hence is gathered that the melancholic by adustion accompany great understanding with much imagination, but they are all weak of memory, for the much adustion much also drieth and hardeneth the brain. These are good preachers, or at least the best that may be found, saving those perfect ones of whom we spoke, for although memory fail them, they enjoy of themselves such invention that the very imagination serveth them instead of memory and remembrance, and ministereth unto them figures and sentences to deliver without that they stand in need of ought besides. Which these cannot bring about who have conned bosom-sermons,° and swerving from that bias are straight set a ground, without having the furniture of any second means, to bring themselves afloat again. And that melancholy by adustion hath this variety of temperature, namely, cold and dry for the understanding, and heat for the imagination, Aristotle declareth in these words: 'Melancholic men are variable and unequal, for the force of choler adust is variable and unequal, as if the same might be greatly both hot and cold.' And as if he had said: 'Melancholic men by adustion are variable and unequal in their complexion, for that choler adust is very unequal, inasmuch as sometimes it is exceeding hot, and sometimes cold beyond measure.'

The signs by which men of this temperature may be known are very manifest: they have the colour of their countenance a dark green, or sallow, their eyes very fiery (of whom it was said: 'He is a man that hath blood in his eyes'[97]), their hair black and bald, their flesh lean, rough and hairy, their veins big. They are of very good conversation and affable, but lecherous, proud, stately, blasphemers, wily, double, injurious, friends of ill doing and desirous of revenge. This is to be understood when melancholy is kindled, but if it be cooled, forthwith there grow in them the contrary virtues: chastity, humility, fear and reverence of God, charity, mercy, and great acknowledgement of their sins, with sighings and tears,[98] for which cause they live in continual war and strife without ever enjoying ease or rest. Sometimes vice prevaileth in them, sometimes virtue, but with all these defects, they are wittiest, and most able for the function of preaching, and for all matters of wisdom which befall in the world, for they have an understanding to know the truth, and a great imagination to be able to persuade the same.

Wherethrough we see that which God did when He would fashion a man in his mother's womb to the end that He might be able to discover to the world the coming of His Son, and have the way to prove and persuade that Christ was the Messiah and promised in the law. For making him of great understanding and of much imagination, it fell out of necessity (keeping the natural order) that He should also make him choleric and adust.[99] And that this is true may easily be understood by him who considereth the great fire and fury with which he persecuted the church, the grief conceived by the synagogues when they saw him converted as they who had forgone a man of high importance and of whom the contrary party had made a gainful purchase. It is also known by the tokens of the reasonable choler with which he spoke and answered the deputy, consuls, and the judges who had arrested him defending his own person and the name of Christ with so great art and readiness as he convinced them all. Yet he had an imperfection in his tongue and was not very prompt of speech,

---

[97] hath blood in his eyes] in sixteenth-century Spain: to be courageous, to have a sense of duty and honour, and the drive to defend it.

[98] 34. *Index*: 'chastity, humility, fear and reverence of God, charity, mercy, and great acknowledgement of their sins, with sighings and tears' [censored].

[99] adust] any of the humours of the body abnormally concentrated and dark in colour, and associated with a pathological state of hotness and dryness of the body.

which Aristotle affirmeth to be a property of the melancholic by adustion. The vices whereto he confessed himself to be subject before his conversion show him to have been of this temperature: he was a blasphemer, a wrong-doer, and a persecutor, all which springeth from abundance of heat. But the most evident sign which showed that he was choleric adust is gathered from that battle which himself confesseth he had within himself betwixt his part superior and inferior, saying: 'I see another law in my members striving against the law of my mind, which leadeth me into the bondage of sin.' And this self contention have we proved (by the mind of Aristotle) to be in the melancholic by adustion.

True it is that some expound very well that this battle groweth from the disorder which original sin made between the spirit and the flesh, albeit being such and so great, I believe also that it springs from the choler adust which he had in his natural constitution. For the royal prophet David participated equally of original sin, and yet complained not so much as did St Paul, but saith that he found the inferior portion accorded with his reason when he would rejoice with God: 'My heart,' saith he, 'and my flesh joined in the living God', and (as we will touch in the last chapter save one) David possessed the best temperature that nature could frame. And hereof we will make proof by the opinion of all the philosophers that the same ordinarily inclineth a man to be virtuous without any great gainstriving° of the flesh. The wits then which are to be sorted out for preachers are first those who unite a great understanding with much imagination and memory, whose signs shall be expressed in the last chapter save one. Where such want, there succeed in their room the melancholic by adustion. Those unite a great understanding with much imagination but suffer defect of memory, wherethrough they are not sorted with copy of words, nor can preach with full store in presence of the people.

In the third rank succeed men of great understanding but defective in their imagination and memory. These shall have but a bad grace in preaching, yet will preach sound doctrine. The last (whom I would not charge with preaching at all) are such as unite much memory with much imagination and have defect of understanding. These draw the auditory after them and hold them in suspense and well pleased, but when they least misdoubt it, they fetch a turn to the Holy House, for by way of their sweet discourses and blessings, they beguile the innocent.

CHAP. XI.

That the theoric of the laws appertaineth to the memory, and pleading and judging (which are their practice) to the understanding, and the governing of a commonwealth to the imagination.

In the Spanish tongue, it is not void of a mystery that this word 'lettered' being a common term for all men of letters and learning, as well divines, as lawyers, physicians, logicians, philosophers, orators, mathematicians, and astrologers, yet in saying that 'such a one is learned', we all understand it by common sense that he maketh profession of the laws, as if this were their proper and peculiar title and not of the residue.

The answer of this doubt, though it be easy, yet to yield the same such as is requisite, it behoveth first to be acquainted what law is, and whereunto they are bound who set themselves to study that profession that afterwards they may employ the same to use when they are judges or pleaders. The law (who so well considereth thereof) is nought else but a reasonable will of the law-maker by which he declareth in what sort he will that the cases which happen daily in the commonwealth be decided for preserving the subjects in peace, and directing them in what sort they are to live, and what things they are to refrain.

I said 'a reasonable will' because it sufficeth not that the king or emperour (who are the efficient cause of the laws), declaring his will in what sort soever, doth thereby make it a law, for if the same be not just and grounded upon reason, it cannot be called a law, neither is it. Even as he cannot be termed a man who wanteth a reasonable soul. Therefore it is a matter established by common accord that kings enact their laws with assent of men very wise and of sound judgement, to the end they may be right, just, and good, and that the subjects may receive them with good will and be the more bound to observe and obey them. The material cause of the law is that it consist of such cases as accustomably befall in the commonwealth, according to the order of nature, and not of things impossible or such as betide very seldom. The final cause is to order the life of man and to direct him what he is to do and what to forbear, to the end that, being conformed to reason, the commonwealth may be preserved in peace. For this cause we see that the laws are written in plain words, not doubtful, nor

40 obscure, nor of double understanding, without ciphers and without abbreviations, and so easy and manifest that whosoever shall read them may readily understand and retain them in memory. And because no man should pretend ignorance, they are publicly proclaimed that whosoever afterward breaketh them may be chastised.

In respect therefore of the care and diligence which the good law-makers use that their laws may be just and plain, they have given in charge to the judges and pleaders that in actions or judgements, none of them follow his own sense, but suffer himself
50 to be guided by the authority of the laws, as if they should say: 'We command that no judge or advocate employ his conceit, nor intermeddle in deciding whether the law be just or unjust, nor yield it any other sense than that that is contained in the text of the letter.' So it followeth that the lawyers are to construe the text of the law, and to take that sense which is gathered out of the construction thereof, and none other.

This doctrine thus presupposed, it falleth out a matter very manifest for what reason the lawyers are termed 'lettered', and other men of learning not so, for this name is derived from the
60 word 'letter', which is to say, 'a man who is not licensed to follow the capacity of his own understanding', but is enforced to ensue the sense of the very letter. And for that the well-practised in this profession have so construed it, they dare not deny or affirm anything which appertaineth to the determination of any case whatsoever, unless they have lying before them some law which in express terms decideth the same. And if sometimes they speak of their own head, interlacing their conceit and reason without grounding upon some law, they do it with fear and bashfulness, for which cause it is a much worn proverb, 'we blush when we
70 speak without law.' Divines cannot call themselves lettered (in this signification), for in the Holy Scripture the letter killeth, and the spirit giveth life; it is full of mysteries, replenished with figures and ciphers, obscure, and not understood by all readers; the vowels and phrases of speech hold a very different signification from that which the vulgar and three-tongued° men do know.[100] Therefore whosoever shall set himself to construe the letter, and

---

[100] Implicitly, the three languages are Latin, Greek and Hebrew.

take the sense which riseth of that grammatical construction, shall fall into many errors.

The physicians also have no letter whereto to submit themselves, for if Hippocrates and Galen, and the other grave authors of this faculty say and affirm one thing, and that experience and reason approve the contrary, they are not bound to follow them. For in physic, experience beareth more sway than reason, and reason more than authority, but in the laws it betideth quite contrary, for their authority and that which they determine is of more force and vigour than all the reasons that may be alleged to the contrary. Which being so, we have the way laid open before us to assign what wit is requisite for the laws. For if a lawyer have his understanding and imagination tied to follow that which the law avouched, without adding or diminishing, it falleth out apparent that this faculty appertaineth to the memory, and that the thing wherein they must labour is to know the number of the laws, and of the rules which are in the text, and to call to remembrance each of them in particular, and to rehearse at large his sentence and determination to the end that when occasion is ministered, we may know there is a law which giveth decision, and in what form and manner. Therefore to my seeming it is a better difference of wit for a lawyer to have much memory and little understanding than much understanding and little memory. For if there fall out no occasion of employing his wit and ability, and that he must have at his fingers' ends so great a number of laws as are extant, and so far different from the other with so many exceptions, limitations, and enlargements, it serves better to know by heart what hath been determined in the laws for every point which shall come in question, than to discourse with the understanding in what sort the same might have been determined: for the one of these is necessary, and the other impertinent,° since none other opinion than the very determination of the law must bear the stroke.

So it falls out for certain that the theoric of the law appertaineth to the memory and not to the understanding, nor to the imagination. For which reason, and for that the laws are so positive,[101] and that because the lawyers have their understanding so tied to the will of the law-maker and cannot intermingle their

---

[101] positive] recognizing or dealing only with matters of scientifically verifiable fact and experience.

own resolution save in case where they rest uncertain of the determination of the law, when any client seeketh their judgement, they have authority and licence to say: 'I will look for the case in my book'; which if the physician should answer when he is asked a remedy for some disease, or the divine in cases of conscience, we would repute them for men but simply seen in the faculty whereof they make profession. And the reason hereof is that those sciences have certain universal principles and definitions under which the particular cases are contained, but in the law-faculty, every law containeth a several particular case without having any affinity with the next though they both be placed under one title. In respect whereof, it is necessary to have a notice of all the laws, and to study each one in particular, and distinctly to lay them up in memory. But here against Plato noteth a thing worthy of great consideration, and that is how in his time a learned man was held in suspicion that he knew many laws by heart, seeing by experience that such were not so skilful judges and pleaders as this their vaunt seemed to pretend. Of which effect it appeareth he could not find out the cause, seeing in a place so convenient he did not report the same, only he saw by experience that lawyers endowed with good memory, being set to defend a cause, or to give a sentence, applied not their reasons so well as was convenient.

The reason of this effect may easily be rendered in my doctrine, presupposing that memory is contrary to the understanding, and that the true interpretation of the laws (to amplify, restrain, and compound them with their contraries and oppositions) is done by distinguishing, concluding, arguing, judging, and choosing, which works we have often said heretofore belong to discourse, and the learned man possessing much memory cannot by possibility enjoy them.

We have also noted heretofore that memory supplieth none other office in the head than faithfully to preserve the figures and fantasies of things; but the understanding and the imagination are those which work therewithal.

And if a learned man have the whole art of memory, and yet want understanding and imagination, he hath no more sufficiency to judge or plead than the very *Code* or *Digest*,[102] which compassing within them all the laws and rules of reason, for all that cannot

---

[102] digest] body of Roman laws compiled by order of the Emperor Justinian.

THE EXAMINATION OF MEN'S WITS

write one letter. Moreover, albeit it be true that the law ought to be such as we have mentioned in his definition, yet it falleth out a miracle to find things with all the perfections which the understanding attributeth unto them: that the law be just and reasonable, and that it proceed fully to all that which may happen, that it be written in plain terms, void of doubt and oppositions, and that it receive not diverse constructions, we see not always accomplished. For in conclusion, it was established by man's counsel and that is not of force sufficient to give order for all that may betide. And this is daily seen by experience, for after a law hath been enacted with great advisement and counsel, the same in short space is abrogated again, for when it is once published and put in practice, a thousand inconveniences discover themselves whereof (when it was persuaded) no man took regard. And therefore kings and emperors are advised by the same laws that they shame not to amend and correct their laws, for, in a word, men they are, and marvel there is none if they commit an error, so much the rather, for that no law can comprehend in words and sentences all the circumstances of the case which it decideth, for the craft of bad people is more wily to find holes than that of good men to foresee how they are to be governed. And therefore it was said: neither the laws nor the resolutions of the Senate can be set down in writing in such sort that all the cases which severally chance may be comprised therein, but it sufficeth to comprehend the things which fall out oftenest, and if other cases succeed afterward, for which no law is enacted, it decideth them in proper terms.

The law-faculty is not so bare of rules and principles but that if the judge or pleader have a good discourse to know how to apply them, they may find their true determination and defence, and whence to gather the same. In sort that if the cases be more in number than the laws, it behoveth that in the judge and in the pleader there be much discourse to make new laws and that not at all adventures, but such as reason (by his consonance) may receive them without contradiction. This the lawyers of much memory cannot do, for if the cases which the law trusteth into their mouth be not squared and chewed to[103] their hands, they are to seek what to do. We are wont to resemble a lawyer who can

---

[103] chewed to] prepared for another.

rehearse many laws by heart, to a regrater[104] or hosier that hath many pairs of hosen ready-made in his shop, who, to deliver you one that may fit you, must make you to assay them all, and if none agree with the buyers' measure, he must send him away hoseless. But a learned man of good understanding is like a good tailor who hath his shears in his hand and his piece a cloth on the table, and taking measure, cutteth his hosen after his stature that demandeth them.

The shears of a good pleader is his sharp understanding, with which he taketh measure of the case, and apparelleth the same with that law which may decide it. And if he find not a whole one that may determine it in express terms, he maketh one of many pieces and therewith useth the best defence that he may. The lawyers who are endowed with such a wit and ability are not to be termed 'lettered', for they construe not the letter, neither bind themselves to the formal words of the law, but it seemeth they are law-makers, or counsellors at law of whom the laws themselves inquire and demand how they shall determine. For if they have power and authority to interpret them, to reave, to add, and to gather out of them exceptions and fallacies, and that they may correct and amend them, it was not unfitly said that they seem to be law-makers.

Of this sort of knowledge it was spoken: 'By the knowledge of the laws it is not meant to con° their words by rote, but to take notice of their force and power.' As if he should say: 'Let no man think that to know the laws is to bear in mind the formal words with which they are written, but to understand how far their forces extend, and what the point is which they may decide, for their reason is subject to many varieties by means of the circumstances as well of time as of person, of place, of manner, of matter, of cause, and of the thing itself.' All which breedeth an alteration in the decision of the law, and if the judge or pleader be not endowed with discourse, to gather out of the law, or to take away or adjoin that which the law self doth not express in words, he shall commit many errors in following the letter. For it hath been said that the words of the law are not to be taken after the Jewish manner, that is, to construe[105] only the letter, and so take the sense thereof.

---

[104] regrater] a person who buys commodities to sell them on at a profit.
[105] to construe] to translate or analyse the grammatical construction of a sentence.

On the things already alleged, we conclude that pleading is a work of discourse and that if the learned in the laws possess much memory, he shall be untoward to judge or plead through the repugnancy of these two powers. And this is the cause for which the learned of so ripe memory (whom Plato mentioneth) could not defend well their clients' causes, nor apply the laws. But in this doctrine there presents itself a doubt, and that (in mine opinion) not of the lightest, for if the discourse be that which putteth the case in the law, and which determineth the same by distinguishing, limiting, amplifying, inferring, and answering the arguments of the contrary party, how is it possible that the discourse may compass all this, if the memory set not down all the laws before it? For (as we have above remembered) it is commanded that no man in actions or judgements shall use his own sense, but leave himself to be guided by the authority of the laws. Conformable hereunto, it behoveth first to know all the laws and rules of the law-faculty, ere we can take hold of that which maketh to the purpose of our case. For albeit we have said that the pleader (of good understanding) is lord of the laws, yet it is requisite that all his reasons and arguments be grounded on the principles of this faculty, without which they are of none effect or valure. And to be able to do this it behoveth to have much memory that may preserve and retain so great a number of laws which are written in the books.

This argument proveth it to be necessary, to the end a pleader may be accomplished, that there be united in him a great discourse and much memory. All which I confess but that which I would say is that, since we cannot find great discourse united with much memory, through the repugnancy which they carry each to other, it is requisite that the pleader have much discourse and little memory, rather than much memory and little discourse, for to the default of memory are found many remedies (as books, tables, alphabets,[106] and other things devised by men), but if discourse fail, there can nothing be found to remedy the same.

Besides this, Aristotle saith that men of great discourse, though they have a feeble memory, yet they have much remembrance by which they retain a certain diffuse notice of things they have seen, heard, and read, whereupon discoursing they call them to

---

[106] alphabets] dictionaries, manuals or compendia arranged alphabetically.

memory. And albeit they had not so many remedies to present unto the understanding the whole body of the civil law, yet the laws are grounded on so great reason as Plato reporteth that the ancients termed the law 'wisdom' and 'reason'. Therefore the judge or pleader of great discourse, though judging or counseling he have not the law before him, yet seldom shall he commit an error, for he hath with him the instrument with which the emperors made the laws. Whence oftentimes it falleth out that a judge of good wit giveth a sentence without knowing the decision of the law, and afterwards findeth the same so ruled in his books. And the like we see sometimes betideth the pleaders when they give their judgement in a case without studying. The laws and rules of reason, whosoever well marketh them, are the fountain and original whence the pleaders gather their arguments and reasons to prove what they undertake. And this work (for certain) is performed by the discourse, which power if the pleader want, he shall never skill to shape an argument though he have the whole civil law at his fingers' ends. This we see plainly to befall in such as study the art of oratory when the aptness thereunto is failing, for though they learn by art the *Topics* of Cicero (being the spring from which flow the arguments that may be invented to prove every problem, both on the affirmative and the negative part), yet they cannot thereout shape a reason. Again, there come others of great wit and towardness who, without looking in book or studying the *Topics*, make 1000 arguments serving for the purpose as occasion requireth.

This self falleth out in the lawyers of good memory, who will recite you a whole text very perfectly, and yet of so great a multitude of laws as are comprised therein cannot collect so much as one argument to prove their intention. And contrariwise, others who have studied simply without books, and without allowance, work miracles in pleading of causes. Hence we know how much it importeth the commonwealth that there may be such an election and examination of wits for the sciences, inasmuch as some without art know and understand what they are to effect, and others, loden with precepts and rules for that they want a convenient towardliness for practice, commit a thousand absurdities which very ill beseem them. So then, if to judge and plead be effected by distinguishing, inferring, arguing, and choosing, it standeth with reason that whosoever setteth himself to study the laws enjoy a good understanding, seeing that such actions appertain to this power and not to the memory or to the imagination. How we may find whether a child be endowed with this difference of wit or no, it would do well to understand. But first it behoveth to lay down what are the qualities of discourse, and how many differences it compriseth in itself, to the end we may likewise know with distinction to which of these the laws appertain. For the first, we must weet° that albeit the understanding be the most noble power, and of greatest dignity in man, yet there is none which is more easily led into error (as touching the truth) than the understanding. This Aristotle attempted to prove when he said that the sense is ever true, but the understanding (for the most part) discourseth badly, the which is plainly seen by experience, for if it were not so amongst the divines, the physicians, the philosophers, and the lawyers, there would not fall out so many weighty dissentions, so diverse opinions, and so many judgements and conceits upon every point, seeing the truth is never more than one. Whence it groweth that the senses hold so great a certainty in their objects, and the understanding is so easily beguiled in his, may well be conceived if we consider that the objects of the five senses, and the spices by which they are known, have their being (real, firm, and stable by nature) before they are known. But that truth which is to be contemplated by the understanding, if itself do not frame and fashion the same, it hath no formal being of his own, but is wholly scattered and loose in his materials, as a house converted into stones, earth, timber and tiles, with which so many errors may be committed in building as there shall men set themselves to build with ill imagination.

The like befalleth in the building which the understanding raiseth when it frameth a trueth, for if the wit be not good, all the residue will work a thousand follies with the selfsame principles. Hence springs it that amongst men there are so sundry opinions touching one self matter, for every one maketh the composition and figure such as is his understanding.

From these errors and opinions are the five senses free, for neither the eyes make the colour, nor the taste the savours, nor the feeling the palpable qualities, but the whole is made and compounded by nature before any of them be acquainted with his object. Men because they carry not regard to this bad operation of the understanding, take hardiness to deliver confidently their own opinion without knowing in certainty of what sort their wit is, and whither it can a fashion a truth well or ill. And if we be not resolved herein, let us ask some of these learned men, who after they have set down in writing and confirmed their opinions with many arguments and reasons, and have another time changed their opinions and conceit, when or how they can assure themselves that, now at last, they have hit the nail on the head. Themselves will not deny but that they erred the first time, seeing they unsay what they said tofore.

Secondly, I avouch that they ought to have the less confidence in their understanding because the power which once ill compoundeth the truth, whilst his patron placed so much assurance in his arguments and reasons, should therefore the sooner take suspect that he may once again slide into error whilst he worketh with the self same instrument of reason. And so much the rather for that it hath been seen by experience that the first opinion hath born most trueth, and afterwards he hath relied upon a worse, and of less probability. They hold it for a sufficient token that the understanding compoundeth well a truth when they see it enamoured of such a figure, and that there are arguments and reasons which move it to conclude in that sort. And verily they miss their cushion, for the same understanding carrieth the same proportion to his false opinions that the inferior powers have each with the differences of their object. For if we demand of the physicians what meat is best and most savoury of all that men accustomably feed upon, I believe they will answer that for men who are distempered and of weak stomach, there is none absolutely good or evil, but such as the stomach is that shall receive it. For there are stomachs, saith Galen, which better brook beef than hens or cracknels, and othersome abhor eggs and milk, and others again have a longing after them. And in the manner of using meats, some like roast, and some boiled; and in roast, some love to have the blood run in the dish, and some to have it brown and burned. And (which is more worthy of consideration) that meat which this day is savourly eaten, and with good appetite, tomorrow will be loathed, and a far worse longed for in his room. All this is understood when the stomach is good and sound, but if it fall into a certain infirmity which the physicians call 'pica',[107] or 'malacia',[108] then arise longings after things which

---
[107] pica] craving to eat substances other than normal foodstuffs.
[108] malacia] abnormal craving for certain kinds of food.

man's nature abhorreth: so as they eat earth, coals, and lime with greater appetite than hens or trouts.

If we pass on to the faculty generative, we shall find as many appetites and varieties, for some men love a foul woman and abhor a fair; others cast better liking to a fool than her that is wise; a fat wench is fulsome, and a lean hath their liking; silks and brave attire offend some men's fancies, who leese° themselves after one that totters in her rags. This is understood when the genital parts are in their soundness, but if they fall into their infirmity of stomach, which is termed 'malacia', they covet detestable beastliness. The same befalleth in the faculty sensitive, for of the palpable qualities hard and soft, rough and smooth, hot and cold, moist and dry, there is none of them which can content every one's feeling: for there are men who take better rest on a hard bed than a soft, and othersome better on a soft than a hard.

All this variety of strange tastes and appetites is found in the compositions framed by the understanding, for if we assemble 100 men of learning and propound a particular question, each of them delivereth a several judgement, and discourseth thereof in different manner. One self argument to one seemeth a sophistical reason, to another probable, and some you shall meet with to whose capacity it concludeth as if it were a demonstration. And this is not only true in diverse understandings, but we see also by experience that one self reason concludeth to one self understanding at one time thuswise, and at another time otherwise. So much that every day men vary in opinion, some by process of time purging their understanding know the default of reason which first swayed them, and others, leesing the good temperature of their brain, abhor the truth and give allowance to a leasing. But if the brain fall into the infirmity which is termed 'malacia', then we shall see strange judgements and compositions, arguments false and weak to prove more forcibly than such as carry strength and truth to good arguments, an answer shaped, and to bad a condescending. From the premises whence a right conclusion may be collected, they father a wrong, and by strange arguments and fond reasons, they prove their bad imaginations. This grave and learned men duly advising labour to deliver their opinion, concealing the reasons whereon they ground (for men persuade themselves that so far a man's authority availeth as the reason is of force on which he buildeth), and the arguments resting so indifferent for concluding through the diversity of understandings, every man giveth a judgement of the reason conformably to the wit which he possesseth. For which cause it is reputed greater gravity to say: 'This is mine opinion, for certain reasons which move me so to think', than to display the arguments whereon he relieth. But if they be enforced to render a reason of their opinion, they overslip not any argument how slight soever for that which they least valued, with some concludeth and work the more effect than the most urgent. Wherein the great misery of our understanding is discovered, which compoundeth and divideth, argueth, and reasoneth, and at last (when it is grown to a conclusion) is void of proof or light, which may make it discern whether his opinion be true or no.

This self uncertainty have the divines in matters which appertain not to the faith, for after they have argued at full they cannot then assure themselves of any infallible proof or evident success that may discover which reasons carried greatest weight, and so every divine casteth how he may best ground himself, and answer with most apparence to the adverse parties' arguments, his own reputation saved, and this is all whereabouts he must bestow his endeavour. But the charge of a physician, and a general in the field, after he hath well discoursed and refuted the grounds of the contrary party is to mark the success which, if it be good, he shall be held for discreet; if bad, all men will know that he relied upon guileful reasons.

In matters of faith propounded by the Church, there can befall none error: for God, best weeting how uncertain men's reasons are, and with how great facility they run headlong to the deceived, consenteth not that matters so high and of so weighty importance should rest upon our only determination.[109] But when two or three are gathered together in his name with the solemnity of the Church, he forthwith entreth into the midst of them as president of the action, and so giveth allowance to that which they say well, and reaveth their errors, and of himself revealeth that to whose notice by humane forces we cannot attain. The proof then which the reasons formed in matters of faith must receive is to advise well whether they prove or infer the same which the Catholic Church saith and declareth. For if they collect ought to the contrary, then, without doubt, they are faulty, but in other

---

[109] 1. *Carew's note in the margin*: 'Take heed you receive no hurt for leaving out the Pope'.

questions where the understanding hath liberty of discourse, there hath not yet any manner been advised to know what reasons conclude, nor when the understanding doth well compound a truth. Only we rely upon the good consonance which they make, and that is an argument which may err, for many false points carry better appearance and likelier proof of truth than the true themselves.

Physicians, and such as command in martial affairs, have success and experience for proof of their reasons. For if ten captains prove by many reasons that it is best to join battle, and so many (on the other side) defend the contrary, that which succeedeth will confirm the one opinion and convince the other. And if two physicians dispute whether the patient shall die or live, after he is cured or deceased, it will appear whose reason was best. But for all this, the success is yet no sufficient proof, for whereas an effect hath many causes, it may very well betide happily for one cause, and yet the reasons perhaps were grounded on a contrary. Aristotle moreover affirmeth that to know what reasons conclude, it is good to ensue the common opinion, or if many wise men say and affirm one self thing, and all conclude with the same reasons, it is a sign (though topical°) that they are conclusive, and that they compound well the truth. But who so taketh this into due consideration shall find it a proof subject also unto beguiling, for in the forces of the understanding, weight is of more pre-eminence than number, for it fareth not in this as in bodily forces that when many join together to list up a weight, they prevail much, and when few, but little. But to attain to the notice of a truth deeply hidden, one high understanding is of more value than a hundred thousand which are not comparable thereunto. And the reason is because the understandings help not each other, neither of many make one, as it falls out in bodily powers. Therefore, well said the wise man: 'Have many peace-makers but take one of a thousand to be thy counsellor.' As if he should say: 'Keep for thyself many friends who may defend thee when thou shalt be driven to come to handstrokes, but to ask counsel, choose only one amongst a thousand.' Which sentence was also expressed by Heraclitus, who said: 'One with me is worth a thousand.'

In contentions and causes, every learned man bethinketh how he may best ground himself on reason, but after he hath well revolved every thing, there is no art which can make him know with assurance whether his understanding have made that composition which in justice is requisite. For if one pleader prove with law in hand that reason standeth on the demandant's side, and another by way also of the law proveth the like for the defendant, what remedy shall we devise to know which of the two pleaders hath formed his reasons best? The sentence of the judge maketh no demonstration of true justice, neither can the same be termed a success, for his sentence also is but an opinion, and he doth none other than cleave to one of the two pleaders. And to increase the number of learned men in one self opinion is no argument to persuade that what they resolve upon is therefore true. For we have already affirmed and proved that many weak capacities (though they join in one to discover some dark conceived truth) shall never arrive to the power and force of someone alone if the same be an understanding of high reach. And that the sentence of the judge maketh no demonstration is plainly seen in that at another higher seat of justice they reserve the same and give a diverse judgement. And, which is worst, it may so fall that the inferior judge was of an abler capacity than the superior, and his opinion more conformable unto reason. And that the sentence of the superior judge is not a sufficient proof of justice neither, it is a matter very manifest, for in the same actions, and from the same judges, without adding or reaving any one jot, we see daily contrary sentences to issue. And he that once is deceived by placing confidence in his own reasons, falleth duly into suspect that he may be deceived of new. Wherethrough we should the less rely upon his opinion, for he that is once naught, sayth the wiseman, chase him from thee.

Pleaders, seeing the great variety of understandings which possess the judges, and that each of them is affectionate to the reason which best squareth with his wit, and that sometime they take satisfaction at one argument, and sometimes assent to the contrary, they thereupon boldly thrust themselves forth to defend every cause in controversy, both on the part affirmative and the negative. And this so much the rather, because they see by experience that in the one manner and the other they have a sentence in their favour, and so that comes very rightly to be verified, which wisdom said: 'The thoughts of mortal men are timorous, and their foresights uncertain.' The remedy then which we have against this, seeing the reasons of the lawyer fail in proof and experience, shall be to make choice of men of great understanding who may be judges and pleaders. For the reasons and arguments of such, sayth Aristotle, are no less certain and firm than experience itself. And by making this choice, it seemeth that the commonwealth resteth assured that her officers shall administer justice. But if they give them all scope to enter without making trial of their wit, as the use is at this day, the inconveniences (which we have noted) will evermore befall.

By what signs it may be known that he who shall study the laws hath the difference of wit requisite to this faculty, heretofore (after a sort) we have expressed. But yet, to renew it to the memory, and to prove the same more at large, we must know that the child who being set to read soon learneth to know his letters, and can pronounce every one with facility, according as they be placed in the ABC, giveth token that he shall be endowed with much memory. For such a work as this for certain is not performed by the understanding nor by the imagination, but it appertaineth unto the office of the memory to preserve the figures of things and to report the natures of each when occasion so requireth, and where much memory dwelleth, we have proved before that default of understanding also raigneth.

To write also with speed, and a faire hand, we said that it bewrayed an imagination, wherethrough the child who in few days will frame his hand, and write his lines right, and his letters even and with good form and figure, yieldeth sign of mean understanding, for this work is performed by the imagination, and these two powers encounter in that contrariety which we have already spoken of and noted.

And it being set to grammar, he learn the same with little labour, and in short time make good Latin and write fine epistles with the well ruled clauses of Cicero. He shall never be good judge nor pleader, for it is a sign that he hath much memory, and (save by great miracle) he will be of slender discourse. But if such a one wax obstinate in plodding at the laws, and spend much time in the schools, he will prove a famous reader, and shall have a stint of many hearers, for the Latin tongue is very gracious in chairs, and to read with great show there are requisite many allegations, and to fardel up° in every law whatsoever hath been written touching the same, and to this purpose, memory is of more necessity than discourse. And albeit it is true that in the chair he be to distinguish, infer, argue, judge, and choose to gather the true sense of the law, yet in the end he putteth the case as best liketh himself: he moves doubts, maketh objections and giveth sentence after his own will without that any gainsay him for which a mean discourse is sufficient. But when one pleader speaketh for the plaintiff, and another for the defendant, and a third lawyer supplieth the judge's place, this is a true controversy, and men cannot speak so at random as when they skirmish without an adversary. And if the child profit slenderly in grammar, we may thereby gather that he hath a good discourse, I say we may so conjecture because it followeth not of necessity that whosoever cannot learn Latin hath therefore straightways a good discourse, seeing we have proved tofore that children of great imagination never greatly profit in the Latin tongue. But that which may best discover this is logic, for this science carrieth the same proportion with the understanding as the touchstone[110] with gold. Wherethrough it falleth out certain that if he who taketh lesson in the arts begin not within a month or two to discourse and to cast doubts, and if there come not in his head arguments and answers in the matter which is treated of, he is void of discourse. But if he prove towardly in his science, it is an infallible argument that he is endowed with a good understanding for the laws, and so he may forthwith addict himself to study them without longer tarrying. Albeit I would hold it better done first to run through the arts, because logic, in respect of the understanding, is nought else than those shackles which we clap on the legs of an untrained mule, which going with them many days, taketh a steady and seemly pace. Such a march doth the understanding make in his disputations when it first bindeth the same with the rules and precepts of logic.

But if this child, whom we go thuswise examining, reap no profit in the Latin tongue, neither can come away with logic as were requisite, it behoveth to try whether he possess a good imagination ere we take him from the laws, for herein is lapped up a very great secret, and it is good that the commonwealth be done to ware thereof, and it is that there are some lawyers who getting up into the chair work miracles in interpreting the texts, and others in pleading. But if you put the stuff of justice into their hands, they have no more ability to govern than as if the laws had never been enacted to any such end. And contrariwise, some other

---

[110] touchstone] smooth, fine-grained, black or dark-coloured variety of quartz or jasper, used for testing the quality of gold and silver alloys by the colour of the streak produced by rubbing them upon it.

there are who with three misunderstood laws, which they have learned at all adventures, being placed in any government, there cannot more be desired at any man's hands than they will perform. At which effect, some curious wits take wonder because they sink not into the depth of the cause from whence it may grow. And the reason is that government appertaineth to the imagination, and not to the understanding nor the memory. And that this is so, the matter may very manifestly be proved, considering that the commonwealth is to be compounded with order and concert, with everything in his due place, which all put together maketh good figure and correspondence. And this (sundry times heretofore) we have proved to be a work of the imagination, and it shall prove nought else to place a great lawyer to be a governor, than to make a deaf man a judge in music. But this is ordinarily to be understood, and not as a universal rule, for we have already proved it is possible that nature can unite great understanding with much imagination, so shall there follow no repugnancy to be a good pleader and a famous governor, and we heretofore discovered that nature being endowed with all the forces which she may possess, and with matter well seasoned, will make a man of great memory and of great understanding and of much imagination, who studying the laws will prove a famous reader, a great pleader, and no less governor. But nature makes so few such, as this cannot pass for a general rule.

CHAP. XII.

How it may be proved that of theorical physic, part appertaineth to the memory, and part to the understanding, and the practice to the imagination.

What time the Arabian physic flourished, there was a physician very famous as well in reading as in writing, arguing, distinguishing, answering, and concluding, who men would think in respect of his profound knowledge were able to revive the dead, and to heal any disease whatsoever, and yet the contrary came to pass, for he never took any patient in cure who miscarried not under his hands. Whereat greatly shaming and quite out of countenance, he went and made himself a friar, complaining on his evil fortune, and not able to conceive the cause how he came so to miss. And because the freshest examples afford surest proof and do most sway the understanding, it was held by many grave physicians that John Argentier,[111] a physician of our time, far surpassed Galen in reducing the art of physic to a better method, and yet for all this it is reported of him that he was so infortunate in practice as no patient of his country durst take physic at his hands, fearing some dismal success. Hereat it seemeth the vulgar have good reason to marvel, seeing by experience (not only in those rehearsed by us, but also in many others with whom men have daily to deal) that if the physician be a great clerk, for the same reason he is unfit to minister.

Of this effect Aristotle procured to render a reason but could not find it out. He thought that the cause why the reasonable physicians of his time failed in curing grew from that such men had only a general notice and knew not every particular complexion, contrary to the empirics, whose principal study bent itself to know the properties of every several person, and let pass the general. But he was void of reason, for both the one and the other exercised themselves about particular cures and endeavoured (so much as in them lay) to know each one's nature singly by itself. The difficulty then consisteth in nothing else than to know for what cause so well learned physicians, though they exercise themselves all their lifelong in curing, yet never grow skilful in practice, and yet other simple souls with three or four rules learned very soon, and the scholars can more skill of ministering than they.

The true answer of this doubt holdeth no little difficulty, seeing that Aristotle could not find it out, nor render (at least in some sort) any part thereof. But grounding on the principles of our doctrine, we will deliver the same, for we must know that the perfection of a physician consisteth in two things no less necessary to attain the end of his art, than two legs are to go without halting. The first is to weet° by way of method the precepts and rules of curing men in general without descending to particulars. The

---

[111] John Argentier] Juan Argenterio/Jean Argentier (1513–1572) was a physician from Piedmont who taught at Pisa, Naples, Rome and Paris. He was the author of books such as *De consultationibus medicis liber* (Florence, 1551); *Commentarii tres in artem medicinalem galeni* (Paris, 1553), *De erroribus veterum medicorum* (Florence, 1553), or *De morbis libri XIV* (Florence, 1558), widely known throughout Europe at the time.

second, to be long-time exercised in practice, and to have visited many patients. For men are not so different each from other but that in diverse things they agree; neither so conjoined but that there rest in them particularities of such condition as they can neither be delivered by speech, nor written, nor taught, nor so collected, as that they may be reduced into art. But to know them is only granted to him who hath often seen and had them in handling. Which may easily be conceived, considering that man's face, being composed of so small a number of parts as are two eyes, a nose, two cheeks, a mouth, and a forehead, nature shapeth yet therein so many compositions and combinations as if you assemble together 100,000 men, each one hath a countenance so different from other and proper to himself, that it falleth out a miracle to find two who do altogether resemble. The like betideth in the four elements, and in the four first qualities (hot, cold, moist, and dry) by the harmony of which the life and health of man is compounded. And of so slender a number of parts, nature maketh so many proportions that if a 100,000 men be begotten, each of them comes to the world with a health so peculiar and proper to himself that if God should on the sudden miraculously change their proportion of these first qualities, they would all become sick except some two or three that by great disposition had the like consonance and proportion. Whence two conclusions are necessarily inferred. The first is that every man who falleth sick ought to be cured conformable to his particular proportion, in sort that if the physician restore him not to his first consonance of humours, he cannot recover. The second that to perform this as it ought is requisite the physician have first seen and dealt with the patient sundry times in his health by feeling his pulse, perusing his stare, and what manner countenance and complexion he is of, to the end that when he shall fall sick, he may judge how far he is from his health, and in ministering unto him may know to what point he is to restore him.

For the first (namely, to weet and understand the theoric and composition of the art) saith Galen, it is necessary to be endowed with great discourse and much memory, for the one part of physic consisteth in reason, and the other in experience and history. To the first is understanding requisite, and to the other memory, and it resting a matter of so great difficulty to unite these two powers in a large degree, it followeth of force that the physician become unapt for the theoric. Wherethrough we behold many physicians learned in the Greek and Latin tongue, and great anatomists and simplicists[112] (all works of the memory) who brought to arguing or disputations, or to find out the cause of any effect (that appertaineth to the understanding), can small skill thereof.

The contrary befalleth in others who show great wit and sufficiency in the logic and philosophy of this art, but being set to the Latin and Greek tongue, touching simples[113] and anatomies can do little because memory in them is wanting. For this cause Galen said very well that it is no marvel if among so great a multitude of men who practise the exercise and study of the art of physic and philosophy, so few are found to profit therein, and yielding the reason, he saith it requires a great toil to find out a wit requisite for this science, or a master who can teach the same with perfection, or can study it with diligence and attention. But with all these reasons Galen goeth groping, for he could not hit the cause whence it comes to pass that few persons profit in physic. Yet in saying it was a great labour to find out a wit requisite for this science, he spoke truth, albeit he did not so far-forth specify the same as we will; namely, for that it is so difficult a matter to unite a great understanding with much memory, no man attaineth to the depth of theorical physic. And for that there is found a repugnancy between the understanding and the imagination (whereunto we will now prove that practice and the skill to cure with certainty appertaineth) it is a miracle to find out a physician who is both a great theorist, and withal a great practitioner, or contrariwise: a great practitioner, and very well seen in theoric. And that the imagination, and not the understanding, is the power whereof the physician is to serve himself in knowing and curing the diseases of particular persons may easily be proved.

First of all presupposing the doctrine of Aristotle, who affirmeth that the understanding cannot know particulars, neither distinguish the one from the other, nor discern the time and place and other particularities which make men different each from other, and that everyone is to be cured after a diverse manner. And the reason is (as the vulgar philosophers avouch) for that the understanding is a spirital power and cannot be altered by the particulars which are replenished with matter. And for this cause

---

[112] simplicist] one who has a knowledge of medicinal simples; a simplist.

[113] simple] plant or herb employed for medical purposes.

Aristotle said that the sense is of particulars and the understanding of universals.

If then medicines are to work in particulars and not in universals (which are unbegotten and uncorruptible) the understanding falleth out to be a power impertinent° for curing. Now the difficulty consisteth in discerning why men of great understanding cannot possess good outward senses for the particulars, they being powers so repugnant. And the reason is very plain, and this is it, that the outward senses cannot well perform their operations unless they be assisted with a good imagination, and this we are to prove by the opinion of Aristotle, who, going about to express what the imagination was, saith it is a motion caused by the outward sense, in sort as the colour (which multiplieth by the thing coloured) doth alter the eye. And so it fareth that this self colour, which is in the crystalline humour, passeth farther into the imagination, and maketh therein the same figure which was in the eye. And if you demand of which of these two kinds the notice of the particular is made, all philosophers avouch (and that very truly) that the second figure is it which altereth the imagination, and by them both is the notice caused, conformable to that so common speech 'from the object and from the power the notice springeth'. But from the first which is in the crystalline humour, and from the sightful° power, groweth no notice if the imagination be not attentive thereunto, which the physicians do plainly prove, saying that if they lance or sear the flesh of a diseased person, who for all that feeleth no pain, it shows a token that his imagination is distracted into some profound contemplation. Whence we see also by experience in the sound, that if they be raught into some imagination, they see not the things before them, nor hear though they be called, nor taste meat savoury or unsavoury, though they have it in their mouth. Wherefore it is a thing certain that not the understanding or outward senses but the imagination is that which maketh the judgement and taketh notice of particular things.

It followeth then that the physician, who is well seen in theoric, for that he is endowed with great understanding or great memory, must of force prove a bad practitioner as having defect in his imagination. And contrariwise, he that proveth a good practitioner, must of force be a bad theorist, for much imagination cannot be united with much understanding and much memory. And this is the cause for which so few are thoroughly seen in physic, or commit but small errors in curing. For, not to halt in the work, it behoveth to know the art and to possess a good imagination for putting the same in practice, and we have proved that these two cannot stick together.

The physician never goeth to know and cure a disease but that secretly to himself he frameth a syllogism in Darii,[114] though he be never so well experienced, and the proof of his first proportion belongeth to the understanding, and of the second to the imagination, for which cause the great theorists do ordinarily err in the *minor*, and the great practitioners in the *major*, as if we should speak after this manner: 'Every fever which springeth from cold and moist humours ought to be cured with medicines hot and dry' (taking the tokening of the cause); 'this fever which the man endureth dependeth on humours cold and moist, therefore the same is to be cured with medicines hot and dry.' The understanding will sufficiently prove the truth of the *major* (because it is a universal) saying that cold and moist require for their temperature hot and dry, for every quality is abated by his contrary. But coming to prove the *minor*, there the understanding is of no value, for that the same is particular and of another jurisdiction whose notice appertaineth to the imagination, borrowing the proper and particular tokens of the disease from the five outward senses.

And if the tokening is to be taken from the fever, or from his cause, the understanding cannot reach thereunto: only it teacheth the tokening is to be taken from that which sheweth greatest peril. But which of those tokenings is greatest is only known to the imagination by counting the damages which the fever produceth with those of the symptoms of the evil, and the cause and the small or much force of the power. To attain this notice, the imagination possesseth certain unutterable properties with which the same cleareth matters that cannot be expressed nor conceived, neither is there found any art to teach them. Wherethrough we see a physician enter to visit a patient and by means of his sight, his hearing, his smelling, and his feeling, he knoweth things which seem impossible. In sort that if we demand of the same physician how he could come by so ready a knowledge, himself cannot tell the reason, for it is a grace which springeth from the fruitfulness

---

[114] Darii] mnemonic word designating the third mood of the first figure of syllogisms.

of the imagination, which by another name is termed 'a readiness of capacity', which by common signs, and by uncertain conjectures, and of small importance, in the twinkling of an eye knoweth 1000 differences of things, wherein the force of curing and prognosticating with certainty consisteth.

This spice° of promptness men of great understanding do want, for that it is a part of the imagination, for which cause, having the tokens before their eyes (which give them notice how the disease fareth) it worketh no manner alteration in their senses, for that they want imagination. A physician once asked me in great secrecy what the cause was that he having studied with much curiosity all the rules and considerations of the art prognosticative,° and being therein throughly instructed, yet could never hit the truth in any prognostication which he made. To whom (I remember) I yielded this answer that the art of physic is learned with one power, and put in execution with another. This man had a very good understanding, but wanted imagination. But in this doctrine there ariseth a difficulty very great, and that is: how physicians of great imagination can learn the art of physic, seeing they want that of understanding? And if it be true that such were better than those who were well learned, to what end serveth it to spend time in the schools? To this may be answered that first to know the art of physic is a matter very important, for in two or three years a man may learn all that which the ancients have been getting in two or three thousand. And if a man should herein ascertain himself by experience, it were requisite that he lived some thousands of years, and in experimenting of medicines, he should kill an infinite number of persons before he could attain to the knowledge of their qualities. From whence we are freed by reading the books of reasonable experienced physicians who give advertisement of that in writing which they found out in the whole course of their lives, to the end that the physicians of these days may minister some receits with assurance and take heed of othersome as venomous.

Besides this, we are to weet that the common and vulgar points of all arts are very plain and easy to learn, and yet the most important of the whole work. And contrariwise, the most curious and subtile are the most obscure and of least necessity for curing. And men of great imagination are not altogether deprived of understanding nor of memory. Wherethrough, by having these two powers in some measure they are able to learn the most necessary points of physic: for that they are plainest, and with the good imagination which they have, can better look into the disease and the cause thereof than the cunningest doctors. Besides that the imagination is it which findeth out the occasion of the remedy that ought to be applied, in which grace the greatest part of practice consisteth, for which cause Galen said that the proper name of a physician was the finder out of occasion.

Now to be able to know the place, the time, and the occasion, for certain is a work of the imagination, since it toucheth figure and correspondence. But the difficulty consisteth in knowing (amongst so many differences as there are of the imagination) to which of them the practice of physic appertaineth, for it is certain that they all agree not in one self particular reason. Which contemplation hath given me much more toil and labour of spirit than all the residue, and yet for all that, I cannot as yet yield the same a fitting name unless it spring from a less degree of heat which partaketh that difference of imagination wherewith verses and songs are indited. Neither do I rely altogether on this, for the reason whereon I ground myself is that such as I have marked to be good practitioners do all piddle somewhat in the art of versifying, and raise not up their contemplation very high, and their verses are not of any rare excellency, which may also betide for that their heat exceedeth that term which is requisite for poetry. And if it so come to pass for this reason, the heat ought to hold such quality as it somewhat dry the substance of the brain, and yet much resolve not the natural heat, albeit (if the same pass further) it breedeth no evil difference of the wit for physic, for it uniteth the understanding to the imagination by adustion. But the imagination is not so good for curing as this which I seek, which inviteth a man to be a witch, superstitious, a magician, a deceiver, a palmister,° a fortune-teller, and a calker,[115] for the diseases of men are so hidden and deliver their motions with so great secrecy that it behoveth always to go calking° what the matter is.

This difference of imagination may hardly be found in Spain, for tofore we have proved that the inhabitants of this region want memory and imagination and have good discourse. Neither yet the imagination of such as dwell towards the north is of avail in physic, for it is very slow and slack, only the same is towardly to

---

[115] calker] astrologer.

make clocks, pictures, puppets, and other ribaldries° which are impertinent for man's service.

Egypt alone is the region which engendereth in his inhabitants this difference of imagination, wherethrough the historians never make an end of telling how great enchanters the Egyptians are, and how ready for obtaining things and finding remedies to their necessities. Josephus to exaggerate the wisdom of Solomon said in this manner: 'So great was the knowledge and wisdom which Solomon received of God, that he outpassed all the ancients, and even the very Egyptians, who were reputed the wisest of all others.' And Plato also said that the Egyptians exceeded all the men of the world in skill how to get their living, which ability appertaineth to the imagination. And that this is true, may plainly appear for that all the sciences belonging to the imagination were first devised in Egypt, as the mathematics, astrology, arithmetic, perspective, judiciary, and the rest. But the argument which most over-ruled me in this behalf is that when Francis of Valois, King of France,[116] was molested by a long infirmity, and saw that the physicians of his household and court could yield him no remedy, he would say every time when his fever increased it was not possible that any Christian physician could cure him, neither at their hands did he ever hope for recovery, wherethrough one time aggrieved to see himself thus vexed with this fever, he dispatched a post into Spain praying the Emperor Charles the fifth that he would send him a Jew physician, the best of his court, touching whom he had understood that he was able to yield him remedy for his sickness if by art it might be effected. At this request the Spaniards made much game, and all of them concluded it was a humorous conceit of a man whose brains were turmoiled with the fever. But for all this, the Emperor gave commandment that such a physician should be sought out if any there were, though to find him they should be driven to send out of his dominions, and when none could be met withal, he sent a physician newly made a Christian[117] supposing that he might serve to satisfy the king's

---

[116] Francis of Valois] Francis I of the House of Valois (1494–1547) was King of France from 1515 until his death. Son of Charles, Count of Angoulême, and Louise of Savoy, Francis succeeded his cousin and father-in-law Louis XII. He engaged in numerous wars against Emperor Charles V, his great rival. In 1530, Francis I married his second wife Eleanor of Austria, sister of Emperor Charles V. They had no children.

humour. But the physician being arrived in France, and brought to the king's presence, there passed between them a gracious discourse in which it appeared that the physician was a Christian, and therefore the king would receive no physic at his hands. The king (with opinion which he had conceived of the physician that he was an Hebrew), by way of passing the time, asked him whether he were not as yet weary in looking for the Messiah promised in the law. The physician answered: 'Sir I expect not any Messiah promised in the Jews' law.' 'You are very wise in that', replied the king, 'for the tokens which were delivered in the divine Scripture whereby to know his coming are all fulfilled many days ago.' 'This number of days', rejoined the physician, 'we Christians do well reckon, for there are now finished 1542 years that he came and conversed in the world thirty-three years, in the end of which he died on the cross, and the third day rose again, and afterwards ascended into heaven, where he now remaineth.' 'Why then', quoth the king, 'you are a Christian?' 'Yea, Sir, by the grace of God, I am a Christian', quoth the physician. 'Then', answered the king, 'return you home to your own dwelling in good time, for in mine own house and court I have Christian physicians very excellent, and I held you for a Jew, who (in mine opinion) are those that have best natural ability to cure my disease.' After this manner he licensed him without once suffering him to feel his pulse, or see his state, or telling him one word of his grief. And forthwith he sent to Constantinople for a Jew, who healed him with the only milk of a she ass.

This imagination of King Francis (as I think) was very true, and I have so conceived it to be, for that in the great hot distemperatures of the brain I have proved tofore how the imagination findeth out that which (the party being found) could never have done. And because it shall not seem that I have spoken in jest, and without relying herein upon a material ground, you shall understand that the varieties of men, as well in the compositions of the body as of the wit and conditions of the soul, spring from their inhabiting countries of different temperature, from drinking diverse waters, and from not using all of them one kind of food. Wherein Plato said: 'Some through variable winds and heats are amongst themselves diverse in manners and kinds;

---

[117] newly made a Christian] a Jew converted to Christianity.

others through the waters and food which spring of the earth, who not only in their bodies, but in their minds also, can skill to do things better and worse.' As if he should say: 'Some men are different from others, either by reason of the contrary air, or through drinking several waters, or for that they feed not all upon one kind of meat, and this difference is discerned not only in the countenance and demeanour of the body, but also in the wit of the soul.'

If I then shall now prove that the people of Israel dwelt many years in Egypt and that departing from thence they did eat and drink waters and meats which are appropriate to make this difference of imagination, I shall then yield a demonstration for the opinion of the King of France, and by consequence we shall understand what wits of men are in Spain to be made choice of for studying the art of physic. As touching the first, we must know that Abraham asking tokens whereby to be assured that he or his descendants should possess the land of promise, the Text sayth that whilst he slept, God made him answer saying: 'Know that thy seed shall be a stranger in a country not his own, and they shall make them underlings in bondage, and afflict them for 400 years. Notwithstanding, I will judge that nation whom they serve, and after this, they shall depart from thence with great substance.' Which prophecy was accomplished, albeit God for certain respects added thereunto thirty years more, for which cause the Scripture sayth: 'But the abode of the children of Israel in Egypt was 430 years, which being finished, that very day the whole army of the Lord departed out of the land of Egypt.' But although this text say manifestly that the people of Israel abode in Egypt 400 years, a gloss declareth that these years were the whole time which Israel went on pilgrimage until he possessed his own country. Inasmuch as he remained in Egypt but 210 years, which declaration agreeth not well with that which St Stephan the protomartyr made in his discourse to the Jews, namely that the people of Israel was 430 years in the bondage of Egypt.

And albeit the abode of 210 years sufficed that the qualities of Egypt might take hold in the people of Israel, yet the time whiles they lived abroad was no lost season in respect of that which appertaineth to the wit, for those who live in bondage, in misery, in affliction, and in strange countries, engender much choler adust, because they want liberty of speech and of revenging their injuries. And this humour, when the same is grown dry, becometh the instrument of subtility, of craft, and of malice, whence we see by experience that if a man rake hell for bad manners and conditions, he cannot find worse than in a slave whose imagination always occupieth itself in devising how to procure damage to his master and freedom to himself. Moreover, the land which the people of Israel walked through was not much estranged nor different from the qualities of Egypt, for in respect of the misery thereof, God promised Abraham to give him another, much more abundant and fruitful. And this is a matter greatly verified, as well in good natural philosophy as in experience, that barren and beggarly regions (not fat, nor plentiful of fruit) engender men of very sharp wit. And contrariwise, abundant and fertile soils bring forth persons big limbed, courageous, and of great bodily forces, but very slow of wit.

Touching Greece, the historians never make an end to recount how appropriate that region is to breed men of great hability, and particularly Galen avoucheth that it is held a miracle for a man to find a fool in Athens. And we must note that this was a city the most miserable, and most barren of all the rest in Greece. Whence we collect that through the qualities of Egypt, and of the provinces where the Hebrew people lived, they grew very quick of capacity. But it behoveth likewise to understand for what cause the temperature of Egypt produceth this difference of imagination. And this will fall out a plain matter when you are done to ware that in this region the sun yieldeth a fervent heat, and therefore the inhabitants have their brain dried and choler adust, the instrument of wiliness and aptness. In which sense, Aristotle demandeth: 'Why the men of Ethiopia and Egypt have their feet crooked, and are commonly curl-pated° and flat nosed?' To which problem he answereth that the much heat of the country rosteth the substance of these members and wrieth them, as it draweth together a piece of leather set by the fire, and for the same cause, their hair curleth and themselves also are wily. And that such as inhabit hot countries are wiser than those who are born in cold regions we have already proved by the opinion of Aristotle, who demandeth whence it grows that men are wiser in hot climates than in cold. But he wist° not to answer this problem, nor make distinction of wisdom, for we have proved heretofore that in man there rest two sorts of wisdom: one whereof Plato said: 'Knowledge which is severed from justice ought rather to be termed craft than wisdom'; another there is found accompanied with justice and simplicity, without doubleness and without wiles, and this is properly called wisdom, for it goeth always guided by justice and duty. They who inhabit very hot countries are wise in the first kind of wisdom, and such are those of Egypt.

Now let us see when the people of Israel was departed out of Egypt and come into the desert, what meat they did eat, what water they drank, and of what temperature the air was where they travailed, that we may know whether upon this occasion the wit with which they issued out of bondage took exchange, or whether the same were more confirmed in them. Forty years, saith the Text, God maintained this people with manna, a meat so delicate and savoury as any might be that ever man tasted in the world. In sort that Moses, seeing the delicacy and goodness thereof, commanded his brother Aaron to fill a vessel and place the same in the Ark of confederacy, to the end the descendants of this people, when they were settled in the land of promise, might see the bread with which God had fed their fathers while they lived in the wilderness, and how bad payment they yielded him in exchange of such cherishments. And to the end that we who have not seen this meat may know of what manner the same was, it will do well that we describe the manna which nature maketh, and so adjoining thereunto the conceit of a great delicacy, we may wholly imagine his goodness.

The material cause of which manna is engendered is a very delicate vapour which the sun, with the force of his heat, draweth up from the earth, the which taking stay aloft is concocted and made perfect. And then the cold of the night coming on, it congealeth, and through his weightiness turneth to fall upon the trees and stones where men gather the same and preserve it in vessels to serve for food. It is called 'dewy' and 'airy honey', through the resemblance which it beareth to the dew, and for that it is made in the air. His colour is white, his savour sweet as honey, his figure like that of coriander, which signs the Holy Scripture placeth also in the manna which the people of Israel did eat. And therefore I carry an imagination that both were semblable in nature. But if that which God created were of more delicate substance, so much the better shall we confirm our opinion. But I am ever of opinion that God applied Himself to natural means when with them He could perform what He meant, and where nature wanted, His omnipotence supplied. This I say because to give them manna to eat in the desert (besides that which hereby

480 He would signify) me seemeth was founded in the self disposition of the earth, which (even at this day) produceth the best manna in the world, through which Galen affirmeth that on Mount Libanus (which is not far distant from this place) there is great and very choice abundance. In sort, that the country people are wont to sing in their pastimes that Jupiter raineth honey in that region. And though it be true that God miraculously created that manna in such quantity, at such time, and on special days, yet it may be that it partaked the same nature with ours, as had also the water which Moses drew forth of the
490 rock, and the fire which Elias[118] with his word caused to rain from heaven; all of them natural things, though miraculously brought to pass.

The manna described by the Holy Scripture it saith was as dew and as the seed of coriander, white, and in taste like honey, which conditions are also in the manna produced by nature. The temperature of this meat, the physicians say, is hot, and consisting of subtile and very delicate parts, which composition the manna eaten by the Jews should also seem to have. Whereon (complaining of his tenderness) they said in this manner: 'Our soul hath
500 a fulsomeness° at this slight meat.' As if they should say that they could no longer endure nor brook so light a meat in their stomach, and the philosophy of this was that their stomachs had been made strong by onions, chibols,[119] and leeks, and coming to eat a meat of so small resistance, it wholly with them turned into choler.[120] And for this cause, Galen gave the charge that men endowed with much natural heat should forbear to eat honey or other light meats, for they would turn to corruption and, instead of digestion, would parch up like soot.

The like hereof befell to the Hebrews as touching manna, which
510 with them wholly turned into choler adust, and therefore they were altogether dry and thin, for this meat had no corpulence to

---

[118] Elias] Elijah. Prophet in Israel during the reign of Ahab (ninth century BC), according to the Books of Kings. He denounced the worship of the god Baal as worship to an idol. Huarte refers here to the episode in which Jehovah's fire descends from the sky.
[119] chibol] species of allium in appearance intermediate between the onion and the leek.
[120] 35. *Index*: 'and coming to eat a meat of so small resistance, it wholly with them turned into choler' [censored].

fatten them.[121] Our soul, said they, is dry, and our eyes see nothing but manna. The water which they drank after this meat was such as they would desire, and if they could not find any such, God showed to Moses a wood of so divine virtue that dipping the same in gross and salt waters it made them to become delicate and of good savour. And when they had no sort of water at all, Moses took the rod with which he had parted the Red Sea, and striking therewith the rocks, there issued springs of waters so delicate and savoury as their taste could desire. In sort, that St Paul saith: 'The rock followed them.' As if he should say: the water of the rock seconded their taste, issuing delicate, sweet, and savoury. And they had accustomed their stomachs before to drink waters thick and brinish, for in Egypt, saith Galen, they boiled them ere they could serve for drink, for that they were naughty and corrupt. So as afterwards drinking waters so delicate, it could not fall out otherwise but that they should turn into choler, for that they found small resistence. Water requireth the same qualities, to digest well in our stomach, saith Galen, and not to corrupt, that the meat hath whereon we accustomably feed. If the stomach be strong, it behoveth to give the same strong meat, which may answer in proportion; if the same be weak and delicate, such also the meat ought to be. The like regard is to be held as touching the water, wherethrough we see by experience that if a man use to drink gross water, he never quencheth his thirst with the purer, neither feeleth it in his stomach. Rather the same increaseth his thirst, for the excessive heat of the stomach burneth and resolveth it so soon as it is received, because therein is no resistance.

The air which they enjoyed in the desert we may also say that it was subtle and delicate, for journeying over mountains and through uninhabited places, they had the same always fresh, cleansed, and without any corruption, for they never made long stay in any one place. So did it always carry a temperature, for by day a cloud was set before the sun which suffered him not to scorch over vehemently, and by night, a pillar of fire which moderated the same. And to enjoy an air of this manner, Aristotle affirmeth, doth much quicken the wit. We may consider then that the men of this folk must needs have a seed very delicate and adust, eating such meat as manna was, and drinking the waters

---

[121] 36. *Index*: 'The like hereof befell [509] [...] no corpulence to fatten them' [censored].

550 before specified, and breathing and enjoying an air so cleansed and pleasant, as also that the Hebrew women bred flowers very subtle and delicate.

Again, let us call to mind that which Aristotle said, that the flowers being subtle and delicate, the child who is bred of them shall be a man of great capacity. How much it importeth that for begetting children of great sufficiency the fathers do feed on delicate meats, we will prove at large in the last chapter of this work. And because all the Hebrews did eat of one self so spiritual and delicate meat, and drank of one self water, all their children
560 and posterity proved sharp and great of wit in matters appertaining to this world.

Now then, when the people of Israel came into the land of promise with so great a wit as we have expressed, there befell unto them afterwards so many travails, dearths, sieges of enemies, subjections, bondages, and ill entreatings that though they had not brought from Egypt and the wilderness that temperature, hot, dry, and adust before specified, they would yet have made it so by this dismal life. For continual sadness and toil uniteth the vital spirits and the arterial blood in the brain, in the liver, and in the
570 heart, and there staying one above another, they grow to dryness and adustion. Wherethrough, often times they procure the fever, and their ordinary is to make melancholy by adustion. Whereof they (in manner) do all partake even to this day, in respect of that (which Hippocrates saith) fear and sadness continuing a long time signifieth melancholy. This choler adust (we said before) to be the instrument of promptness, craftiness, sharpness, subtility, and maliciousness.[122] And this is applied to the conjectures of physic, and by the same a man getteth notice of the diseases, their causes and remedies. Wherefore King Francis understood this
580 marvellous well, and it was no lightness of the brain or invention of the devil which he uttered. But through his great fever, lasting so many days, and with the sadness to find himself sick and without remedy, his brain grew dry and his imagination rose to such a point of which we made proof tofore, that if it have the temperature behooful,° a man will on a sudden deliver that which he never learned. But there presents itself a difficulty very great against all these things rehearsed by us, and that is, that if the

---

[122] 37. *Index*: 'sharpness, subtility, and maliciousness' [censored].

children or nephews of those who were in Egypt and enjoyed manna, the waters, and the subtle air of the wilderness, had been made choice of for physicians, it might seem that King Francis's opinion were in some part probable for the reasons by us reported. But that their posterity should preserve till our days those dispositions of the manna, the water, the air, the afflictions, and the travails which their ancestors endured in the prison of Babylon, it is a matter hard to be conceived, for if in 430 years, during which the people of Israel lived in Egypt, and forty in the desert, their seed could purchase those dispositions of ability better, and with more facility could they leese° it again in 2000 years whilst they have been absent. And specially sithence their coming into Spain, a region so contrary to Egypt, and where they have fed upon different meats and drunk waters of nothing so good temperature and substance as those other.

This is agreeable to the nature of man, and whatso other living creature and plant which forthwith partaketh the conditions of the earth where they live, and leese those which they brought with them from elsewhere. And whatsoever instance they can allege, the like will betide it within few days beyond all gainsaying.

Hippocrates recounteth of a certain sort of men who to be different from the vulgar chose for a token of their nobility to have their head like a sugar-loaf. And to shape this figure by art, when the child was born the midwives took care to bind their heads with swaths and bands until they were fashioned to the form. And this artificialness grew to such force as it was converted into nature. For in process of time, all the children that were born of nobility had their head sharp from their mothers' womb. So from thenceforth, the art and diligence of the midwives herein became superfluous. But so soon as they left nature to her liberty and her own ordering, without oppressing her any longer with art, she turned by little and little to recover again the figure which she had before.

In like sort might it befall the children of Israel, who notwithstanding the region of Egypt, the manna, the delicate waters, and their sorrowfulness, wrought those dispositions of wit in that seed. Yet those reasons and respects surceasing, and other contrary growing on, it is certain that by little and little the qualities of the manna would have worn away, and other far different therefrom have grown on conformable to the country where they inhabited, to the meats which they fed upon, to the waters which they drank, and to the air which they breathed. This doubt in natural philosophy holdeth little difficulty, for there are some accidents to be found which are brought in at a moment, and afterwards endure forever in the subject without possibility of corrupting. Others there are which waste as much time in undoing as they occupied in engrafting, and some more, some less, according to the action of the agent and the disposition of the patient. For example, of the first we must know that a certain man, through a great fear whereinto he was driven, rested so transformed and changed in colour that he seemed dead. And the same lasted not only during all the time of his own life, but also the children which he begot had the same colour, without that he could find any remedy to take it away. Conformable hereunto, it may be that in 430 years, whilst the people of Israel led their lives in Egypt, forty in the wilderness, and sixty in the bondage of Babylon, there needed more than 3000 years that this seed of Abraham should take a full loss of their disposition of wit, occasioned by this manna, seeing to reform the bad colour, settled upon a sudden through fear, more than one hundred years were requisite. But because the truth of this doctrine may be understood from the root, it behoveth to resolve two doubts which serve to the purpose, and as yet I have not cleared. The first is whence it commeth that meats, by how much the more delicate and savoury they are (as hens and partridge), so much the sooner the stomach doth abhor and loathe them. And contrariwise, we see that a man eateth beef all the year long without receiving any annoyance thereby, and if he eat hens' flesh but three or four days together,[123] the fifth he cannot abide the savour thereof but that it will turn his stomach upside down. The second is whence it commeth that bread of wheat and flesh of mutton, not being of substance so good and savoury as hen and partridge, yet the stomach never loatheth them though we feed thereon all our lives long. But wanting bread we cannot eat other meats, neither do they content us.

He that can shape an answer to these two doubts shall easily understand for what cause the descendants of the people of Israel have not yet lost the dispositions and accidents which manna brought into that seed. Neither will the promptness of wit, and

---

[123] together] in a row.

subtlety whereof they then possessed themselves so soon take an end. Two certain and very true principles there are in natural philosophy on which the answer and resolution of these doubts dependeth. The first is that all powers whatsoever which govern man are naked and deprived of the conditions and qualities which rest in their object, to the end that they may know and give judgement of all the differences. The eyes partake this property, who being to receive into themselves all figures and colours, it was of necessity utterly to deprive them of figures and colours. For if they were pale, as in those who are overcome with the yellow jaundice,[124] all things whereon they looked would appear to them of the same colour. So the tongue, which is the instrument of taste, ought to be void of all savours, and if the same be sweet or bitter, we know by experience that whatsoever we eat or drink hath the like taste. And the same may be avouched of hearing, of smelling, and of feeling. The second principle is that all things created naturally covet their preservation and labour to endure forever, and that the being which God and nature have given them may never take end, notwithstanding that afterward they are to possess a better nature. By this principle, all natural things endowed with knowledge and sense abhor and fly from that which altereth and corrupteth their natural composition.

The stomach is naked and deprived of the substance and qualities of all meats in the world, as the eye is of colours and figures. And when we eat ought, though the stomach overcome it, yet the meat turneth against the stomach (for that the same is of a contrary principle) and altereth and corrupteth his temperature and substance, for no agent is of such force but that in doing, it also suffereth. Meats that are very delicate and pleasing do much alter the stomach; first, because it digesteth and embraceth them with great appetite and liking, and then, through their being so subtle and void of excrements, they pierce into the substance of the stomach, from whence they cannot depart again. The stomach then feeling that this meat altereth his nature and taketh away the proportion which he carrieth to other meats, groweth to abhor the same, and if he must needs feed thereon, it behoveth to use many salads and seasonings, thereby to beguile him.

---

[124] jaundice] morbid condition caused by obstruction of the bile, characterized by yellowness of the conjunctiva, skin, fluids, and tissues, and by constipation, loss of appetite, and weakness.

All this, manna had even from the beginning, for though the same were a meat of such delicacy and pleasing relish, yet in the end the people of Israel found it fulsome,[125] and therefore said: 'Our soul loatheth this over light meat.' A complaint far unworthy of a people so specially favoured by God, who had pretended a remedy in that behalf, which was that manna had those relishes and tastes which well agreed with them to the end they might eat thereof. 'Thou sentest them bread from heaven which had in it all pleasingness', for which cause many amongst them fed thereon with good appetite, for they had their bones, their sinews, and their flesh so imbued with manna and his qualities that by means of the resemblance from each to other they longed after nothing else. The like befalleth in bread of wheat, and weathers flesh, whereon we accustomably feed.

Gross meats and of good substance (as beef) have much excrements, and the stomach receiveth them not with such desire as those that are delicate and of good relish, and therefore is longer ere the same take alteration by them. Hence commeth it that to corrupt the alteration which manna made in one day, it behoveth to feed a whole month upon contrary meats. And, after this reckoning, to deface the qualities that manna brought into the seed in the space of forty years, there need 4000 and upward. And if any man will not herewith rest satisfied, let us say that as God brought out of Egypt the twelve tribes of Israel, so he had taken then twelve male and twelve female Moors of Ethiopia, and had placed them in our country, in how many years think we would these Moors and their posterity linger to leave their native colour, not mixing themselves the while with white persons? To me it seemeth a long space of years would be requisite. For though 200 years have passed over our heads sithence the first Egyptians came out of Egypt into Spain,[126] yet their posterity have not forlorn that their delicacy of wit and promptness, nor yet that roasted colour which their ancestors brought with them from

---

[125] fulsome] of a sickly or sickening taste.
[126] 200 years have passed over our heads sithence the first Egyptians came out of Egypt into Spain] actually, the first written record on the arrival of a gipsy tribe in Spain (more precisely, in Saragossa) is from May 1425, when King Alfonso V of Aragon, the Magnanimous, signed a three-month safe-conduct to a group of gipsies led by a Juan de Egipto Menor, allowing them to travel through his kingdom and continue their pilgrimage to Santiago de Compostela.

Egypt. Such is the force of man's seed when it receiveth thereinto any well rooted quality. And as in Spain the Moors communicate the colour of their elders by means of their seed, though they be out of Ethiopia, so also the people of Israel coming from thence may communicate to their descendants their sharpness of wit without remaining in Egypt or eating manna. For to be ignorant or wise is as well an accident in man as to be black or white. True it is that they are not now so quick and prompt as they were a thousand years since, for from the time that they left to eat manna, their posterity have ever lessened hitherto because they used contrary meats and inhabited countries different from Egypt, neither drank waters of such delicacy as in the wilderness, as also by mingling with those who descended from the gentiles, who wanted this difference of wit. But that which cannot be denied them is that as yet they have not lost it altogether.

CHAP. XIII.

By what means it may be showed to what difference of ability the art of warfare appertaineth, and by what signs the man may be known who is endowed with this manner of wit.

What is the cause, saith Aristotle, that seeing fortitude is not the greatest of all virtues, but justice and prudence are greater than it, yet the commonwealth, and in a manner all men with a common consent do make greater account, and within themselves do more honour a valiant man than either the just of wise though placed in never so high callings or offices? To this problem Aristotle answereth saying there is no king in the world who doth not either make war, or maintain war against some other, and for so much as the valiant procure them glory and empire, take revenge on their enemies, and preserve their estate, they yield chiefest honour not to the principal virtue, which is justice, but to that by which they reap most profit and advantage. For if they did not in this wise entreat the valiant, how were it possible that kings should find captains and soldiers who would willingly jeopard their lives to defend their goods and estates?

Of the Asiaticans it is recounted that there was a people inhabiting a part thereof who bore themselves very courageously, and being asked why they had neither king nor law, they made answer that laws made men cowards, and seeing it was necessary to undergo the hazard of the wars for depriving another of his estate, they made choice to fight for their own behoof,° and themselves to reap the benefit of the victory. But this was an answer rather of barbarous men than reasonable people, who well know that without a king, without a commonwealth, and without laws, it is impossible to preserve men in peace. That which Aristotle said serveth very well to the purpose, though there be a better answer to be framed, namely, that when Rome honoured her captains with those triumphs and solemnities, she did not only reward the courage of their triumpher, but also the justice with which he maintained his army in peace and concord, the wisdom with which he performed his enterprises, and their temperancy used in abstaining from wine, women, and meat, which trouble the judgement and turn counsels into error. Yea, wisdom is more highly to be regarded and rewarded in a general than courage and manliness, for as Vegetius[127] well said, few over-courageous captains bring their enterprises to lucky pass. Which groweth for that wisdom is more necessary in war than courage in bickering. But Vegetius could never attain to the notice what manner of wisdom this is, neither could plot down with what difference of wit he ought to be endowed who taketh charge in war. Neither do I ought marvel thereat, for the manner of philosophy whereon this dependeth was not then devised. True it is that to verify this point answereth not our first intent, which purporteth to make choice of apt wits for learning. But martial affairs are so dangerous, and of so deep counsel, and it falleth out a matter so important for a king to know well unto whom he credit his power and state, that we shall perform no less thanks-worthy a part of service to the commonwealth to teach this difference of wit and his signs, than in the other which we have already described. For which cause we must note that *malitia* and *militia*, viz. martial matters, and malice, have as it were one self name, and likewise, one self definition. For changing *a* into *i*, of *malitia* you make *militia*, and of *militia*, *malitia*, with great facility. What the nature and property of malice is Cicero teacheth saying: 'Malice is a way

---

[127] Vegetius] Vegetius was a writer of the Later Roman Empire. His two surviving works are *Digesta Artis Mulomedicinae*, a guide to veterinary medicine, and *Epitoma rei militaris*, a work on Roman military organization, institutions and behaviour in war.

of hurting, crafty, and full of guile.' In war, likewise, nothing falleth so much into consideration as how to offend the enemy and defend ourselves from his entrappings. Therefore, the best property whereof a general can be possessed is to be malicious with his enemy and never to construe any his demeanour to a good sense, but to the worst that may be, and to stand on his guard.

'Believe not', sayth *Ecclesiasticus*, 'thine enemy, with his lips he sweetneth° and in his heart he betrayeth thee to make thee fall into the dike; he weepeth with his eyes, and if he light upon a fit occasion, he will not be satisfied with thy blood.' Hereof we find a manifest example in the Holy Scripture, for the people of Israel being besieged in Bethulia, and straightened with hunger and thirst, that famous lady Judith[128] issued out with a resolution to kill Holofernes,[129] and going towards the army of the Assyrians, she was taken by the sentinels and guards, and being asked whither she was bound, made answer with a two-fold mind: 'I am a daughter of the Hebrews, whom you hold besieged, and fly unto you, for I have learned that they shall fall into your hands and that you shall evil entreat them because they would not yield themselves to your mercy. Therefore, I determined to fly unto Holofernes and to discover unto him the secrets of this obstinate people, showing him how he may enter without the loss of any one soldier.'

So Judith being brought to Holofernes' presence, threw herself down to the ground, and with closed hands began to worship him and utter words full of deceit, the most craftily that might be, in sort that Holofernes and all his counsel verily believed she said nothing but truth. But she not forgetful what in heart she had purposed, found a convenient occasion and chopped off his head.

Contrary hereunto are the conditions of a friend, and therefore it behoveth ever to yield him credit. Wherethrough Holofernes

---

[128] Judith] Judith is the heroine of the Book of Judith in Apocrypha. She is a courageous widow who delivered her countrymen from the assault of Holofernes, general of Nebuchadnezzar. Her actions resulted in the cutting off of the head of Holofernes.

[129] Holofernes] According to the Book of Judith, Holofernes was the chief captain of Nebuchadnezzar, king of the Assyrians. Holofernes became a persecutor of the Jews with the mission of making Nebuchadnezzar the object of universal human worship. He was slain by Judith during the siege of Bethulia.

should have done better to believe Achior,[130] seeing he was his friend, and on zeal that he should not leave the siege with dishonour, said unto him: 'Sir, first inform yourself whether his people have sinned against God, for if it be so, Himself will deliver them into your hands without that you shall need to conquer them; but if He hold them in grace, know for certain that He will defend them, and we shall not be able to vanquish them.' Holofernes conceived displeasure at this advertisement, as a man confident, lascivious and a wine-bibber, which three things turn topsy-turvy that counsel which is requisite for the art of war. For which cause Plato said he liked very well of a law which the Carthaginians had by which they commanded that the general, whilst he had charge of the army, should drink no wine, for this liquor, as Aristotle affirmeth, maketh a man of wit be quite burned up with choler (as Holofernes showed in those so furious words which he spoke to Achior).

Now that wit which is requisite for ambushes and stratagems, as well to prepare them as to perceive them, and to find out such remedy as appertaineth, Cicero describeth drawing his descent from this known *versutia*, which he saith is derived from this verb *versor*, for those who are winding,° crafty, double, and cavillers, upon a sudden contrive their wiles and employ their conceit with facility. And so the same Cicero exemplified it, saying: 'Chrisippus[131] a man doubtless winding and crafty. I call those winding whose mind is suddenly winded about.' This property to attain suddenly the means is *solertia* (quickness) and appertaineth to the imagination, for the powers which consist in heat perform speedily their work. And for this cause men of great understanding are little worth for the war, for this power is very slow in his operation and a friend of uprightness, of plainness, of simplicity and mercy, all which is wont to breed much damage in war. These are good to treat with friends, with whom the wisdom

---

[130] Achior] General of the Ammonites who spoke in behalf of Israel before Holofernes, the Assyrian general, and afterwards converted to Judaism.

[131] Chrisippus] Chrysippus of Soli (*c.* 279 BC–*c.* 206 BC) was a Greek Stoic philosopher. He was a pupil of Cleanthes in the Stoic school, and after Cleanthes' death, he became the third head of the school. In his many writings Chrysippus expanded the doctrines of Zeno of Citium, the founder of Stoicism. Only fragments of his works have survived. Some of these are included in the works of later authors like Cicero, Seneca, and Galen, among others.

of the imagination is not needful, but only the rightfulness and singleness of the understanding which admitteth no doubleness, nor doth any wrong. Therefore, with the enemy it booteth nothing, for he always studieth to offend with wiles, and such wit is requisite wherewith to counter guard ourselves. And so Christ our redeemer advised his disciples saying: 'Behold I send you as sheep amongst wolves, be you therefore wise as serpents and simple as doves.' With our enemies we must practise wisdom, and with our friends plainness and simplicity.

Now if the captain be not to give credit to his enemy, but is always to misdoubt that he will go beyond him, it is necessary that he hold a difference of imagination, forecastful, wary, and which can skill to discern the wiles which come vailed with any coverture, for the self power which finds them out can only devise the remedies which are behooful° in that behalf. That seemeth to be another difference of the imagination which deviseth the engines and war-like instruments, whereby invincible fortresses are won, which pitcheth the camp and marshalleth every squadron in his due place, and which knoweth the occasions of joining and retiring, which plotteth treaties, consortments and capitulations with the enemy. For all which the understanding is impertinent° as are the ears to see withal. And therefore I nothing doubt but that the art of war appertaineth to the imagination, for all whatsoever a good captain is to perform importeth consonance, figure and correspondence.

Now the difficulty resteth to set down with what difference of the imagination in particular war is to be managed. And in this I cannot resolve with certainty because the knowing thereof is very nice. Yet I conjecture that it requireth a degree more of heat than the practice of physic, and that it allay choler but not utterly quench it.

This is very manifest, for those captains who are full of promptness and subtlety are not very courageous, nor desirous of bickering, neither covet to come to handystrokes, but by stratagems and fetches without adventuring a broken pate do bring their purposes to pass. Which property better pleased Vegetius than any other. 'Good captains', saith he, 'not by open war in which the peril is common, but by secret practices, ever assay with the safety of their own soldiers to cut their enemies in pieces, or at least to make them afraid.' The fruit of this manner of wit the Roman Senate very wisely looked into, for though they had many famous captains who achieved sundry wars, yet returning to Rome to receive the triumph and glory due to their enterprise, so great were the plaints which the parents made for their children, the children for the parents, the wives for their husbands, and brothers for brethren, that through the sorrow for them who perished in the wars they could take little pleasure in the sports and pastimes. Wherefore the Senate took a resolution not to seek out so courageous captains wholly desirous to come to handstrokes, but men somewhat timorous and very ready, as Q. Fabius, of whom it is written that it was a wonder to see him offer a pitched battle in the open field, and specially when he was far from Rome, whereby in ill successes he could not readily be relieved, and he did nought else but give way to the enemy and devise stratagems and wiles with which he exploited great enterprises and obtained many victories without the loss of any one soldier. He was received into Rome with great joy of all men, for if he carried forth 100,000 soldiers, he returned with as many, unless some perhaps miscarried by sickness. The shout which the people gave at his return was, as Ennius reporteth, of this tenour:

One man by lingering, only us relieved.

As if they had said: 'This man with giving way to our enemies hath made us lords of the world, and brought back our soldiers to their houses in safety.' Some captains have since that time endeavoured to imitate him, but because they wanted his wit and readiness, they sundry times let slip many fit occasions of fighting, whence greater damages and inconveniences arose than if they had speedily joined battle. We may also take example of that famous Carthaginian captain of whom Plutarch writeth these words: 'Hannibal,[132] after he had attained this so great a victory, commanded that many Italian prisoners should freely be set at liberty without ransom to the end the same of his courtesy and pardoning might be dispersed among the people. Albeit of

---

[132] Hannibal] Hannibal (247–c.180 BC) was a Punic Carthaginian military commander, son of Hamilcar Barca, leading Carthaginian commander during the First Punic War, and brother of Mago and Hasdrubal. At the outbreak of the Second Punic War he marched an army from Iberia into northern Italy. Hannibal occupied much of Italy for 15 years until he had to return to Carthage to fight a Roman counter-invasion of North Africa. There he was defeated by Scipio Africanus at the Battle of Zama.

disposition he were very wide from this virtue, for of his own nature he was fell and unmerciful, and in such sort was trained up from the tender years of his youth, that he never learned laws or civil conditions but wars, slaughters, and betrayings of the enemy. Wherethrough he grew to be a captain very cruel and malicious in beguiling men, and always devising how he might entrap his enemy. And when he saw he could not prevail by open war, he sought to get the upper hand by policies, as was plainly seen in this deed of arms by us rehearsed, and by the battle which he fought against Sempronius[133] near the river Trebia.'

The tokens to know a man that is possessed of this difference of wit are very strange and well worthy of contemplation. Wherethrough Plato saith that the man who is very wise (in this sort of ability which we trace out) cannot be courageous nor well conditioned, for Aristotle saith that wisdom consisteth in cold, and stomach and manliness in heat. Therefore these two qualities being repugnant and contrary, it is impossible that a man be very full of hardiness and also of wisdom therewithal. For which cause it is necessary that choler be burned and become choler adust, to the end that a man may prove wise. But where this spice° of melancholy is found, inasmuch as the same is cold, fear and cowardice are straightways entertained. In sort, that craft and readiness require heat, for that the same is a work of the imagination, but not in such degree as courage, wherethrough they repugn each to other in extension. But herein befalleth a matter worth the noting, that of the four moral virtues (justice, prudence, fortitude, and temperance), the two first require a wit and good temperature to the end that they may be put in practice. For if a judge be not endowed with understanding to make himself capable of the point of justice, little avails it that he carry a good will to render every man his due, since this his good meaning may wander out of the way and wrong the true proprietary. The like is to be understood of wisdom, for if the only will sufficed to set things in good order, then in no work, good or evil, should any error be committed. There is no thief whatsoever who seeketh not to rob in such manner as he may not be espied,

---

[133] Sempronius] Tiberius Sempronius Longus (*c.* 260 BC–210 BC) was a Roman consul during the Second Punic War. Sempronius captured Malta from the Carthaginians, and later led an attack at the Battle of the Trebia against the Carthaginian army of Hannibal's brother, Mago.

and there is no captain who desireth not to be owner of so much wisdom as may serve to vanquish his enemy. But a thief that is not his craftsmaster in filching, soon falleth to be discovered, and the captain that wanteth imagination, ere long is overcome.

Fortitude and temperance are two virtues which men carry in their fist though they want a natural disposition. For if a man be disposed to set little of his life, and show hardiness, he may well do it. But if he be courageous of his own natural disposition, Aristotle and Plato affirm very truly, it is not possible that he can be wise though he would. In sort, that by this reason, there groweth no repugnancy to unite the wisdom of the mind with courage, for a wise and skilful man hath the understanding to hazard his honour in respect of his soul, and his life in respect of his honour, and his goods in respect of his life, and so he doth. Hence it comes that gentlemen, for that they are so much honoured, are so courageous, and there is none who will endure more hardness in the wars (for that they are brought up in so many pleasures) to the end they may not be termed ribalds. Hereon is that byword grounded: 'God keep me from a gent. by day and a thief by night', for the one, because he is seen, and the other that he may not be known, do fight with double resolution. On this self reason is the religion of Malta grounded, who, knowing how much it importeth nobility to be a man of valure, have a firm law that all those of their order shall be issued from gentility, both on the father's side and the mother's, for so each of them must in the combat show himself worth two of a baser progeny. But if a gentleman had the charge given him to encamp an army, and the order whereby he should put the enemy in rout, if he had not a wit appropriate hereunto, he would commit and utter a thousand disorders, for wisdom lieth not in men's disposition. But if there were recommended unto him the guard of a gate, they might soundly sleep on his eyes, although by nature he were a baggage." The sentence of Plato is to be construed when a wise man followeth his own natural inclination and doth not correct the same by reason. And in that sort it is true that a very wise man cannot of his natural disposition be courageous, for choler adust (which maketh him wise) maketh him also, saith Hippocrates, timorous and fearful.

The second property, wherewith a man possessed of this difference of wit cannot be endowed, is to be pleasant and of quaint behaviour, for with his imagination he frameth many plots

THE EXAMINATION OF MEN'S WITS

and weeteth° that whatsoever error or negligence are the way to cast away an army, wherethrough he ever carrieth an eye to the main chance. But people of little worth call carefulness a toil; chastisement, cruelty; and mercy, softness; suffering and dissembling of lewd parts, a good disposition. And this verily springeth because men are sots° who pierce not into the true value of things, nor in what sort they ought to be managed. But the wise and skilful cannot hold patience, nor bear to see matters ill handled though they nothing appertain unto themselves, and therefore live a small while and with much trouble of spirit. Whence Solomon said: 'I gave also my mind to understand wisdom, doctrine, errors, and folly, and found that in these also there is weariness and affliction of spirit. For into much wisdom entereth much displeasure, and who so attaineth science getteth sorrow.' In which words it seemeth that Solomon gave us to understand that he lived better contented being ignorant, than after he had received wisdom. And so verily it came to pass, for the ignorant live most careless inasmuch as nothing giveth them pain nor vexation, and they little reck who have a better capcase[134] than themselves. The vulgar accustometh to call such the 'angels of heaven', for they see how they take nothing at heart, neither find fault with anything ill done, but let all pass. But if they considered the wisdom and condition of the angels, they should see it were a word that carried evil consonance, and a case for the Inquisition House, for from the day when we receive the use of reason, until that of our death, they do nought else save reprove us for all our evil doings, and advise us to that which we ought to do. And if as they speak to us in their spiritual language (by moving our imagination), so they should deliver us their opinion in material words, we would hold them importunate and unmannerly brought up. And he that believeth not this, let him mark that the angel of whom St Matthew maketh mention, seemed such a one to Herod and to the wise of his brother Philip, seeing (because they would not hear his fault-findings) they faire and well chopped off his head. Better were it that these men, who by the vulgar are fondly termed 'angels of heaven', were called

---

[134] In Spanish this reads: 'ni piensan que en saber nadie les hace ventaja' ('nor do they think that in knowing someone else is better than them'). According to the OED, cap-case refers to a receptacle (like a box, chest or case). So, cap-case could then mean figuratively 'head', and, by extension, knowledge.

310 asses of the earth, for amongst brute beasts, saith Galen, there is none more blunt or of less wit than the ass, although in memory he outreach all the rest. He refuseth no burden, he goeth whither he is driven without any gainstriving,° he winceth not, he biteth not, he is not fugitive, not jadish° conditioned; if he be laboured[135] with a cudgel, he setteth not by it, he is wholly made to the well-liking and service of him that is to use him. These self properties do those men partake whom the vulgar term angels of heaven, which sport-making springeth in them, for that they are block-heads and void of imagination, and have their wrathful power
320 very remiss, which tokeneth a great defect in a man and argueth that he is ill compounded.

There was never angel nor man in the world better conditioned than Christ our redeemer, and he entering one day into the temple, belaboured well favouredly those whom he found there selling of merchandise, and this he did because the irascible° is the chastice° giver, and sword of reason, and the man who reproveth not things ill done, either showeth himself but a fool, or is deprived of the wrathful power. In sort, that it falls out a miracle to see a wise man of that gentleness or conditions
330 which are best liking to lewd men's fancies. Wherethrough such as set down in writing the actions of Julius Caesar, marvelled to see how his soldiers could support a man so rough and severe, and this grew in him because he lighted upon a wit requisite for the wars.

The third property of those who are endowed with this difference of wit is to be reckless touching the attiring of their person, and in a manner all of them are slovenly, homely, with their hosen hanging about their heels, full of wrinkles, their cap sitting upon the one side with some threadbare gaberdine° on their
340 back, and never long to change suits.

This property, Lucius Florus[136] recounteth, had that famous captain Viriatus,[137] by nation a Portuguese, of whom (exaggerating his great humility) he saith and affirmeth that he despised so

---

[135] laboured] beaten.
[136] Lucius Florus] Lucius Annaeus Florus (*c.* 74 AD–*c.* 130 AD) was a Roman historian from the time of Trajan and Hadrian. He is the author of *Epitome de T. Livio Bellorum omnium annorum DCC Libri duo*. The book is a summary of Roman history from the foundation of Rome to the closing of the temple of Janus by Augustus (25 BC).

much all ornament of his person as there was no private soldier in his army that went worse apparelled than himself. And verily this was no virtue, neither did he the same artificially, but it is a natural effect of those who are possessed with that difference of imagination after which we inquire.

This recklessness in Julius Caesar greatly deceived Cicero, for being asked after the battle the cause which moved him to follow the party of Pompey, he answered, as Macrobius[138] recounteth: 'His girding deceived me.' As if he had said: 'It was my beholding of Julius Caesar to be a man somewhat slovenly, and who never wore his girdle handsomely', whom his soldiers in scoff called 'Loosecoat'. But this should have moved and made him to know that he was endowed with a wit requisite to the counsel of war. Rightly did Sulla[139] hit the nail on the head who, as Suetonius Tranquillus[140] reporteth, seeing the recklessness of Julius Caesar in his apparelling himself when he was a boy, advertised the Romans, saying: 'Take heed of this ill girded young fellow.' The historians busy themselves much in recounting how carelessly Hannibal bore him touching his apparel, and how little he recked to go neat and handsome. To grow in great dislike at motes on the cape, to take much care that his stockings sit clean and his cloak handsome, without plaits, appertaineth to a difference of the imagination of very base alloy,° and gainsaith the understanding and that imagination which the war requireth.

The fourth sign is to have a bald head, and the reason hereof may soon be learned: for this difference of imagination resideth in the forepart of the head, as do all the rest, and excessive heat

---

[137] Viriatus] Viriatus (d. 138 BC) led the resistance of the Lusitanian people to the Roman expansion into what would later become the Roman province of Lusitania, roughly present-day Portugal and Galicia. Despite being a victorious leader, he was eventually betrayed to the Romans and killed.

[138] Macrobius] Macrobius was a Roman author from the early fifth century who penned the still extant books *Saturnalia* and the *Commentarii in Somnium Scipionis* (Commentary on the Dream of Scipio).

[139] Sulla] Lucius Cornelius Sulla Felix (c. 138 BC–78 BC) was a Roman general and statesman. He held the office of consul twice. Sulla noted in his memoirs Caesar's notorious ambition.

[140] Suetonius Tranquillus] Suetonius (c. 69–after 122 AD) was a Roman historian of the times of the Roman Empire. *De Vita Caesarum* (*The Twelve Caesars*), his only extant full work is a collective biography of twelve leaders of the Roman Empire, beginning with Julius Caesar and ending with Domitian.

burneth the skin of the head, and closeth the pores through which the hair is to pass. Besides, that the matter whereof the hair is engendered (as the physicians avouch) are those excrements which the brain expelleth in time of his nourishing, and by the great fire that there is, they are consumed and burned up, and so the matter faileth whereof they may breed. And if Julius Caesar had been seen in this point of philosophy, he would not so much have shamed at his bald-head, as that to cover the same, he caused the hinder part of his hair (which should hang down on his neck) to be featly turned towards his forehead. And Suetonius maketh mention that nothing so much contented him as when the Senate enacted that he might wear a laurel garland on his head, and that on none other ground than because thereby he might cover his baldness. Another sort of baldness groweth from having the hair hard and earthly, and of a gross composition, but that betokeneth a man void of understanding, imagination and memory.

The fifth sign, whereby those are known who have this difference of imagination is that such are spare in words and full of sentences, and the reason importeth because the brain being hard, it followeth of necessity that they suffer a defect in memory, to which copy of words appertaineth. To find much what to say, springeth from a conjunction which the memory maketh with the imagination in his first degree of heat. Such as have this conjoining of both powers are ordinarily great liars, and never want words and tales though you stand hearkening unto them a whole day together.

The sixth property of those who have this difference of imagination is to be honest and to take great dislike at filthy and bawdy talk, and therefore Cicero saith that men very reasonable do imitate the honesty of nature, who hath hidden the unseemly and shameful parts which she made to provide for the necessity of mankind and not to adorn it, and she consenteth not to fasten the eyes on those, nor that the ears should once hear them named. This we might well attribute to the imagination and say that the same resteth offended at the evil representation of these parts, but in the last chapter we rendered a reason of this effect and reduced the same to the understanding, and we adjudged him defective in this power, who took not offence at such dishonesty. And because to the difference of imagination appurtenant to the art military there is joined this discourse, therefore are good captains very honest. Wherethrough, in the history of Julius Caesar, we find an

THE EXAMINATION OF MEN'S WITS

action of the greatest[141] honesty that might be,[142] and that is, whilst they murdered him with daggers in the Senate-house, he (perceiving it was impossible to escape death) gave himself to fall to the ground, and so sitted his imperial robe about him, that after his death they found him couched with great honesty, with his legs and other parts covered that might any way offend the sight.

The seventh property and of greatest importance is that the general have good fortune and be lucky, by which sign we shall perfectly find that he is seized of the wit and hability behooful for the art martial. For in substance and truth, there is nothing which ordinarily maketh men unfortunate, and that their enterprises do not always take success after their desire, save that they are deprived of wisdom and lay not hold on the convenient means for achieving their exploits. For that Julius Caesar showed such wisdom in the affairs which he managed, he bare away the bell (in respect of fortunateness) from all other captains of the world, so as in perils of importance, he encouraged his soldiers saying: 'Fear not, for you have Caesar's good fortune to fight on your party.'

The Stoics held opinion that as there was a first cause, everlasting, almighty, and of infinite wisdom, known by the order and concert of his marvellous works, so also there was another unwise and unconcerted° whose works proved without order, without reason, and void of discretion. For with an affection no way reasonable, it giveth and reaveth from men riches, dignity, and honour. This they termed 'fortune', seeing her a friend to men who perform their business by haphazard without forecasting, without wisdom, and without submitting themselves to the government of reason. They portrayed her (the better to make her manners and malice known) in form of a woman, a royal sceptre in her hand, her eyes vailed, her feet upon a round ball, accompanied with persons sottish and void of all trade of living. By painting her like a woman, they noted her great lightness and little discretion; by her royal sceptre, they acknowledged her sovereignty over riches and honour; her veiled eyes gave to understand the ill fashion which she held in distributing her gifts; her feet standing on the round ball, betokened the small firmness

---

[141] 38. *Index*: 'greatest' [censored].

[142] 39. *Index*: 'that might be' [censored]. In Spanish, the censored sentence is 'que ha hecho hombre en el mundo'.

in the favours which she imparted, for she snatcheth them away with the like facility that she reacheth them forth, without keeping steadfasteness in ought whatsoever. But the worst part they found in her was that she favoureth the wicked and persecuteth the virtuous; loveth the foolish, and abhorreth the wise; abaseth the noble, and advanceth the base; what is foul pleaseth her, and what is fair worketh her annoyance. Many men, placing confidence in these properties, because they know their own good fortune, take hardiness to undertake fond and headlong enterprises which yet prosper with them very luckily, and yet other men, very wise and advised, dare not adventure to execute those enterprises which they have begun with great discretion, finding by experience that such find worst success. How great a friend fortune showeth herself to bad people Aristotle maketh known by this problem: 'Whence groweth it that riches (for the most part) are possessed rather by the wicked than by men of worth?' Whereto he shapeth answer: 'Perhaps because fortune being blind cannot know nor make choice of what is best.' But this is an answer unworthy of so great a philosopher, for it is not fortune that bestoweth wealth on men, and though it were, yet he yieldeth no reason why she always cherisheth the bad and abandoneth the good. The true solution of this demand is that the lewd sort are very witty and have a gallant imagination to beguile in buying and selling, and can profit in bargaining, and employing their stock where occasion of gain is offered. But honest men want this imagination, many of whom have endeavoured to imitate these bad fellows, and by trafficking and trucking,° within few days have lost their principal.

This, Christ our redeemer pointed at, considering the sufficiency of that steward whom his master called to account, who reserving a good portion of the goods to his own behoof, salved up all his reckonings and got his *quietus est*. Which wisdom (though it were faulty) yet God commended saying: 'The children of this world are more wise in their kind than the children of light', for these ordinarily enjoy a good understanding with which power they place their affection on their law, and have want of imagination, whereto the knowledge how to live in this world appertaineth, wherethrough many are morally good, because they lack the wit how to be naught. This manner of answering is more easy and apparent. The natural philosophers, because they could not reach so far, devised so fond and ill jointed a cause, as lady

490 fortune, to whose power they might impute good and bad success, and not to the unskilfulness and little knowledge of men.

Four sorts of people there are in every commonwealth (if a man list to mark them). For some men are wise, and seem not so; others seem so, and are not; others, neither are, nor seem; and some both are, and seem so. Some men there are silent, slow in speech, staid in answering, not curious nor copious of words. Yet they retain hidden within them a natural power appertaining to the imagination, whereby they know the fit time and occasion to bring their purpose to pass, and how they are therein to demean
500 themselves without communicating or imparting their mind to any other. These by the vulgar are called happy and lucky, them seeming that with little knowledge and less wit everything falleth into their lap.

Others, contrariwise, are of much eloquence in words and discourse, great conversers,° men that take upon them to govern the whole world, who go about hunting how with small expense they may reap great gains, and therein (after the vulgars' conceit) no man in judgement can step an ace beyond them, and yet, coming to the effect, all falleth to the ground between their hands.
510 These cry out upon fortune and call her blind buzzard and jade,[143] for the matters which they design and work with much wisdom she suffereth not to take good effect. But if there were a fortune who might plead her own defence, she would tell them: 'Yourselves are the buzzards, the sots, and the do-noughts,° whom you speak of, that being unskilful hold yourselves wise, and using unfit means, would yet reap good successes.' This sort of people have a kind of imagination which decketh up° and setteth forth their words and reasons, and maketh them seem to be what indeed they are not. Whereon I conclude that the general
520 who is endowed with a wit requisite for the art military, and doth duly forecast what he is to exploit, shall be fortunate and happy. Otherwise, it is lost labour to look that he ever prevail to victory, unless God do fight for him as he did for the armies of Israel, and yet withal, they chose the wisest and skilfulest amongst them to be commanders, for we must not leave all upon God's hands, neither yet may a man wholly affy[144] on his own wit and

---

[143] jade] applied to Fortune, Nature, personified.
[144] affy] rely on.

sufficiency, but it will do best to join both together, for there is no other fortune save God and a man's own good endeavour.

He who first devised chess-play[145] made a model of the art military, representing therein all the occurrents and contemplations of war without leaving any one behind. And as in this game fortune beareth no stroke, neither can the player who beateth the adverse party be termed fortunate, nor he who is beaten unfortunate. So the captain that overcometh ought to be called wise, and the vanquished, ignorant; and not the one happy, or the other unhappy. The first thing which he ordained in this play was that, when the king is mated, the contrary party is vanquisher, thereby to let us understand that the chief force of an army consisteth in a good commander to govern and direct the same, and for proof hereof, he lotted as many chief men to the one side as to the other, to the end that whosoever lost might be ascertained it so fell out through default of his own knowledge and not of fortune. And this is more apparently seen if we consider that a skilful player will spare half his men to the other party, and yet for all that get the game. And this was it which Vegetius noted: that often few soldiers and weak vanquish many and valiant if they be governed by a general who can skill in ambushes and stratagems. He ordained also that the pawns might not turn back, thereby to advise the commander that he duly forecast all chances ere he send forth his soldiers to the service, because if any mischance alight, it behoves rather that they be cut in pieces where they were placed than to turn their backs, for the soldier is not to know when time serveth to fly or to fight save by direction of his captain, and therefore so long as his life lasteth, he is to keep his place under pain of becoming infamous.

Hereunto he adjoined another law that the pawn which had made seven draughts° without being taken should be made a queen and might make any draught at pleasure, and be placed next the king as one set at liberty and endowed with nobility; whereby he gave us to understand how in the war it importeth greatly for making the soldier valiant to proclaim advantages, free camps, and preferments, for such as shall have done any special piece of service. And principally, that the honour and profit pass to their posterity, for then, they will exploit with greater courage

---

[145] Chess-play] set of materials for the game, chess-board and chess-men.

and gallantness. For which cause Aristotle affirmeth that a man maketh more reck to be chief of his linage than of his own proper life. This Saul well perceived when he caused to be proclaimed in the army: 'Whosoever shall strike that man', meaning kill the giant Goliath, 'shall be made rich by the king and shall have his daughter to wife, and his house shall be enfranchised° in Israel from all manner tribute.' Conformable unto this proclamation, there was a court in Spain which ordained that whatsoever soldier by his good usage deserved to receive for his pay 500 soldi[146] (this was the greatest stipend allowed in the wars) should himself and his posterity be discharged forever from all taxes and services.

The Moors (as they are great players at chess) have in their plays set seven degrees in imitation of the seven draughts, which the pawn must make to be a queen, and so they enlarge the play from one to the second, and from the second to the third, until they arrive to seven, answerable to the proof that the soldier shall give of himself. And if she be so gallant as to enlarge his pay to the seventh, they yield him the same, and for this cause they are termed 'septerniers' or 'sevenstears'.[147] These have large liberties and exemptions, as in Spain those gentlemen who are called *hidalgos*.[148] The reason hereof in natural philosophy is very plain, for there is no faculty of all those that govern man which will willingly work unless there be some interest to move the same, which Aristotle proveth in the generative power and the self reason swayeth in the residue. The object of the wrathful faculty (as we have above specified) is honour and advantage, and if this cease, straightways courage and stomach decay. By all this may be conceived the great signification which it carrieth to make that pawn a queen who hath made seven draughts without taking, for

---

[146] soldi] Plural of 'soldo', Italian coin and money of account, formerly the twentieth part of a lira.

[147] 'septerniers' or 'sevenstears'] in Spanish, 'septenarios o matasiete' (1575, fol. 229ᵛ); 'septenarios' refers to something made up of seven parts or elements, 'matasiete' (literally, 'killer of seven') to a loudmouthed rogue who boasts of his courage. In Italian, the phrase becomes 'settennarii o ammazza sette' (1586, Q1ᵛ) — 'ammazza sette' also literally means 'killer of seven'. 'Sevenstears' might be a compound of 'seven' and 'stir', the latter with the meaning of 'to wield (a weapon); to brandish, flourish'.

[148] hidalgo] in Spain one of the lower nobility; a gentleman by birth; baser form of rural nobility exempt from paying taxes.

whatsoever the greatest nobility in the world that hath been or shall be hath sprung and shall spring from pawns and private men, who by the valour of their person have done such exploits as they deserved for themselves and their posterity the title of gentlemen, knights, noblemen, earls, marquises, dukes and kings. True it is that some are so ignorant and void of consideration as they will not grant that their nobility had a beginning, but that the same is everlasting and grown into their blood not by the grace of some particular king but by the supernatural and divine reason. To the bent of this purpose (though we shall thereby somewhat lengthen our matter) I cannot but recount a very witty discourse which passed between our Lord the Prince Don Carlos,[149] and the doctor Suárez of Toledo, who was judge of the Court in Alcalá of Henares.

Prince: Doctor, what think you of this people?

Doctor: Very well, my Lord, for here is the best air and the best soil of any place in Spain.

P: For such the physicians made choice of to recover my health. Have you seen the university?

D: No my L.

P: See it then, for it is very special, and where they tell me the sciences are very learnedly read.

D: Verily, for a college and particular study, it carrieth great fame, and should be such in effect as your highness speaketh of.

P: Where did you study?

D: In Salamanca, my lord.

P: And did you proceed doctor in Salamanca?

D: My lord, no.

P: That me seemeth was evil done to study in one university and take degree in another.

D: May it please your highness that the charges of taking degrees in Salamanca are excessive, and therefore we poor men fly the same and get us to some other university, knowing that we receive our sufficiency and learning not from the degree but from our study and pains, albeit my parents were not so poor but if them listed might have born the charge of my proceeding in

---

[149] Prince Don Carlos] Don Carlos (1545–1568) was the eldest son of King Philip II of Spain and Maria Manuela of Portugal. Don Carlos had multiple physical and psychological afflictions and was mentally unstable. King Philip II imprisoned him in 1568; he died months after that.

Salamanca. But your highness well knoweth that the doctors of this university have the like franchises as the gentlemen of Spain, and to us who are such by nature, this exemption doth harm, at least to our posterity.

P: Which of the kings mine ancestors gave this nobility to your lineage?

D. None. And to this end your highness must understand there are two sorts of gentlemen in Spain: some of blood, and some by privilege. Those in blood, as myself, have not received their nobility at the king's hand, but those by privilege have.

P: This matter is very hard for me to conceive, and I would gladly that you expressed it in plainer terms, for if my blood royal (reckoning from myself to my father, and from him to my grandfather, and so by order from each to other) cometh to finish in Pelagius,[150] to whom by the death of the king Don Roderick[151] the kingdom was given, before which time he was not king, if we reckon up after this sort your pedigree, shall we not come at last to end in one who was no gentleman?

D: This discourse cannot be denied, for all things have had a beginning.

P: I ask you then, from whence that first man had his nobility, who gave beginning to your nobility? He could not enfranchise himself, nor pluck out his own neck from the yoke of tributes and services which before time he paid to the kings my predecessors, for this were a kind of theft, and a prefering himself by force with the king's patrimony, and it foundeth not with reason that gentlemen of blood should have so bad an original as this. Therefore it falleth out plain that the king gave him freedom and yielded him the grace of that nobility. Now tell me from whom he had it.

D: Your highness concludeth very well, and it is true that there is no true nobility save of the king's grant. But we term those noble of blood of whose original there is no memory, neither is it specified by writing, when the same began, nor what king yielded them this favour. And this obscureness is received in the

---

[150] Pelagius] Pelayo (*c.* 685–737) was a Visigothic nobleman credited with beginning in Covadonga, Asturias, the Christian conquest of the Iberian Peninsula from Muslim rule (the *Reconquista*).

[151] Don Roderick] Don Rodrigo was the Visigothic King of Hispania between 710 and 711, when he lost the battle of Guadalete to the Moors.

commonwealth for more honourable than distinctly to know the contrary.

    The commonwealth also maketh gentlemen, for when a man groweth valorous, of great virtue, and rich, it dareth not to challenge such a one as seeming thereby to do him wrong, and that it is fit a man of that worth do live in all franchise. This reputation passing to the children and to the nephews groweth to nobility, and so they get a pretence against the king. These are not therefore gentlemen because they receive 500 soldi of pay, but when the contrary cannot be proved, they pass for such.

    That Spaniard who devised this name of a gentleman, *hijodalgos*, gave very well to understand this doctrine which we have set down, for by his opinion, men have two kinds of birth: the one natural, in which all are equal; the other spiritual. When a man performeth any heroical enterprise, or any virtue or extraordinary work, then is he now born, and procureth for himself other new parents, and leeseth° that being which he had tofore. Yesterday he was called the son of Peter, and nephew of Sanchius, and now he is named the son of his own actions. Hence had that Castilian proverb his original which saith 'every man is the son of his own works'. And because the good and virtuous works are in the Holy Scripture termed 'somewhat' and in the Spanish tongue it signifieth *algo*, and vices and sins 'nothing', which in the Spanish is termed *nada*. This Spaniard compounded this word *hijodalgo* thereof, which importeth nought else but that such a one is descended of him who performed some notorious and virtuous action for which he deserved to be rewarded by the king or commonwealth together with all his posterity forever. The law of the *Partida*[152] saith that *hijodalgo* signifieth the son of goods. But if we understand the same of temporal goods, the reason was not good, for there are infinite gentlemen poor, and infinite rich men who are no gentlemen. But if he mean the son of goods, that is to say, of good qualities, it carrieth the same sense which we before expressed.

---

[152] *Partida*] The *Partidas* (or *Siete Partidas*, Seven-Part Code) were a statutory code compiled under the reign of Alfonso X of Castile (1252–1284), el Sabio ('the Wise'), in order to provide a uniform body of normative rules for his kingdom. The name alludes to the seven sections into which the code is divided; the fragment to which Huarte particularly refers here is included in the Second Partida, Title XXI, Law II.

700　　Of the second birth which men ought to have besides their natural, there is afforded us a natural example in the Scripture, where Christ our redeemer reprehendeth Nicodemus because he (being a doctor of the law) wist° not yet, it was necessary that a man should be born of new, thereby to obtain a better being and more honourable parents than his natural. For which cause, all the time that a man performeth no heroical enterprise, in this sense he is called *hijo de nada*, to weet the son of nothing, although by his ancestors he bear the name of *hijodalgo*, that is the son of somewhat, or a gentleman. To the purpose of this doctrine, I will
710　　recite unto you a discourse which passed between a very honourable captain and a cavalier who stood much on the pantofles of his gentility. Whereby shall be discovered in what the honour of this second birth consisteth. This captain then falling in company with a knot of cavaliers, and discoursing of the largesse and liberty which soldiers enjoy in Italy, in a certain demand which one of them made him, he gave him the *you* because he was native of that place, and the son of mean parents, born in a village of some few houses. But the captain (aggrieved thereat) answered saying: 'Signore,° your signory shall understand
720　　that soldiers who have enjoyed the liberty of Italy cannot content themselves to make abode in Spain because of the many laws which are here enacted against such as set hand to their sword.'

The other cavaliers, hearing him use the term of *signoria*, could not forbear laughter. The cavalier blushing hereat used these words: 'Your *mercedi* may weet that in Italy, to say *signoria*, importeth so much as in Spain to say *mercede*, and this Signor Capitano, being accustomed to the use and manner of that country, giveth the term of *signoria* where he should do that of *mercede*.' Hereto the captain answered saying: 'Let not your
730　　signory hold me to be a man so simple but that I know when I am in Italy to apply myself to the language of Italy, and in Spain, to that of Spain. But he that in Spain talking with me may give me the *you*, it behoveth at least that he have a signory in Spain, and yet so I can scarce take it well.'[153] The cavalier, somewhat

---

[153] *Vuestra merced* was the most generalized courtesy formula in sixteenth-century Spain, while *señoría* was the treatment given to the nobility with titles. In contrast, in Italy *señoría* was the ordinary treatment, the equivalent to *vuestra merced* in Spanish. 'The *you*' in Carew's text translates the Spanish *vos*, used to address inferiors in rank, and therefore fairly disrespectful at the time.

affronted, made reply saying: 'Why Signor Capitano are you not native in such a place, and son to such a man? And know you not again who I am, and what mine ancestors have been?' 'Signore', answered the captain, 'I know right well that your signory is a good cavalier, and such have been your elders, but I and my right arm (which now I acknowledge for my father) are better than you and all your lineage.' This captain meant to allude to the second birth when he said 'I and my right arm, which now I acknowledge to be my father', and that not unduly, for with his right arm and with his sword he had performed such actions as the valour of his person was equal to the nobility of that cavalier.

For the most part, the laws and nature, saith Plato, are contrary, for a man sometimes issueth out of nature's hands with a mind very wise, excellent, noble, frank, and with a wit apt to command a whole world. Yet because his hap was to be born in the house of Amiclas, a base peasant, by the laws he remaineth deprived of that honour and liberty wherein nature placed him. And contrariwise, we see others whose wit and fashions were ordained to be slaves, and yet for that they were born in noble houses, they come by force of the laws to be great Lords. But one thing hath been noted many ages ago which is worthy of consideration, that those who are born in villages and thatched houses prove more sufficient men, and of greater towardness for the sciences and arms, than such as have great cities for their birthplace. Yet is the vulgar so subject to ignorance as they gather a consequence to the contrary, from birth in mean places. Hereof the Sacred Scripture affordeth as an example where it is read that, the people of Israel much wondering at the great works of our saviour Christ, said: 'Is it possible that out of Nazareth can come ought that is good?' But to return to the wit of this captain of whom we have discoursed, he ought to be endowed with much understanding and with the difference of imagination which is requisite for the art of war. Wherethrough, in this treatise we deliver much doctrine whence we may gather wherein the valour of men consisteth that they may reap estimation in the commonwealth.

Six things me seemeth a man ought to have to the end he may be termed honourable, and which of them soever want, his being is thereby impaired. But yet all of them are not placed in one self degree, nor partake a like value or the self qualities.

The first and principal is the valour of a man's own person as touching his wisdom, justice, mind, and courage. This maketh riches and birthright, from hence grow honourable titles; from this beginning all the nobility in the world fetcheth his original. And if any be settled in a contrary opinion, let him go to the great houses in Spain, and he shall find that they all derive their original from particular men who by the valour of their persons attained to that which now by their successions is possessed.

The second thing which honoureth a man, next to the valour of his person, is substance, without which we find not that any man carrieth estimation in the commonwealth.

The third is the nobility and antiquity of his ancestors to be well born, and of honourable blood, is a thing very precious. But yet retaineth in itself a great defect, for by itself alone it yieldeth a slender avail, as well in regard of the gentleman himself as of others who stand in need thereof. For a man can neither eat nor drink the same, nor apparel himself therewithal, nor give nor bestow the same, but it maketh a man to live as dying by depriving him of the remedies which he might otherwise procure to supply his necessities. But let him unite the same with riches, and by no degree of honour it can be countervailed. Some are wont to resemble nobility to a cipher in numbering, which of itself beareth no value, but united with another number, multiplieth the same.

The fourth point which maketh a man to be of account is to have some dignity or honourable office, and contrariwise nothing so much abaseth a man as to get his living by some handy-craft.

The fifth thing which honoureth a man is to be called by a good surname and a gracious Christian name, which may deliver a pleasing consonance° to the ear, and not to be termed *pasty* or *pestel*, as some that I know. We read in the general history of Spain that there came two ambassadors out of France unto King Alfonse the ninth to demand one of his daughters in marriage for their sovereign King Philip. One of which ladies was very fair and named *Urraca*,[154] the other nothing so gracious and called *Blanche*.[155] They both coming in presence of the ambassadors, all

---

[154] Urraca] Urraca of Castile (1187–1220) was a daughter of Alfonso VIII, king of Castile, and Eleanor of England (daughter of Henry II of England and Eleanor of Aquitaine). In 1206 Urraca married Afonso II of Portugal and together had five children.

[155] Blanche] Blanche of Castile (1188–1252). She was sister (of the same parents) of Urraca and became the wife of Louis VIII of France.

810 men held it as a matter resolved that the choice would light upon Urraca, as the elder, and fairer, and better adorned. But the ambassadors, inquiring each of their names, took offence at the name of Urraca,[156] and made choice of the lady Blanche, saying that her name would be better received in France than the other.

The sixth thing which honoureth a man is the seemly ornament of his person, and his going well apparelled and attended with many waiters. The good descent of the Spanish nobility is of such as through the valour of their person, and through their honourable enterprises achieved, grew in the wars to the pay of 820 500 soldi. The original whereof our late writers cannot verify, for if they find not their matter laid down in writing, and expressed to their hands by others, they are unable to supply the same with any invention of their own. The difference which Aristotle placeth betwixt memory and remembrance is that if the memory have lost any of those things which at first it knew, it cannot call the same to mind without new learning thereof, but remembrance enjoyeth this special grace that if it forget ought by stopping a while to discourse thereupon, it turneth to find out that which was before lost. Which may be the Court that speaketh in favour of good 830 soldiers, we find at this day recorded neither in books, nor in the memory of men, but there are left as relics these words, *hijodalgo*, in those that receive 500 soldi of pay after the Court of Spain, and their known wages. By making discourse and arguing whereon, it will fall out an easy matter to find out their associates. Antony of Lebrissa,[157] giving the signification of this verb, *vendico, cas*, saith the same signifieth 'to draw unto it that which is due for pay, or by reason', as we say nowadays by a new phrase of speech 'to take pay from the king'. And it is a thing so used in Castile the Old to say 'such a one hath well impaid° his travail' when he is 840 well paid, that amongst the civiler sort there is no manner of speech more ordinary.

From this signification, the word *vindicare* fetched his original, namely, when any one would stir at the wrong offered him by another, for injury metaphorically is termed debt. After this sort, when we now say 'such a one is *hijodalgo, de vengar quincentos sueldos*' (that is, a gentleman of the pay of 500 soldi), we mean

---

[156] Urraca means 'magpie' in Spanish. Blanche's name in Spanish was Blanca, literally 'white'.
[157] Antonio de Nebrija or de Lebrija. See Chapter X, footnote 94.

that he is descended from a soldier so valiant as for his prowess he deserved to receive so large a pay as is that of 500 soldi, who by the court of Spain was (with all his posterity) enfranchised from paying any tallages° or services to the king. This known pay is nought else save the entrance which such a soldier made into the number of those whose stipend was 500 soldi. For then were registered in the king's book the name of the soldier, the country where he was born, and who were his parents and progenitors, for the more certainty to him who received this benefit and stipend. Even as at this day we read in the book of *Bezerro*, which is kept at Salamanca, where are found written the beginning of well-near all the Spanish nobility. The semblable diligence used Saul when David slew Goliath, for forthwith he sent Abner his captain to take information of what stock the young man was descended. Anciently they termed *solaro* the house of the villain, as well as of the gentleman.

But sithence we have stepped aside into this digression, it behoveth to make return to our purpose from whence we parted, and to know whence it groweth that in play at chess, which we termed a counterfeit of war, a man shameth more to lose than at any other game, albeit the same turn him to no damage, neither is the play for money. And whence it may spring that the lookers on see more draughts than the players themselves, though they are less seen in the play? And that which most importeth is that some gamsters play best fasting, and some better after meat.

The first doubt holdeth like difficulty, for we have avouched that in war and in chess-play fortune hath nought to do, neither may we be allowed to say 'who would ever have thought this?'. But all is ignorance and carelessness in him that leeseth, and wisdom and cunning in him that getteth. And when a man is overcome in matters of wit and sufficiency, and is cut off from all allegations of excuse or pretence other than his own ignorance, it followeth a matter of necessity that he wax ashamed. For man is reasonable and a friend to his reputation, and cannot brook that in the works of this power any other should step a foot before him. For which cause Aristotle demandeth what the reason may be why the ancients consented not that special rewards should be assigned to those who surpassed the rest in the sciences, and yet ordained some for the best leaper, runner, thrower of the bar and wrestler. To which he frameth answer that in wrestling and bodily contentions, it is tolerated that there be judges assigned who shall censure how far one man exceedeth another, to the end they may justly yield prize to the vanquisher, it falling out a matter of no difficulty for the eye to discern who leapeth most ground, or runneth with greatest swiftness. But in matters of science it proveth very hard to try by the understanding, which exceedeth other, for that it is a thing appertaining to the spirit, and of much quaintness, and if the judge list to give the prize maliciously, all men cannot look thereunto for it is a judgement much estranged from the sense of the beholders. Besides this answer, Aristotle giveth another which is better, saying that men make no great reck to be overcome in throwing, wrestling, running, and leaping, for that they are graces wherein the very brute beasts outpass us. But that which we cannot endure with patience is to have another adjudged more wise and advised than ourselves, wherethrough they grow in hatred with the judges, and seek to be revenged of them, thinking that of malice they went about to shame them. Therefore, to shun these inconveniences, they would not yield consent that in works appertaining to the reasonable part, men should be allowed either judges or rewards. Whence is gathered that the universities do ill who assign judges and rewards of the first, second, and third degree in licensing those that prove best at the examinations.

For besides that the inconveniences alleged by Aristotle do betide, it is repugnant to the doctrine of the Gospel[158] that men grow into contention who should be chief. And that this is true, we see manifestly for that the disciples of our saviour Christ, coming one day from a certain voyage, treated amongst themselves who should be the greatest, and being now arrived at the lodging, their master asked them whereof they had reasoned upon the way. But they (though somewhat blunt) well understood how this question was not allowable, wherethrough the Text saith that they durst not tell him. But because from God nothing can be concealed, He spoke unto them in this manner. 'If any will be chief amongst you, he shall be the last of all, and servant to the rest.' The Pharisees were abhorred by Christ our redeemer because they loved the highest seats at feasts and the principal chairs in the synagogues. The chief reason thereon they rely who bestow degrees after this manner is that when scholars know each

---

[158] 40. *Index*: 'it is repugnant to the doctrine of the Gospel' [censored].

of them shall be rewarded according to the trial which they shall give of themselves, they will scantly afford themselves time from their study to sleep or eat. Which would cease were there not a reward for him that taketh pains, or chastisement for him that addicted himself to looseness and loitering. But this is a slender reason, and so only in appearance, and presupposeth a great falsehood, which is that knowledge may be gotten by continual plodding at the book, and by hearing of good masters, and never leesing a lesson. And they mark not that if a scholar want the wit and ability requisite for the learning which he applieth, it falleth out a lost labour to beat his head day and night at his books. And the error is such, that if differences of wits so far distant as these do enter into competency,° the one through his quick capacity without studying or poring in books getteth learning in a trice, and the other, for that he is blockheaded and dull, after he hath toiled all his life long can small skill in the matter.

Now the judges come (as men) to give the first price to him who was enabled by nature and took no travail, and the last to him who was born void of capacity, yet never gave over studying. As if the one had gotten learning by turning over his books, and the other lost the same through his own sluggishness. And it fareth as if they ordained prices for two horses of which the one had his legs sound and nimble, and the other halted downright. If the universities did admit to the study of the sciences none but such as had a wit capable thereof and were all equal, it should seem a thing well done to ordain reward and punishment: for whosoever knew most, it would thereby appear that he pained himself most, and who knew least, had given himself more to his ease.

To the second doubt we answer that, as the eyes stand in need of light and clearness to see figures and colours, so the imagination hath need of light in the brain to see the fantasies which are in the memory. This clearness the sun giveth not, nor any lamp or candle, but the vital spirits which are bred in the heart and dispersed throughout the body. Herewithal it is requisite to know that fear gathereth all the vital spirits to the heart, and leaveth the brain dark, and all the other parts of the body cold. Whereupon Aristotle maketh this demand: 'Whence commeth it that who so feareth, his voice, his hands, and his nether lip do tremble?' Whereto he answereth that through this fear, the natural heat hieth to the heart and leaveth all the residue of the body acold, and the cold (as is before touched) by Galen's mind, hindereth all the powers and faculties of the soul and suffereth not them to work.

Hence beginneth the answer of this second doubt, and it is that those who play at chess conceive fear to lose because the game standeth upon terms of reputation and disgrace, and for that fortune hath no stroke therein. So the vital spirits assembling to the heart, the imagination is foreslowed by the cold, and the fantasms in the dark, for which two reasons, he who playeth cannot bring his purpose to effect. But the lookers on, inasmuch as this no way importeth them, neither stand in fear of looking through want of skill, do behold more draughts for that their imagination retaineth his heat, and his figures are enlightened by the light of the vital spirits. True it is that much light reaveth also the light of the imagination, and it befalleth what time the player waxeth ashamed and out of countenance to see his adversary beat him. Then through this aggrievedness,[159] the natural heat increased and enlighteneth more than is requisite of all which he that standeth by is devoid. From hence issueth an effect very usual in the world, that what time a man endeavoureth to make the best muster of himself and his learning, and sufficiency most known, it proveth worst with him. With others again the contrary betideth, who being brought to their trial make a great show, and passed out of the lists appear of little worth. And of all this, the reason is very manifest, for he whose head is filled with much natural heat, if you appoint him to do an exercise of learning or disputation, within four and twenty hours after, a part of that excessive heat which he hath flieth to the heart and so the brain remaineth temperate. And in this disposition (as we will prove in the chapter ensuing) many points worth the utterance present themselves to a man's remembrance. But he who is very wise and endowed with a great understanding, being brought to trial, by means of fear cannot retain the natural heat in his head, whereon through default of light, he findeth not in his memory what to deliver.

If this fell into their consideration, who take upon them to control the generals of armies, blaming their actions, and the order which they set down in the field, they should discern how great a difference resteth between the giving a looking on the sight

---

[159] aggrievedness] feeling of injury causing grief.

out at a window, or the breaking of a lance therein, and the fear to leese an army whose charge their sovereign hath committed to their hands.

No less damage doth fear procure the physician in curing, for his practice (as we have proved heretofore) appertaineth to the imagination, which resteth more annoyed by cold than any other power, for that his operation consisteth in heat. Whence we see by experience that physicians can sooner cure the vulgar sort than princes and great personages. A counsellor at law one day asked me (knowing that I handled this matter) what the cause might be that in the affairs where he was well paid many cases and points of learning came to his memory, but with such as yielded not to his travail what was due, it seemed that all his knowledge was shrunk out of his brain. Whom I answered that matters of interest appertained to the wrathful faculty which maketh his residence in the heart, and if the same receive not contentment, it doth not willingly send forth the vital spirits by whose light the figures which rest in the memory may be discerned. But when that findeth satisfaction, it cheerfully affordeth natural heat. Wherethrough the reasonable soul obtaineth sufficient clearness to see whatsoever is written in the head. This defect do men of great understanding partake who are pinching, and rely much on their interest, and in such is the property of that counsellor best discerned. But who so falleth into due consideration hereof shall observe it to be an action of justice that he who laboureth in another man's vineyard be well paid his wages.

The like reason is current for the physicians, to whom (when they are well hired) many remedies present themselves. Otherwise, the art (as well in them as the lawyer) slippeth out of their fingers. But here a matter very important is to be noted, namely, that the good imagination of the physician discovereth on a sudden what is necessary to be done. And if he take leisure and farther consideration, a thousand inconveniences come into his fancy, which hold him in suspense, and this while the occasion of the remedy passeth away. Therefore it is never good to advise the physician to consider well what he hath in hand, but that he forthwith execute what first he purposed. For we have proved heretofore that much speculation maketh the natural heat to avoid out of the head, and again the same may increase so farforth as to turmoil the imagination. But the physician in whom it is slack shall not do amiss to use long contemplation, for the heat advancing itself up to the brain shall come to attain that point which to this power is behooful.°

The third doubt in the matters already rehearsed hath his answer very manifest, for the difference of the imagination with which we play at chess requireth a certain point of heat to see the draughts, and he that playeth well fasting hath then the degree of heat requisite thereunto. But through the heat of the meat, the same exceedeth that point which was necessary, and so he playeth worse. The contrary befalleth to such as play well after meals, for the heat rising up together with the meat and the wine, arriveth to the point which wanted whiles he was fasting. It is therefore needful to amend a place in Plato, who saith that nature hath with great wisdom disjoined the liver from the brain to the end the meat with his vapours should not trouble the contemplation of the reasonable soul. But here if he mean those operations which appertain to the understanding, he speaketh very well, but it can take no place in any of the differences of the imagination. Which is seen by experience in feasts and banquets, for when the guests are come to mid meal, they begin to tell pleasant tales, merriments, and similitudes, where at the beginning none had a word to say, but at the end of the feast, their tongue faileth them, for the heat is passed beyond the bound requisite for the imagination. Such as need to eat and drink a little to the end the imagination may lift up itself are melancholic by adustion, for such have their brain like hot lime, which taken up into your hand is cold and dry in feeling, but if you bath the same in any liquor, you cannot endure the heat which groweth thereof.

We must also correct that law of the Carthaginians which Plato allegeth, whereby they forbade their captains to drink wine when they went to their wars, and likewise their governors during the year of their office. And albeit Plato held the same for a very just law, and never maketh an end of commending the same, yet it behoveth to make a distinction. We have alleged heretofore that the work of judging appertaineth to discourse, and that this power abhorreth heat, and therefore receiveth much damage by wine. But to govern a commonwealth (which is a distinct matter from taking into your hand a process and giving sentence thereupon) belongeth to the imagination, and that requireth heat. And the governor not arriving to the point which is requisite may well drink a little wine so to attain the same. The like may be said touching the general of an army whose counsel partaketh also with the imagination. And if the natural heat be by any hot thing to be advanced, none performeth it so well as wine, but it is requisite that the same be temperately taken, for there is no nourishment which so giveth and reaveth a man's wit as this liquor. Wherefore it behoveth the general to know the manner of his imagination, whether the same be of those which need meat and drink to supply the heat that wanteth, or to abide fasting. For in this only consisteth how to manage his affairs well or evil.

CHAP. XIV.

How we may know to what difference of ability the office of a king appertaineth, and what signs he ought to have who enjoyeth this manner of wit.

When Solomon was chosen king and head of so great and numberful° a people as that of Israel, the Text saith that for governing and ruling them he craved wisdom from Heaven and nothing besides. Which demand so much pleased God as in reward of having asked so well, He made him the wisest king of the world, and not so contented, He gave him great riches and glory, evermore holding his request in better price. Whence is manifestly gathered that the greatest wisdom and knowledge which may possibly be in the world is that foundation upon which the office of a king relieth. Which conclusion is so certain and true as it were but lost labour to spend time in the proof thereof. Only it behoveth to show to what difference of wit the art of being a king, and such a one as is requisite for the commonwealth, appertaineth, and to unfold the tokens whereby the man may be known who is endowed with this wit and ability. Wherethrough it is certain that as the office of a king exceedeth all the arts in the world, so the same requireth a perfection of wit in the largest measure that nature can devise. What the same is we have not as yet defined, for we have been occupied in distributing to the other arts their differences and manners. But since we now have the same in handling, it must be understood that of nine temperatures which are in mankind, one only, saith Galen, maketh a man so surpassing wise as by nature he can be. Wherein the first qualities are in such weight and measure that the heat exceedeth not the cold, nor the moist the dry, but are found in such equality and conformity as if really they were not contraries, nor had any natural opposition. Whence resulteth an instrument so appliable to the operations of the reasonable soul, that man commeth to possess a perfect memory of things passed, and a great imagination to see what is to come, and a great understanding to distinguish, infer, argue, judge and make choice. The other differences of wit by us recounted have not any one amongst them of sound perfection, for if a man possess great understanding, he cannot (by means of much dryness) comprise the sciences which appertain to the imagination and the memory; and if he be of great imagination, by reason of much heat, he remaineth insufficient for the sciences of the understanding and the memory; and if he enjoy a great memory, we have tofore expressed how unable those of much memory (through their excessive moisture) do prove for all the other sciences. Only this difference of wit which we now are a searching is that which answereth all the arts in proportion. How much damage the unableness of adjoining the rest breedeth to any one knowledge, Plato noteth, saying that the perfection of each in particular dependeth on the notice and knowledge of them all in general.

No sort of knowledge is found so distinctly and severed from another but that the skill in the one much aideth to the others' perfection. But how shall we do if having sought for this difference of wit with great diligence in all Spain, I can find but one such?[160] Whereby I conceive that Galen said very well that out of Greece nature not so much as in a dream maketh any man temperate or with a wit requisite for the sciences. And the same Galen allegeth the reason hereof, saying that Greece is the most temperate region of the world, where the heat of the air exceedeth not the cold, nor the moist the dry. Which temperature maketh men very wise and able for all the sciences as appeareth, considering the great number of famous men who thence have issued, as Socrates, Plato, Aristotle, Hippocrates, Galen, Theophrastus,[161] Demosthenes,

---

[160] 2. *Carew's note in the margin:* 'No doubt your own king'.

[161] Theophrastus] Theophrastus (*c.* 371–*c.* 287 BC) was a Greek philosopher who studied in Plato's school and later joined Aristotle's. He eventually became Aristotle's successor at the Peripatetic school, which he presided over for more than thirty years.

Homer, Thales Milesius,[162] Diogenes Cynicus,[163] Solon,[164] and infinite other wise men mentioned in histories whose works we find replenished with all sciences. Not as the writers of other provinces, who if they treat of physic, or any other science, it proves a miracle for them to allege any other sort of science in their aid or favour. All of them are beggerly and without furniture, as wanting a wit capable of all the arts. But which we may most marvel at in Greece is that, whereas the wit of women is found so repugnant unto learning (as hereafter we will prove), yet there have been so many the Greeks so specially seen in the sciences, as they have grown into competence[165] with the sufficientest men, as namely Leontia,[166] a most wise woman, who wrote against Theophrastus, the greatest philosopher of his time, reproving him for many errors in philosophy. But if we look into other provinces of the world, hardly shall we find sprung up any one wit that was notable, which groweth for that they inhabit places distempered where men become brutish, slow of capacity, and ill conditioned.

For this cause Aristotle moveth a doubt saying: 'What meaneth it that those who inhabit a country, either over cold or over hot, are fierce and fell in countenance and conditions?' To which problem he answereth very well saying that a good temperature not only maketh a good grace in the body, but also aideth the wit and ability. And as the excesses of heat and cold do hinder nature that she cannot shape a man in good figure, so (also for the like reason) the harmony of the soul is turned topsy-turvy, and the wit proveth slow and dull.

This the Greeks well wist° inasmuch as they termed all the nations of the world barbarians, considering their slender sufficiency and little knowledge. Whence we see that of so many

---

[162] Thales Milesius] Thales of Miletus (c. 624–c. 546 BC) was a pre-Socratic Greek philosopher and mathematician from Asia Minor.

[163] Diogenes Cynicus] Diogenes of Sinope or the Cynic (late fifth and early fourth century BC) was a one of the founders of Cynic philosophy. None of his many writings have survived.

[164] Solon] Solon (c. 638 BC–558 BC) was an Athenian statesman and lawmaker. He was involved in building the political foundations for democracy in Athens.

[165] competence] rivalry in dignity or relative position, vying.

[166] Leontia] Leontium or Leontia was an Athenian woman, originally a courtesan, although afterwards the wife of either Metrodorus or the philosopher Epicurus (341–270 BC). She was in any case disciple of Epicurus, and the author of an essay against Theophrastus and in defence of Epicurean philosophy.

that are born and study out of Greece, if they be philosophers, none of them arriveth to the perfection of Plato and Aristotle; if physicians, to Hippocrates and Galen; if orators, to Demosthenes; if poets, to Homer. And so in the residue of the sciences and arts the Greeks have ever held the foremost rank beyond all contradiction. At least the problem of Aristotle is very well verified in the Greeks, for verily they are the men of most sufficiency and loftiest capacity in the world. Were it not that they live in disgrace, oppressed by force of arms in bondage, and all hardly entreated by the coming of the Turks, who banished all learning and caused the University of Athens to pass unto Paris in France, where at this day the same continueth. And (thus through want of manurance[167]) so many gallant wits (as we have before reported) are utterly perished. In the other regions out of Greece, though schools and exercise of learning are planted, yet no man hath proved in them of any rare excellency.

    The physician holdeth he hath waded very far if with his wit he can attain to that which Hippocrates and Galen delivered, and the natural philosopher reckoneth himself so full of knowledge as he can be capable of no more, if he once grow to the understanding of Aristotle. But this notwithstanding, it goeth not for a universal rule that all such as have Greece for their birth-place must of force be temperate and wise, and all the residue distemperate and ignorant. For the same Galen recounteth of Anacharsis,[168] who was born in Scythia, that he carried the reputation of a rare wit amongst the Grecians though himself a Barbarian. A philosopher, born in Athens, falling in contention with him, said unto him: 'Get thee hence, thou Barbarian.' Then Anacharsis answered: 'My country is to me a shame, and so art thou to thine: for Scythia, being a region so distemperate, and where so many ignorant persons live, myself am grown to knowledge, and thou being born in Athens, a place of wit and wisdom, wert never other than an ass.' In sort that we need not utterly despair in regard of the temperature, neither think it a case of impossibility to meet herewithal out of Greece, and especially in Spain, a region not very distemperate. For as I have found one

---

[167] manurance] cultivation or training of the character or faculties.
[168] Anacharsis] Scythian philosopher from the early sixth century BC, traditionally considered a forerunner of the Cynics. None of his works have survived.

of these differences in Spain, so it may well be that there are many others not yet come to knowledge and which I have not been able to find out. It shall do well therefore to entreat the tokens by which a temperate man may be discerned, to the end where such a one is, he may not be hidden.

Many signs have the physicians laid down to discover this difference of wit, but the most principal, and which afford best notice, are these following.

The first, saith Galen, is to have his hair auburn,° a colour between white and red, and that passing from age to age, they ever become more golden. And the reason is very clear, for the material cause whereof they [*sic*] hair consisteth, the physicians say, is a gross vapour which ariseth from the digestion that the brain maketh at the time of his nourishment, and look what colour is of the member such also is that of his excrements. If the brain in his composition partake much of phlegm, the hair in growth is white; if much choler, saffron coloured. But if these two humours rest equally mingled, the brain becometh temperate, hot, cold, moist, and dry, and the hair auburn, partaking both the extremes. True it is, Hippocrates saith, that this colour in men who live under the north (as are the English, Flemish, and Almains) springeth, for that their whiteness is parched up with much cold, and not for the reason by us alleged. Wherefore in this token it behoveth to be well advised, otherwise we may soon slip into error.

The second token which a man who shall be endowed with this difference of wit must have is, saith Galen, to be well shaped, of good countenance, of seemly grace, and cheerful. In sort, that the sight may take delight to behold him as a figure of rare perfection. And the reason is very plain, for if nature have much force and a seed well seasoned, she always formeth of things possible the best and most perfect in his kind, but being purveyed of forces, mostly she placeth her study in fashioning the brain, for that amongst all other parts of the body the same is the principal seat of the reasonable soul, whence we see many men to be great and foul and yet of an excellent wit.

The quantity of body which a temperate man ought to have, saith Galen, is not resolutely determined by nature, for he may be long, short, and of mean stature, conformable to the quantity of the temperate seed which it had when it was shaped. But as touching that which appertaineth to the wit in temperate persons, a mean stature is better than either a great or little. And if we must lean to either of the extremes, it is better to incline to the little than to the great, for the bones and superfluous flesh (as we have proved heretofore by the opinion of Plato and Aristotle) bring great damage to the wit. Agreeable hereunto, the natural philosophers are wont to demand whence it proceedeth that men of small stature are ordinarily more wise than those of long stature. And for proof hereof, they cite Homer, who saith that Ulysses was very wise and little of body, and contrariwise Ajax very foolish and in stature tall.[169] To this question they make very simple answer saying that the reasonable soul gathered into a narrow room hath thereby more force to work, conformably to that old saw: 'virtue is of more force united than dispersed.' And contrariwise, making abode in a body long and large, it wanteth sufficient virtue to move and animate the same. But this is not the reason thereof, for we should rather say that long men have much moisture in their composition, which extendeth out their flesh and ableth the same to that increase which the natural heat doth ever procure. The contrary betideth in little bodies, for through their much dryness, the flesh cannot take his course, nor the natural heat enlarge or stretch it out, and therefore they remain of short stature. And we have erst proved that amongst the first qualities, none bringeth so great damage to the operations of the reasonable soul as much moisture, and that none so far quickeneth the understanding as dryness.

 The third sign, saith Galen, by which a temperate man may be known is that he be virtuous and of good conditions. For if he be lewd and vicious, Plato affirmeth it groweth for that in man there is some distemperate quality which urgeth him to offend, and if such a one will practise that which is agreeable to virtue, it behoveth that first he renounce his own natural inclination. But whosoever is absolutely temperate standeth not in need of any such diligence, for the inferior powers require nothing at his hands that is contrary to reason. Therefore Galen saith that to a man who is possessed of this temperature, we need prescribe no diet what he shall eat and drink, for he never exceedeth the quantity and measure which physic would assign him. And Galen

---

[169] Ajax] Mythological Greek hero, son of Telamon and Periboea, and king of Salamis. Homer's *Iliad* describes him as of great stature and strength, and a prominent warrior in Agamemnon's army.

contenteth not himself to term them most temperate, but moreover avoucheth that it is not necessary to moderate their other passions of the soul, for his anger, his sadness, his pleasure, and his mirth, are always measured by reason. Whence it followeth that they are evermore healthful and never diseased, and this is the fourth figure.

But herein Galen swerveth from reason, for it is impossible to frame a man that shall be perfect in all his powers, as the body is temperate, and that his wrathful and concupiscential power get not the sovereignty over reason and incite him to sin. For it is not fitting to suffer any man (how temperate soever) to follow always his own natural inclination without gainsetting and correcting him by reason. This is easily understood considering the temperature which the brain ought to have to the end the same may be made a convenient instrument for the reasonable faculty. And that which the heart should hold, to the end the wrathful power may covet glory, empire, victory, and sovereignty over all, and that which the liver ought to have for digesting the meats, and that which ought to rest in the cods° to be able to preserve mankind and to increase the same. Of the brain, we have said sundry times tofore that it should retain moisture for memory, dryness for discourse, and heat for the imagination. But for all this, his natural temperature is cold and moist, and by reason of the more or less of these two qualities, sometimes we term it hot, and sometimes cold; now moist, then dry; but the cold and moist grow to predominate. The liver, wherein the faculty of concupiscence resideth, hath for his natural temperature heat and moisture to predominate; and from this, it never altereth so long as a man liveth. And if sometimes we say it is cold, it groweth, for that the same hath not all the degrees of heat requisite to his own operations. As touching the heart, which is the instrument of the wrathful faculty, Galen affirmeth it of his own nature to be so hot as if (while a creature liveth) we put our finger into his hollowness, it will grow impossible to hold the same there one moment without burning. And albeit sometime we term it cold, yet we may not conceive that the same doth predominate, for this is a case impossible, but that the same consisteth not in such degree of heat as to his operations is behooful.°

In the cods, where the other part of the concupiscible maketh abode, the like reason taketh place, for the predomination of his natural temperature is hot and dry. And if sometimes we say that a man's cods are cold, we must not absolutely so understand the same, neither to predomination, but that the degree of heat requisite for the generative virtue is wanting. Hereon we plainly infer that if a man be well compounded and instrumentalized, it behoveth of force that he have excessive heat in his heart, for otherwise the wrathful faculty would grow very remiss, and if the liver be not exceeding hot, it cannot digest the meat, nor make blood for nourishment, and if the cods have not more heat than cold, a man will prove impotent and without power of begetting. Wherefore these two members (being of such force as we have said) it followeth of necessity that the brain take alteration through much heat, which is one of the qualities that most paineth reason, and which is worst, the will being free, inciteth and inclineth itself to condescend to the appetites of the lower portion.

By this reckoning it appeareth that nature cannot fashion such a man as may be perfect in all his powers, nor produce him inclined to virtue. How repugnant it is unto the nature of man that he become inclined to virtue is easily proved considering the composition of the first man, which though the most perfect that ever mankind enjoyed saving that of Christ our redeemer, and shaped by the hands of so great an artificer, yet if God had not infused into him a supernatural quality, which might keep down his inferior part, it was impossible (abiding in the principles of his own nature) that he should not be inclined to evil.[170] And that God made Adam of a perfect power to wrath and concupiscence is well to be understood in that He said and commanded him: 'Increase and multiply and to replenish the earth.' It is certain that He gave them an able power for procreation and made them not of a cold complexion, inasmuch as He commanded him that he should people the earth with men, which work cannot be accomplished without abundance of heat. And no less heat did He bestow upon the faculty nutritive with which He was to restore his consumed substance and renew another in lieu thereof. Seeing that He said to the man and the woman: 'Behold, I have given you every herb that bringeth forth seed upon the earth, and whatsoever trees have seed of their kind to the end they may serve you for food.' For if God had given them a stomach and liver, cold and of little heat, for certain they could not have digested

---

[170] 41. *Index*: 'How repugnant it is unto the nature of man [263] [...] that he should not be inclined to evil' [censored].

their meat, nor preserve themselves 900 years alive in the world. He fortified also the heart, and gave the same a wrathful faculty which might yield him apt to be a king and lord, and to command the whole world, and said unto them: 'Do you subdue the earth and command over the fishes of the sea, and the fowls of the air, and all the beasts that move on the face of the earth.' But if He had not given them much heat, they had not partaken so much vivacity nor authority of sovereignty of commandment, of glory, of majesty, and of honour.

How much it endamageth a prince to have his wrathful power remiss cannot sufficiently be expressed, for through this only cause it befalleth that he is not feared nor obeyed, nor reverenced by his subjects. After having fortified the wrathful and concupiscible powers, giving unto the forementioned members so much heat, He passed to the faculty reasonable, and shaped for the same a brain cold and moist in such degree, and of a substance so delicate, that the soul might with the same discourse and philosophize, and use his infused knowledge. For we have already avouched and heretofore proved that God to bestow a supernatural knowledge upon men, first ordereth their wit and maketh them capable by way of the natural dispositions delivered by His hand that they may receive the same. For which cause, the text of the Holy Scripture affirmeth that He gave them a heart to conceive and replenished them with the discipline of understanding. The wrathful and concupiscential powers being then so mighty through great heat, and the reasonable so weak and remiss to resist, God made provision of a supernatural quality, and this is termed by the divines 'original justice', by which they come to repress the brunts of the inferior portion, and the part reasonable remaineth superior and inclined to virtue. But when our first parents offended, they lost this quality, and the irascible° and concupiscible remained in their nature, and superior to reason, in respect of the strength of the three members that we spoke of, and man rested ready, even from his youth, unto evil.

Adam was created in the age of youth, which (after the physicians) is the most temperate of all the residue, and from that age forth, he was inclined to evilness, saving that little time whilst he preserved himself in grace by original justice. From this doctrine we gather in good natural philosophy that if a man be to perform any action of virtue to the gainsaying of the flesh, it is impossible that he can put the same in execution without outward aid of grace, for the qualities with which the inferior power worketh are of greater efficacy. I said 'with gainsaying of the flesh' because there are many virtues in man which grow for that he hath his powers of wrath and concupiscence feeble (as chastity in a cold person), but this is rather an impotency of operation than a virtue, for which cause, had not the Catholic church taught us that without the special aid of God we could not have overcome our own nature, philosophy natural would so have learned us,[171] namely, that grace comforteth our will. That then which Galen would have said was that a temperate man exceedeth in virtue all others who want this good temperature, for the same is less provoked by the inferior part.

The fifth property which those of this temperature possess is to be very long lived, for they are strong to resist the causes and occasions which engender diseases, and this was that which the royal prophet David meant: 'The days of our age in themselves are seventy years, but if in the potentates there be eighty or more, it is their pain and sorrow.' As if he should say: 'The number of years which men ordinarily do live arrive unto seventy, and if potentates reach unto eighty, those once passed, they are dead on their feet.' He termeth those men 'potentates' who are of this temperature, for more than any other they resist the causes which abridge the life. Galen layeth down the last token saying that they are very wise, of great memory for things passed, of great imagination to foresee those to come, and of great understanding to find out the truth of all matters. They are not malicious, not wily, not cavillers, for these spring from a temperature that is vicious. Such a wit as this assuredly was not framed by nature to addict itself unto the study of the Latin tongue, logic, philosophy, physic, divinity, or the laws, for put case he might easily attain these sciences, yet none of them can fully replenish his capacity, only the office of a king is in proportion answerable thereunto, and in ruling and governing ought the same solely to be employed. This shall easily be seen if you run over the tokens and properties of a temperate man, which we have laid down by taking into consideration how fitly each of them squareth with the royal sceptre, and how impertinent° they show for the other arts and sciences.

---

[171] 42. *Index*: 'For we have already avouched [302] [...] philosophy natural would so have learned us' [censored].

That a king be fair and gracious is one of the things which most inviteth his subjects to love him and wish him well. For the object of love, saith Plato, is beauty and a seemly proportion, and if a king be hardly favoured and badly shaped, it is impossible that his subjects can bear him affection. Rather, they reck it a shame that a man unperfect and void of the gifts of nature should have sway and commandment over them. To be virtuous and of good conditions, easily may we gather how greatly it importeth, for he who ought to order the lives of his subjects and deliver unto them rules and laws to live conformably to reason, it is requisite that he perform the same also in his own person; for as the king is, such are the great, the mean, and the inferior persons.

Moreover, by this means he shall make his commandments the more authentical and with the better title may chastise such as do not observe them. To enjoy a perfection in all the powers which govern man (namely, the generative, nutritive, wrathful, and reasonable) is more necessary in a king than any artist whatsoever. For, as Plato delivereth, in a well ordered commonwealth, there should be appointed certain surveyors who might with skill look into the qualities of such persons as are to be married, and give to him a wife answerable unto him in proportion, and to every wife a convenient husband. Through this diligence, the principal end of matrimony should not become vain, for we see by experience that a woman who could not conceive of her first husband, marrying another, straightways beareth children, and many men have no children by their first wife, taking another, speedily come to be fathers.

Now this skill, saith Plato, is principally behooful in the marriage of kings, for it being a matter of such importance for the peace and quiet of the kingdom that the prince have lawful children to succeed in the state, it may so fall that the king, marrying at all adventures, shall take a barren woman to wife, with whom he shall be cumbered all days of his life without hope of issue. And if he decease without heirs of his body, straightways it must be decided by civil wars who shall command next after him. But Hippocrates saith this art is necessary for men that are distemperate and not for those who partake this perfect temperature by us described. These need no special choice in their wife, nor to search out which may answer them in proportion, for whomsoever they marry withal, saith Galen, forthwith they beget issue. But this is understood when the wife is found and of the age wherein women by order of nature may conceive and bring forth. In sort, that fruitfulness is more requisite in a king than in any artist whatsoever for the reasons tofore alleged.

The nutritive power, saith Galen, if the same be gluttonous, greedy, and bibbing,° it springeth for that the liver and stomach want the temperature which is requisite for their operations, and for this cause, men become riotous and short lived. But if these members possess their due temperature and composition, the self Galen affirmeth that they covet no greater quantity of meat and drink than is convenient for preservation of life. Which property is of so great importance for a king that God holdeth that land for blessed to whose lot such a prince befalleth. 'Blessed is the land', saith He in *Ecclesiasticus*, 'whose king is noble and whose princes feed in due times for their refreshment and not for riotousness.' Of the wrathful faculty, if the same be extended or remiss, it is a token, saith Galen, that the heart is ill composed and partaketh not that temperature which is requisite for his operations. From which two extremes a king ought to be farther distant than any other artist. For to join wrathfulness with much power maketh smally for the subjects' avail. And as illy fitteth it for a king to have his wrathful power remiss, for if he slightly slip over bad parts and attempts in his kingdom, he groweth out of awe and reverence amongst his subjects, whence great damages, and very difficult to be remedied, do accustomably arise in the commonwealth. But the man who is temperate groweth displeased upon good ground and can pacify himself as is requisite, which property is as necessary to be settled in a king as any of those which we have before remembered. How much it importeth that the faculty reasonable (the imagination, the memory, and the understanding) be of greater perfection in a king than in any other is easily to be proved, for the other arts and sciences (as it seemeth) may be obtained and put in practice by the force of man's wit. But to govern a kingdom, and to preserve the same in peace and concord, not only requireth that the king be endowed with a natural wisdom to execute the same, but it is also necessary that God particularly assist him with his understanding and aid him in governing, whence it was well noted in the Scripture 'the heart of the king is in the hand of God'. To live also many years and to enjoy continual health is a property more convenient for a good king than for any other artisan. For his industry and travail breedeth a universal good to all, and if he fail to hold out in healthfulness, the commonwealth falleth to ruin.

All this doctrine here laid down by us will be evidently confirmed if we can find in any history that at any time there was any king chosen in whom any of those tokens and conditions by us recited were not wanting. And truth hath this as peculiar to her nature that she never lacketh arguments whereby to be confirmed.

The divine Scripture recounteth that God, falling in dislike with Saul for that he had spared Amaleck's life,[172] commanded Samuel[173] that he should go to Bethlehem and anoint for king of Israel one of the eight sons of Jesse.[174] Now the holy man, presuming that God had a liking to Eliah[175] for that he was tall of stature, demanded of him: 'Is this man here in the presence of my Lord, his Christ?' To which question he was answered in this manner: 'Take not regard to his countenance, nor to the tallness of his stature, for I have refused him. I judge not man by his look, for man seeth the things outwardly apparent but the Lord discerneth the heart.' As if God should say: 'Mark not, oh Samuel, the high stature of Eliah, nor that manly countenance which thou beholdest, for I have tried that in Saul. You men judge by the outward signs, but I cast mine eye upon the judgement and wisdom wherewith a people is to be governed.'

Samuel, mistrusting his own skill in choosing, passed on farther in the charge which was commanded him, asking still of God upon every one which of them he should anoint for king. And because God held himself contented with none of them, He said unto Jesse: 'Hast thou yet no more sons but those who stand before us?' Who answered saying that he had yet one more who kept his beasts, but he was of little growth, him seeming that therefore he was not sufficient to wield the royal sceptre. But Samuel, now wisted that a great stature was no sure token, caused

---

[172] Amaleck] Amalek. Agag was the king of the Amalekites (a people considered Amalek's descendants) during the reign of Saul. Saul spared Agag's life contrary to Jehovah's manifest will. Saul's disobedience infuriated God, who then commissioned Samuel to cut Agag in pieces.

[173] Samuel] Prophet who anointed the first two kings of the Kingdom of Israel: Saul and David.

[174] Jesse] Jesse was a man of wealth and position at Bethlehem, a son of Obed and grandson of Boaz and Ruth. Jesse was the father of eight sons, the youngest of whom was David, who became the king of the Israelites.

[175] Eliah] Elijah. See note 118.

him to be sent for. And it is a point worth the noting that the Holy Scripture, before it expressed how he was anointed king, saith in this manner: 'But he was auburn haired, and of a faire countenance, and a visage well shaped; arise and anoint him, for this is he.' In sort, that David had the two first tokens of those which we recounted: auburn haired, handsome shaped, and of a mean stature. To be virtuous and well conditioned, which is the third sign easily we may conceive, that he was therewithal endowed, seeing that God said 'I have found a man after my heart', for albeit he sinned sundry times, yet for all that, he lost not the name and habit of virtue.[176] Even as one by habit vicious, though he perform some good moral works, doth not therefore leese° the name of lewd and vicious. That he led all the course of his life in health, it should seem may be proved, because in his whole history mention is made of his sickness but once (and this is a natural disposition, of all such as are long lived). Now because his natural heat was resolved, and that he could not take heat in his bed, to remedy this, they couched a very fair lady by his side who might foster him with heat. And herethrough he lived so many years that the Text saith he deceased in a good age, full of days, of riches, and of glory, as if it should say: 'David died in a good old age, full of days, of riches, and of glory, having endured so many travails in the wars, and undergone great penance for his transgressions.' And this grew for that he was temperate and of a good complexion, for he refused the occasions which accustomably breed infirmity and shortening of man's life. His great wisdom and knowledge was noted by that servant of Saul when he said: 'My lord, I know a cunning musician, the son of Jesse, born in Bethlehem, courageous in fight, wise in discourse, and of seemly countenance.' By which tokens (above specified) it is manifest that David was a temperate man, and to such is the royal sceptre belonging, for his wit is of the best mould that nature could fashion.

But there presenteth itself a very great difficulty against this doctrine, namely, seeing God knew all the wits and abilities of Israel, and likewise wist that temperate men are seized of the wisdom and knowledge requisite to the calling of a king, for what cause in the first election that He made, He sought not out a man

---

[176] 43. *Index*: 'for albeit he sinned sundry times, yet for all that, he lost not the name and habit of virtue' [censored].

of this sort? Nay the Text avoucheth that Saul was so tall of stature as he passed all the residue of Israel by the head and shoulders, and this sign is not only an evil token of wit in natural philosophy, but even God himself (as we have proved) reproved Samuel because (moved by the high growth of Eliah) he thereupon would have made him king. But this doubt declareth that to be true which Galen said that out of Greece we shall not (so much as in a dream) find out a temperate man, seeing in a people so large (as that of Israel) God could not find one to choose for a king, but it behooved him to tarry till David was grown up and the whiles made choice of Saul. For the Text saith that he was the best of Israel, but verily it seemed he had more good nature than wisdom, and that was not sufficient to rule and govern.[177] 'Teach me', saith the Psalm, 'goodness, discipline, and knowledge.' And this the royal prophet David spoke seeing that it availeth not for a king to be good and virtuous unless he join wisdom and knowledge therewithal. By this example of King David it seemeth we have sufficiently approved our opinion.

But there was also another king born in Israel of whom it was said: 'Where is he that is born king of the Jews?' And if we can prove that he was auburn haired, towardly, of mean bigness, virtuous, healthful, and of great wisdom and knowledge, it will be no way damageable to this our doctrine. The Evangelists busied not themselves to report the disposition of Christ our redeemer, for it served not to the purpose of that which they handled, but is a matter which may easily be understood, supposing that for a man to be temperate, as is requisite, compriseth all the perfection wherewith naturally he can be endowed. And seeing that the Holy Spirit compounded and instrumentalized him, it is certain that as touching the material cause of which He formed him, the distemperature of Nazareth could not resist him, nor make him err in his work (as do the other natural agents). But he performed what him best pleased, for he wanted neither force, knowledge, nor will, to frame a man most perfect and without any defect. And that so much the rather, for that his coming (as himself affirmed) was to endure travails for man's sake and to teach him the truth. And this temperature (as

---

[177] 3. *Carew's note in the margin*: 'A weak reason, rather God chose Saul as a carnal man fit for the Jews obstinate asking, and David as a spiritual man, the instrument of his mercy'.

we have before proved) is the best natural instrument that can be found for these two things. Wherethrough I hold that relation for true, which Publius Lentulus,[178] vice-consul, wrote from Jerusalem unto the Roman Senate after this manner:[179]

'There hath been seen in our time, a man who yet liveth, of great virtue, called Jesus Christ, who by the gentiles is termed the prophet of truth, and his disciples say that he is the son of God. He raiseth the deceased and healeth the diseased, is a man of mean and proportionable stature, and of very fair countenance. His look carrieth such a majesty as those who behold him are enforced both to love and fear him. He hath his hair coloured like a nut full ripe, reaching down to his ears, and from his ears to his shoulders; they are of wax colour, but more bright. He hath in the middle of his forehead a lock after the manner of Nazareth. His forehead is plain, but very pleasing; his face void of spot or wrinkle, accompanied with a moderate colour. His nostrils and mouth cannot by any with reason be reproved; his beard thick and resembling his hair, not long, but forked. His countenance very gracious and grave, his eyes graceful and clear; and when he rebuketh, he daunteth; and when he admonisheth, he pleaseth. He maketh himself to be beloved, and is cheerful with gravity; he hath never been seen to laugh, but to weep diverse times. His hands and arms are very fair, in his conversation he contenteth very greatly, but is seldom in company; but being in company is very modest. In his countenance and port he is the seemliest man that may be imagined.' In this relation are contained three or four tokens of a temperate person.

The first that he had, his hair and beard of the colour of a nut fully ripe, which to him that considereth it well appeareth to be a brown auburn, which colour God commanded they [sic] heifer should have which was to be sacrificed as a figure of Christ.[180] And when he entered into heaven with that triumph and majesty

---

[178] Publius Lentulus] Publius Lentulus supposedly was a Roman Consul during the reign of Augustus (63 BC–14 AD), and Governor of Judea before Pontius Pilate. The 'Letter of Lentulus' is an epistle allegedly written by Publius Lentulus offering a physical description of Christ. It is regarded as apocryphal.

[179] 4. *Carew's note in the margin*: 'And I hold it untrue, because the phrase utterly differeth from the Latin tongue as *speciosus valde interfilios bomimum* [very fair amongst the son of men]'.

[180] 5. *Carew's note in the margin*: 'Unwritten varities'.

which was requisite for such a prince, some angels who had not been informed of his incarnation said: 'Who is this that commeth from Edon, with his garments dyed in Bozra?' As if they had said: 'Who is he that commeth from the red land with his garment stained in the same dye, in respect of his hair and his red beard, and of the blood with which he was tainted?' The same letter also reporteth him to be the fairest man that ever was seen, and this is the second token of a temperate person, and so was it prophesied by the Holy Scripture as a sign whereby to know him: 'Of fair shape above all the children of men.' And in another place he saith: 'His eyes are fairer than the wine, and his teeth whiter than milk' Which beauty and good disposition of body imported much to effect that all men should bear him affection and that there might be nothing in him worthy to be abhorred. For which cause, the letter delivereth that all men were enforced to love him. It reciteth also that he was mean of personage, and that not because the Holy Ghost wanted matter to make him greater, if so it had seemed good, but (as we tofore have proved by the opinion of Plato and Aristotle) because when the reasonable soul is burdened with much bones and flesh, the same incurreth great damage in his wit.

The third sign, namely, to be virtuous and well conditioned, is likewise expressed in this letter, and the Jews themselves with all their false witnesses could not prove the contrary, nor reply when he demanded of them: 'Which of you can reprove me of sin?' And Josephus (through the faithfulness which he owed to his history) affirmed of him that he partaked of another nature above man in respect of his goodness and wisdom. Only long life could not be verified of Christ our redeemer because they put him to death being young, whereas if they had permitted him to finish his natural course, the same would have reached to eighty years and upwards. For he who could abide in a wilderness forty days and forty nights without meat or drink, and not be sick nor dead therewithal, could better have defended himself from other lighter things which had power to breed alteration or offence. Howbeit this action was reputed miraculous and a matter which could not light within the compass of nature.

These two examples of kings which we have alleged, sufficeth to make understood that the sceptre royal is due to men that are temperate, and that such are endowed with the wit and wisdom requisite for that office. But there was also another man made by the proper hands of God to the end he should be king and lord of all things created, and He made him fair, virtuous, sound, of long life, and very wise. And to prove this shall not be amiss for our purpose. Plato holdeth it for a matter impossible that God or nature can make a man temperate in a country distemperate, wherethrough he affirmeth that God, to create a man of great wisdom and temperature, sought out a place where the heat of the air should not exceed the cold, nor the moist the dry. And the divine Scripture, whence he borrowed this sentence, saith not that God created Adam in the earthly paradise, which was that most temperate place whereof He speaketh, but that after He had shaped him there He placed him. 'Then our Lord God', saith he, 'took man and set him in the Paradise of pleasure to the end he might there work and take it in charge.' For the power of God being infinite, and his knowledge beyond measure, when He had a will to give him all the natural perfection that might be in mankind, we must think that neither the piece of earth of which he was framed, nor the distemperature of the soil of Damascus where he was created, could so gainsay him but that He made him temperate. The opinion of Plato, of Aristotle, and of Galen take place in the works of nature, and even she also can sometimes (even in distemperate regions) engender a person that shall be temperate. But that Adam had his hair and his beard auburn, which is the first token of a temperate man, manifestly appeareth. For in respect of this so notorious sign, he had that name, Adam, which is to say (as St Jerome interpreteth it) a red man.

That he was fair and well fashioned, which is the second token, cannot in him be denied, for when God created him, the Text saith: 'God saw all things which He had made and they were very good.' Then it falleth out certain that he issued not from the hands of God foul and ill shaped, for the works of God are perfect. And so much the more for that the trees (as the Text saith) were fair to behold. Then, what may we think of Adam, whom God created to this principal end that he might be lord and president of the world? That he was virtuous, wise, and well conditioned (which are the third and sixth signs) is gathered out of these words: 'Let us make a man after our own image and likeness.' For by the ancient philosophers, the foundation on which the resemblance that man hath with God is grounded are virtue and wisdom. Therefore, Plato avoucheth that one of the greatest contentments which God receiveth in heaven is to see a virtuous and wise man praised and magnified upon earth, for such a one is his lively portraiture. And contrariwise, He groweth displeased when ignorant and vicious persons are held in estimation and honour, which springeth from the unlikeness between God and them.

That he lived healthful and a long space (which are the fourth and fifth tokens) is nothing difficult to prove, inasmuch as his days were 930 years. Wherethrough I may now conclude that the man who is auburn haired, fair, of mean stature, virtuous, healthful, and long lived, must necessarily be very wise and endowed with a wit requisite for the sceptre royal.[181]

We have also (as by the way) disclosed in what sort great understanding may be united with much imagination and much memory, albeit this may also come to pass and yet the man not be temperate. But nature shapeth so few after this model that I could never find but two amongst all the wits that I have tried.[182] But how it can come to pass that great understanding may unite with much imagination and much memory in a man not temperate is a thing which easily may be conceived, if you presuppose the opinion of some physicians, who affirm that the imagination resideth in the forepart of the brain, the memory in the hinder part, and the understanding in that of the middle. And the like may be said in our imagination, but it is a work of great labour that the brain, being (when nature createth the same) of the bigness of a grain of pepper, it should make one ventricle of seed very hot, another very moist, and the middlemost of very dry, but in fine this is no impossible case.

CHAP. XV.

In what manner parents may beget wise children and of a wit fit for learning.

It falleth out a matter worthy of marvel that nature, being such as we all know her (wise, witty, and of great art, judgement and force), and mankind, a work of so special regard, yet for one whom she maketh skilful and wise, she produceth infinite deprived of wit. Of which effect myself searching the reason and

---

[181] 6. *Carew's note in the margin:* 'And such a one if you mistake not is your king Philip'.

[182] 7. *Carew's note in the margin:* 'Your king and yourself'.

natural causes have found (in my judgement) that parents apply not themselves to the act of generation with that order and concert which is by nature established, neither know the conditions which ought to be observed to the end their children may prove of wisdom and judgement. For by the same reason for which in any temperate or distemperate region a man should be born very witty (having always regard to the self order of causes) there will 100,000 prove of slender capacity. Now, if by art we may procure a remedy for this, we shall have brought to the commonwealth the greatest benefit that she can receive. But the knot of this matter consisteth in that we cannot entreat hereof with terms so seemly and modest as to the natural shamefastness of man is requisite. And if for this reason I should forbear to note any part or contemplation that is necessary, for certain the whole matter would be marred, in sort that diverse grave philosophers hold opinion how wise men ordinarily beget foolish children because in the act of copulation, for honesty's sake, they abstain from certain diligences which are of importance that the son may partake of his father's wisdom. Some ancient philosophers have laboured to search out the natural reason of this natural shame which the eyes conceive when the instruments of generation are set before them, and why the ears take offence to hear them named. And they marvel to see that nature hath framed those parts with such diligence and carefulness, and for an end of such importance as the immortalizing of mankind, and yet the wiser a man is, the more he groweth in dislike to behold or hear them spoken of. Shame and honesty, saith Aristotle, is the proper passion of the understanding, and who so resteth not offended at those terms and actions of generation giveth a sure token of his wanting that power, as if we should say that he is blockish who putting his hand into the fire doth not feel the same to burn. By this token, Cato the elder discovered that Manilius (a noble man) was deprived of understanding because it was told him that the other kissed his wife in presence of his daughter, for which cause he displaced him out of the Senate, and Manilius could never obtain at his hands to be restored.[183]

Out of this contemplation, Aristotle frameth a problem

---

[183] Manilius] The conservative Roman statesman Cato the Elder, or Cato the Censor (234–149 BC), during his censorship from 184 BC to 182 BC, degraded Manius Manilius for embracing his wife in broad daylight.

demanding whence it grew that men who desire to satisfy their venerous lusts, do yet greatly shame to confess it, and yet coveting to live, to eat, or to perform any other such action, they stagger not to acknowledge it. To which problem he shapeth a very untoward answer saying: 'Perhaps it commeth because the covetings of diverse things are necessary, and some of them kill if they be not accomplished, but the lust of venerous acts floweth from excess and is token of abundance.' But in effect this problem is false and the answer none other, for a man not only shameth to manifest the desire he carrieth to company with a woman, but also to eat, to drink, and to sleep, and if a will take him to send forth any excrement, he dares not say it or do it but with cumber and shamefastness, and so gets him to some secret place out of sight. Yea, we find men so shamefast, as though they have a great will to make water yet cannot do it if any look upon them, whereas if we leave them alone, straightways the urine taketh his issue. And these are the appetites to send forth the superfluous things of the body, which if they were not effected, men should die, and that much sooner than with forbearing meat or drink. And if there be any, saith Hippocrates, who speaketh or actuateth this in the presence of another, he is not master of his sound judgement. Galen affirmeth that the seed holdeth the semblable proportion with the seed-vessels° as the urine doth with the bladder, for as much urine annoyeth the bladder so much seed endamageth the seed-vessels. And the opinion which Aristotle held in denying that man and woman incur no infirmity or death by retaining of seed is contrary to the judgement of all physicians, and especially of Galen, who saith and avoucheth that many women remaining widows in their youth have therethrough lost their sense, motion, breathing, and finally their life. And the self Aristotle reckoneth up many diseases whereunto continent persons are subject in that behalf.

The true answer of this problem cannot be yielded in natural philosophy because it is not marshalled under her jurisdiction, for it behoveth to pass to a higher, namely, metaphysic, wherein Aristotle saith that the reasonable soul is the lowest of all the intelligences, and for that it partaketh of the same general nature with the angels, it shameth to behold itself placed in a body which hath fellowship with brute beasts. Wherethrough the divine Scripture noteth it as a mystery that the first man being naked was not ashamed, but so soon as he saw himself to be so, forthwith he got a covering. At which time he knew that through his own fault he had lost immortality, and that his body was become subject to alteration and corruption, and those instruments and parts given him for that of necessity he must die and leave another in his room, and that to preserve himself in life that small space which rested, it behoved him to eat and drink and to expel those noisome and corrupt excrements. And principally he shamed seeing that the angels, with whom he had competence, were immortal and stood not in need of eating, drinking, or sleeping for preservation of their life; neither had the instruments of generation but were created all at once without matter and without fear of corrupting.[184] Of all these points were the eyes and the ears naturally done to ware. Wherethrough, the reasonable soul groweth, displeased and ashamed, that these things given man to make him mortal and corruptible are thus brought to his memory. And that this is a well fitting answer we evidently perceive, for God to content the soul after the universal judgement, and to bestow upon him entire glory, will cause that his body shall partake the properties of an angel, bestowing thereupon subtleness, lightness, immortality, and brightness, for which reason he shall not stand in need to eat or drink as the brute beasts.[185] And when men shall thuswise dwell in heaven, they will not shame to behold themselves clothed with flesh,[186] even as Christ our redeemer and his mother nothing shamed thereat. But it will breed an accidental glory to see that the use of those parts which were wont to offend the hearing and the eyes is now surceased. I therefore making due reckoning of this natural modesty of the ear have endeavoured to salve the hard and rough terms of this matter and to fetch certain not ill-pleasing biases of speech. And where I cannot throughly perform it, the honest reader shall afford me pardon. For to reduce to a perfect manner the art which must be observed to the end men may prove of rare capacities is one of the things most requisite for the commonwealth. Besides that, by the same reason they shall prove virtuous, prompt, sound, and long-lived.

I have thought good to sever the matter of this chapter into

---

[184] 8. *Carew's note in the margin*: 'A high speculation'.
[185] 9. *Carew's note in the margin*: 'Note here a sign which showeth the immortality of the soul'.
[186] clothed with flesh] naked.

four principal parts, that thereby I may make plain what shall be delivered, and that the reader may not rest in confusion. The first is to show the natural qualities and temperature which man and woman ought to possess to the end they may use generation. The second, what diligence the parents ought to employ that their children may be male and not female. The third, how they may become wise and not fools. The fourth, how they are to be dealt withal after their birth for preservation of their wit.

To come then to the first point we have already alleged that Plato layeth down how in a well-ordered commonwealth there ought to be assigned certain surveyors of marriages who by art might skill to look into the qualities of the persons that are to be married, and to give each one the wife which answereth him in proportion, and to every wife her convenient husband. In which matter, Hippocrates and Galen began to take some pains and prescribed certain precepts and rules to know what woman is fruitful and who can bear no children; and what man is unable for generation, and who able and likely to beget issue. But touching all this, they uttered very little and that not with such distinction as was behooful,° at least for the purpose which I have in hand. Therefore it falleth out necessary to begin the art even from his principles, and briefly to give the same his due order and concert, that we so may make plain and apparent from what union of parents wise children issue, and from what, fools and do-noughts.° To which end it behoveth first to know a particular point of philosophy, which although in regard of the practices of the art it be very manifest and true, yet the vulgar make little reck thereof. And from the notice of this dependeth all that which as touching this first point is to be delivered. And that is that man (though it seem otherwise in the composition which we see) is different from a woman in nought else, saith Galen, than only in having his genital members without his body.[187] For if we make anatomy of a woman, we shall find that she hath within her two stones,° two vessels for seed, and her belly of the same frame as a man's member, without that any one part is therein wanting.

And this is so very true that if when nature hath finished to form a man in all perfection, she would convert him into a woman, there needeth nought else to be done save only to turn

---

[187] 10. *Carew's note in the margin*: 'This is no chapter for maids to read in sight of others'.

his instruments of generation inwards. And if she have shaped a woman, and would make a man of her, by taking forth her belly and her cods,° it would quickly be performed. This hath chanced many times in nature, as well whiles the creature hath been in the mother's womb, as after the same was born, whereof the histories are full. But some have held them only for fables because this is mentioned in the poets, yet the thing carrieth mere truth, for diverse times nature hath made a female child, and she hath so remained in her mother's belly for the space of one or two months, and afterwards, plenty of heat growing in the genital members, upon some occasion they have issued forth and she become a male. To whom this transformation hath befallen in the mother's womb is afterwards plainly discovered by certain motions which they retain unfitting for the masculine sex, being altogether womanish, and their voice shrill and sweet. And such persons are inclined to perform women's actions, and fall ordinarily into uncouth offences.[188] Contrariwise, nature hath sundry times made a male with his genitors outward, and cold growing on, they have turned inward and it became female. This is known after she is born, for she retaineth a mannish fashion as well in her words as in all her motions and workings.

This may seem difficult to be proved, but considering that which many authentical historians affirm, it is a matter not hard to be credited. And that women have been turned into men after they were born, the very vulgar do not much marvel to hear spoke of, for besides that which sundry our elders have laid down for truth, it befell in Spain but few years since, and that whereof we find experience is not to be called in question or argument. What then the cause may be that the genital members are engendered within or without, and the creature becometh male or female, will fall out a plain case if we once know that heat extendeth and enlargeth all things, and cold retaineth and closeth them up. Wherethrough, it is a conclusion of all philosophers and physicians that if the seed be cold and moist, a woman is begotten and not a man; and if the same be hot and dry, a man is begotten and not a woman. Whence we apparently gather that there is no man who in respect of a woman may be termed cold; nor woman hot, in respect of a man.

Aristotle saith it is necessary for a woman to be cold and moist

---

[188] uncouth offences] sodomy.

that she may be likewise fruitful, for if she were not so, it would fall out impossible that her monthly course should flow, or she have milk to preserve the child nine months in her belly, and two years after it is born, but that the same would soon waste and consume.

All philosophers and physicians avouch that the belly holdeth the same proportion with man's seed that the earth doth with corn and with any other grain. And we see that if the earth want coldness and moisture, the husbandman dareth not sow therein, neither will the seed prosper. But of soils, those are most fruitful and fertile in rendering fruit which partake most of cold and moist, as we see by experience in the regions towards the north (as England, Flanders, and Almaine) whose abundance of all fruits worketh astonishment in such as know not the reason thereof. And in such countries as these, no married woman was ever childless, neither can they there tell what barrenness meaneth, but are all fruitful and breed children through their abundance of coldness and moisture.[189] But though it is true that the woman should be cold and moist for conception, yet she may abound so much therein that it may choke the seed, even as we see excess of rain spoileth the corn, which cannot ripen in overmuch coldness. Whereon we must conceive that these two qualities ought to keep a certain measurableness, which when they exceed or reach not unto, the fruitfulness is spoiled. Hippocrates holdeth that woman for fruitful whose womb is tempered in such sort as the heat exceedeth not the cold, nor the moist the dry. Wherethrough he saith that those women who have their belly cold cannot conceive, no more than such as are very moist or very cold and dry. But so, for the same reason that a woman and her genital parts should be temperate, it were impossible that she could conceive or be a woman. For if the seed of which she was first formed had been temperate, the genital members would have issued forth, and she have been a man. So should a beard grow on her chin and her flowers[190] surcease, and she become as perfect a man as nature could produce. Likewise, the womb in a woman cannot be predominately hot, for if the seed whereof she was engendered had been of that temperature, she should have been born a man, and not a woman. This is past all exception that the qualities

---

[189] 11. *Carew's note in the margin*: 'You are much mistaken'.
[190] flower] menstrual discharge.

which yield a woman fruitful are cold and moisture, for the nature of man standeth in need of much nourishment that he may be able to use procreation and continue his kind. Wherethrough we see that amongst all the females of brute beasts, none have their monthly courses as a woman. Therefore it was requisite to make her altogether cold and moist, and that in such a degree as that she might breed much phlegmatic blood, and not be able to waste or consume the same. I said phlegmatic blood because this is serviceable to the breeding of milk, by which, Hippocrates and Galen avouch, the creature is relieved all the time it remaineth in the mother's belly. Now if the same should be temperate, it would produce much blood unfit for the engendering of milk, and would wholly resolve, as it doth in a temperate man, and so nothing be left for nourishing the babe.

Therefore I hold it for certain, and verily it is impossible that a woman can be temperate or hot, but they are all cold and moist. And if this be not so, let the philosopher or physician tell me for what cause all women are beardless and have their sickness whiles they are healthful, and for what cause the seed of which she was formed, being temperate or hot, she was born a woman and not a man? Howbeit, though it be true that they are all cold and moist, yet it followeth not that they are all in one degree of coldness and moisture. For some are in the first, some in the second, and some in the third, and in each of these they may conceive if a man answer them in proportion of heat, as shall hereafter be expressed. By what tokens we may know these three degrees of coldness and moisture in a woman, and likewise weet° who is in the first, who is in the second, and who in the third, there is no philosopher or physician that as yet hath unfolded. But considering the effects which these qualities do work in women, we may part them, by reason of their being extended, and so we shall easily get notice hereof. The first, by the wit and hability of the woman; the second, by her manners and conditions; the third, by her voice big or small; the fourth, by her flesh, much, or little; the fifth, by her colour; the sixth by her hair; the seventh, by her fairness or foulness.

As touching the first, we may know that though it be true (as tofore we have proved) that the wit and ability of a woman followeth the temperature of the brain and of none other member, yet her womb and cods are of so great force and vigour to alter the whole body that if these be hot and dry, or cold and moist, or of whatsoever other temperature, the other parts, saith Galen, will be of the same tenour. But the member which most partaketh the alterations of the belly, all physicians say, is the brain, though they have not set down the reason whereon they ground this correspondency. True it is Galen proveth by experience that by spaying a sow she becometh fair and fat, and her flesh very savoury, and if she have her cods, she tasteth little better than dog's flesh. Whereby we conceive that the belly and the cods carry great efficacy to communicate their temperature to all the other parts of the body, especially to the brain, for that the same is cold and moist like themselves, between which (through the resemblance) the passage is easy.

Now if we conclude that cold and moist are the qualities which work an impairment in the reasonable part, and that his contraries, namely hot and dry, give the same perfection and increasement, we shall find that the woman who showeth much wit and sufficiency partaketh of cold and moist in the first degree, and if she be very simple, it yieldeth a sign that she is in the third, the partaking between which two extremes argueth the second degree. For to think that a woman can be hot and dry, or endowed with a wit and ability conformable to these two qualities, is a very great error, because if the seed of which she was formed had been hot and dry in their domination, she should have been born a man and not a woman, but in that it was cold and moist, she was born a woman and not a man. The truth of this doctrine may clearly be discerned if you consider the wit of the first woman who lived in the world: for God having fashioned her with his own hands, and that very accomplished and perfect in her sex, it is a conclusion infallibly true that she was possessed of much less knowledge than Adam, which the devil well weeting, got him to tempt her and durst not fall in disputation with the man, fearing his great wit and wisdom. Now to say that Eve for her offence was rest that knowledge which she wanted cannot be avouched, for as yet she had not offended.

So then this defect of wit in the first woman grew for that she was by God created cold and moist, which temperature is necessary to make a woman fruitful and apt for childbirth, but enemy to knowledge. And if he had made her temperate like Adam, she should have been very wise, but nothing fruitful, nor subject to her monthly courses save by some supernatural means. On this nature, St Paul grounded himself when he said: 'Let a woman learn in silence with all subjection.' Neither would he allow the woman to teach or govern the man, but to keep silence. But this is true when a woman hath not a spirit or greater grace than her own natural disposition, but if she obtain any gift from above, she may well teach and speak. For we know that the people of Israel, being oppressed and besieged by the Assyrians, Judith (a very wise woman) sent for the priests of the Cabeits and Carmits, and reproved them saying: 'How can it be endured that Osias[191] should say if within five days there come no succour he will yield the people of Israel to the Assyrians? See you not that these words rather provoke God to wrath than to mercy? How may it be that men should point out a limited time for the mercy of God and in their mind assign a day at which He must succour and deliver them?' And in the conclusion of this reproof she told them in what sort they might please God and obtain their demand. And no less, Elbora[192] (a woman of no less wisdom) taught the people of Israel how they should render thanks unto God for the great victories which she had attained against their enemies. But whilst a woman abideth in her natural disposition, all sorts of learning and wisdom carrieth a kind of repugnancy to her wit. And for this cause, the Catholic Church, upon great reason hath forbidden that no woman do preach, confess, or instruct, for their sex admitteth neither wisdom nor discipline.

It is discovered also by the manners of a woman and by her condition, in what degree of cold and moist her temperature consisteth. For if with a sharp wit she be froward, cursed, and wayward, she is in the first degree of cold and moist; it being true (as we have proved tofore) that an ill condition evermore accompanieth a good imagination, she who partaketh this degree of cold and moist suffereth nothing to escape her hands, noteth all things, findeth fault with all things, and so is insupportable. Such are accustomably of amiable conversation and fear not to look men in the face, nor hold him ill mannered who maketh love

---

[191] Osias] Hosea (eight century BC) was the son of Beeri, and author of the Book of Hosea. He was the only prophet of Israel who left any written prophecy. He is often regarded as a prophet of doom.

[192] Elbora] Deborah. A prophetess, wife of Lapidoth. Deborah roused the people before the degrading subjection in which Jabin, king of Hazor, had held Israel for twenty years. She was a key agent in the victorious attack of the Hebrew host against the army of Jabin.

unto them. But on the other side, to be a woman of good conditions, and to be aggrieved at nothing, to laugh upon every small occasion, to let things pass as they come, and to sleep soundly, descrieth the third degree of cold and moist, for much pleasantness of conceit is ordinarily accompanied with little wit. She who partaketh of these two extremes standeth in the second degree. A voice hoarse, big, and sharp, saith Galen, is a token of much heat and drought, and we have also proved it heretofore by the opinion of Aristotle, wherethrough we may gain this notice that if a woman have a voice like a man, she is cold and moist in the first degree, and if very delicate, in the third. And partaking betwixt both the extremes, she shall have the natural voice of a woman and be in the second degree.

How much the voice dependeth on the temperature of the cods shall shortly hereafter be proved, where we entreat of the tokens appertaining to a man. Much flesh also in women is a sign of much cold and moist, for to be fat and big, say the physicians, groweth in living creatures from this occasion. And contrariwise, to be lean and dry is a token of little coldness and moisture. To be meanly fleshed, that is, neither overmuch nor very little, giveth evidence that a woman holdeth herself in the second degree of cold and moist. Their pleasantness and courtesies showeth the degrees of these two qualities: much moisture maketh their flesh supple and little, rough and hard. The mean is the commendable part. The colour also of the face and of the other parts of the body discovereth the extended or remiss degrees of these two qualities. When the woman is very white, it bodeth, saith Galen, much cold and moist, and contrariwise, she that is swart and brown is in the first degree thereof, of which two extremes is framed the second degree of white and well coloured.

To have much hair and a little show of a beard is an evident sign to know the first degree of cold and moist: for all physicians affirm that the hair and beard are engendered of heat and dryness, and if they be black it greatly purporteth the same. A contrary temperature is betokened when a woman is without hair. Now she whose complexion consisteth in the second degree of cold and moist hath some hair, but the same reddish and golden. Foulness moreover and fairness help us to judge the degrees of cold and moist in women. It is a miracle to see a woman of the first degree very fair, for the seed whereof she was formed, being dry, hindereth that she cannot be fairly countenanced.° It behoveth that clay be seasoned with convenient moisture to the end vessels may be well framed and serve to use. But when that same is hard and dry, the vessel is foul and unhandsome.

Aristotle farther avoucheth that overmuch cold and moist maketh women by nature foul, for if the seed be cold and very moist, it can take no good figure because the same standeth not together, as we see that of over soft clay ill shaped vessels are fashioned. In the second degree of cold and moist, women prove very fair, for they were formed of a substance well seasoned and pleasant to nature, which token, of itself alone affordeth an evident argument that the woman is fruitful, for it is certain that nature could do it, and we may judge that she gave her a temperature and composition fit for bearing of children. Wherethrough she answers in proportion well-near to all men, and all men do desire to have her.

In man there is no power which hath tokens or signs to descry the goodness or malice of his object. The stomach knoweth the meat by way of taste, of smelling, and of sight, wherethrough the divine Scripture saith that Eve fixed her eyes on the tree forbidden, and her seemed that it was sweet in taste. The faculty of generation holdeth for a token of fruitfulness a woman's beauty, and if she be foul, it abhorreth her, conceiving by this sign that nature erred and gave her not a fit temperature for bearing of children.

By what signs we may know in what degree of hot and dry every man resteth.
1.

Man hath not his temperature so limited as a woman, for he may be hot and dry (which temperature, Aristotle and Galen held, was that which best agreed with his sex) as also hot and moist and temperate. But cold and moist, and cold and dry, they would not admit whilst a man was found and without impairment, for as you shall find no woman hot and dry, nor hot and moist, or temperate, so shall you find no man cold and moist, nor cold and dry, in comparison of women, unless in case I shall now express. A man hot and dry, and hot and moist, and temperate, holdeth the same degrees in his temperature as doth a woman in cold and moist, and so it behoveth to have certain tokens whereby to

THE EXAMINATION OF MEN'S WITS

discern what man is in what degree that we may assign him a wife answerable unto him in proportion. We must therefore weet that from the same principles of which we gathered understanding what woman is hot and dry, and in what degree, from the self we must also make use to understand what man is hot and dry, and in what degree. And because we said that from the wit and manners of a man we conjecture the temperature of his cods, it is requisite that we take notice of a notable point mentioned by Galen, namely, that to make us understand the great virtue which a man's cods possess, to give firmness and temperature to all the parts of the body, he affirmeth that they are of more importance than the heart. And he rendereth a reason saying that this member is the beginning of life, and nought else, but the cods are the beginning of living soundly and without infirmities.

How much it endamageth a man to be deprived of those parts (though so small) there need not many reasons to prove, seeing we see by experience that forthwith the hair and the beard pill away, and the big and shrill voice becometh small, and herewithal a man leeseth° his forces and natural heat, and resteth in far worse and more miserable condition than if he had been a woman. But the matter most worth the noting is that if a man before his gelding had much wit and hability, so soon as his stones be cut away, he groweth to leese the same, so far-forth as if he had received some notable damage in his very brain. And this is a manifest token that the cods give and reave the temperature from all the other parts of the body, and he that will not yield credit hereunto, let him consider (as myself have done oftentimes) that of 1000 such capons[193] who addict themselves to their book, none attaineth to any perfection, and even in music (which is their ordinary profession) we manifestly see how blockish they are, which springeth because music is a work of the imagination, and this power requireth much heat, whereas they are cold and moist. So it falleth out a matter certain that from the wit and hability we may gather the temperature of the cods, for which cause, the man who showeth himself prompt in the works of the imagination should be hot and dry in the third degree. And if a man be of no great reach, it tokeneth that with his heat much moisture is united, which always endamageth the reasonable part, and this is the

---

[193] capon] eunuch.

470 more confirmed if he be good of memory. The ordinary conditions of men hot and dry in the third degree are courage, pride, liberality, audacity, and cheerfulness, with a good grace and pleasantness; and in matter of women such a one hath no bridle nor ho.[194] The hot and moist are merry, given to laughter, lovers of pastime, fair conditioned, very courteous, shamefast, and not much addicted to women.

The voice and speech much discovereth the temperature of the cods. That which is big and somewhat sharp giveth token that a man is hot and dry in the third degree, and if the same be pleasant, 480 amiable, and very delicate, it purporteth little heat and much moisture, as appeareth in the gelded. A man who hath moist united with heat will have the same high, but pleasant and shrill. Who so is hot and dry in the third degree is slender, hard and rough fleshed, the same composed of sinews and arteries, and his veins big. Contrariwise, to have much flesh smooth and tender is show of much moisture, by means whereof it extendeth and enlargeth out the natural heat. The colour of the skin, if the same be brown, burned, blackish green, and like ashes, yieldeth sign that a man is in the third degree of hot and dry. But if the flesh 490 appeareth white and well coloured, it argueth little heat and much moisture. The hair and beard are a mark also not to be overslipped, for these two approach very near to the temperature of the cods. And if the hair be very black and big, and specially from the ribs down to the navel, it delivereth an infallible token that the cods partake much of hot and dry. And if there grow some hair also upon the shoulders, the same is so much the more confirmed. But when the hair and beard are of chestnut colour, soft, delicate, and thin, it inferreth not so great plenty of heat and dryness in the cods.

500 Men very hot and dry are never fair, save by miracle, but rather hard-favoured and ill shaped, for the heat and dryness (as Aristotle affirmeth of the Ethiopians) wrieth the proportion of the face, and so they become disfigured. Contrariwise, to be seemly and gracious proveth a measurable hot and moist, for which cause, the matter yielded itself obedient whereto nature would employ it. Whence it is manifest that much beauty in a man is no token of much heat. Touching the signs of a temperate man,

---

[194] ho] cessation, moderation.

we have sufficiently discoursed in the chapter foregoing, and therefore it shall not be needful to reply the same again. It sufficeth only to note that as the physicians place in every degree of heat three degrees of extension, so also in a temperate man we are to set down the largeness and ampleness of three other. And he who standeth in the third, next to cold and moist, shall be reputed cold and moist, for when a degree passeth the mean, it resembleth the other. And that this is true we manifestly find, for the signs which Galen delivereth us to know a man cold and moist are the self same of the temperate man but somewhat more remiss. So is he wise, of good conditions, and virtuous, he hath his voice clear and sweet, is white skinned, of flesh good and supple, and without hair, and if it have any, the same is little and yellow; such are very well favoured and fair of countenance, but Galen affirmeth that their seed is moist and unfit for generation. These are no great friends to women, nor women unto them.

*What women ought to marry with what
man that they may have children.
2.*

To a woman who beareth not children when she is married, Hippocrates commandeth that two points of diligence be used: to know whether it be her defect, or that it grow because the seed of her husband is unable for generation. The first is to make her suffumigations with incense or storax, with a garment close wrapped about her which may hang down on the ground, in sort that no vapour or fume may issue out, and if within a while after she feel the savour of the incense in her mouth, it yieldeth a certain token that the barrenness cometh not through her defect, inasmuch as the same found the passages of the belly open, wherethrough it pierced up to the nostrils and the mouth.

The second is to take a garlic-head clean pilled and put the same into the belly what time the woman goeth to sleep, and if the next day she feel in her mouth the scent of the garlic, she is of herself fruitful without any default.

But albeit these two proofs perform the effect which Hippocrates speaketh of, namely, that the vapour pierce from the inner part up to the mouth, yet the same argueth not an absolute barrenness in the husband, nor an entire fruitfulness in the wife, but an unapt correspondence of both, wherethrough she proveth as barren for him as he for her, which we see to fall out in daily experience, for the man taking another wife begetteth children. And, which increaseth the marvel, in such as are not seen in that point of natural philosophy, is that if these two separate each from other upon pretence of impotency, and so he take another wife, and she another husband, it hath been found that both have had children. And this groweth because there are some men whose generative faculty is unable and not alterable for one woman, and yet for another is apt and begetteth issue. Even as we see by experience in the stomach that to one kind of meat a man hath great appetite, and to another (though better) it is as dead. What the correspondence should be which the man and wife ought to bear each to other to the end they may bring forth children is expressed by Hippocrates in these words: 'If the hot answer not the cold and the dry the moist with measure and equality there can be no generation.' As if he should say that if there unite not in the woman's womb two seeds, the one hot, and the other cold, and the one moist and the other dry, extended in equal degree, they cannot beget children. For a work so marvellous as is the shaping of a man standeth in need of a temperature where the hot may not exceed the cold, nor the moist the dry. For if a man's seed be hot, and the woman's seed hot likewise, there will no engendering succeed.

This doctrine thus presupposed, let us now fit by way of example a woman cold and moist in the first degree (whose signs we said were to be wily, ill conditioned, shrill voiced, spare fleshed, and black and green coloured, hairy and evil favoured), she shall easily conceive by a man that is ignorant, of good conditions, who hath a well sounding and sweet voice, much white and supple flesh, little hair and well coloured, and fair of countenance. She may also be given for wife to a temperate man whose seed (following the opinion of Galen) we said was most fruitful and answerable to whatsoever woman, provided that she be sound and of age convenient. But yet with all their incidents, it is very difficult for her to conceive child, and being conceived, saith Hippocrates, within two months the same miscarrieth, for she wanteth blood wherewith to maintain herself and the babe during the nine months. Howbeit this will find an easy remedy if the woman do bath herself before she company with her husband, and the bain must consist of water fresh and warm, the which (by

Hippocrates) righteth her temperature to a good sort, for it looseneth and moisteneth her flesh even as the earth ought to be alike disposed that the grain may therein fasten itself and gather root.

Moreover, it worketh a farther effect, for it increaseth the appetite to meat, it restraineth resolution,[195] and causeth a greater quantity of natural heat. Wherethrough plenty of phlegmatic blood is increased by which the little creature may those nine months have sustenance. The tokens of a woman cold and moist in the third degree are to be dull witted, well conditioned, to have a very delicate voice, much flesh, and the same soft and white, to want hair and down, and not to be over fair. Such a one should be wedded to a man hot and dry in the third degree, for his seed is of such fury and fervency as it behoveth the same to fall into a place very cold and moist that it may take hold and root. This man is of the quality of cresses, which will not grow save in the water, and if he partaked less hot and dry, his sowing in so cold a belly were nought else than to cast grain into a pool.

Hippocrates giveth counsel that a woman of this sort should first lessen herself and lay aside her flesh and her fat before she marry, but then she need not to take to husband a man so hot and dry, for such a temperature would not serve, nor she conceive. A woman cold and moist in the second degree retaineth a mean in all the tokens which I have specified save only in beauty, which she enjoyeth in a high degree. Which yieldeth an evident sign that she will be fruitful and bear children, and prove gracious and cheerful. She answereth in proportion well-near to all men.

First to the hot and dry in the second degree, and next to the temperate, and lastly to the hot and moist. From all these unions and conjoinings of men and women which we have here laid down may issue wise children, but from the first are the most ordinary. For put case that the seed of a man incline to cold and moist, yet the continual dryness of the mother, and the giving her so little meat, correcteth and amendeth the defect of the father. For that this manner of philosophizing° never heretofore came to light, it was not possible that all the natural philosophers could shape an answer to this problem, which asketh whence proceedeth it that many fools have begotten wise children. Whereto they answer

---

[195] resolution] decomposition, disintegration.

that sottish persons apply themselves affectionately to the carnal act and are not carried away to any other contemplation. But contrarily, men very wise, even in the copulation go imagining upon matters nothing pertinent to that they have in hand, and therethrough weaken the seed and make their children defective as well in the powers reasonal[196] as in the natural. In the other conjoinings it is requisite to take heed that the woman be cleansed, and dried by a ripe age, and marry not over young, for hence it cometh that children prove simple and of little wit. The seed of young parents is very moist, for it is but a while since they were born, and if a man be formed of a matter endowed with excessive moisture, it followeth of force that he prove dull of capacity.

What diligence ought to be used that children male and not female may be born.
3.

Those parents who seek the comfort of having wise children and such as are towards for learning, must endeavour that they may be born male, for the female, through the cold and moist of their sex, cannot be endowed with any profound judgement. Only we see that they talk with some appearance of knowledge in slight and easy matters with terms ordinary and long studied, but being set to learning, they reach no farther than to some smack of the Latin tongue, and this only through the help of memory. For which dullness themselves are not in blame, but that cold and moist which made them women, and these self qualities (we have proved heretofore) gainsay the wit and ability. Solomon considering how great scarcity there was of wise men, and that no woman came to the world with a wit apt for knowledge, said in this manner: 'I found one man amongst 1000, but I have not found one woman amongst the whole rout.' As if he should say that of 1000 men he had found one wise, but throughout the race of women, he could never light upon one that had judgement. Therefore we are to shun this sex and to procure that the child be born male, for in such only resteth a wit capable of learning. It behoveth therefore first to take into consideration what

---

[196] reasonal] rational.

instruments were ordained by nature in man's body to this effect, and what order of causes is to be observed that we may obtain the end which we seek for. We must then understand that amongst many excrements and humours which reside in a man's body, nature, saith Galen, useth only the service of one to work that mankind may be preserved. This is a certain excrement which is termed 'whey', or 'wheyish blood', whose engendering is wrought in the liver and in the veins at such time as the four humours (blood, phlegm, choler, and melancholy) do take the form and substance which they ought to have.

Of such a liquor as this doth nature serve herself to resolve the meat and to work that the same may pass through the veins and through the strait passages, carrying nourishment to all the parts of the body. This work being finished, the same nature provideth the veins, whose office is nought else but to draw unto them this whey, and to send it through their passages to the bladder, and from thence out of the body. And this to free man from the offence which an excrement might breed him. But she, advising that he had certain qualities convenient for generation, provided two veins which should carry part thereof to the cods and vessels of seed, together with some small quantity of blood, whereby such seed might be formed as was requisite for mankind. Wherethrough she planted one vein in the reins° on the right side, which endeth in the right cod, and of the same is the right seed-vessel framed, and another on the left side which likewise taketh his issue at the left cod, and of that is shaped the left seed-vessel.

The requisite qualities of this excrement that the same may be a convenient matter for engendering of seed are, saith Galen, a certain tartness and biting which groweth, for that the same is salt, wherethrough it stirreth up the seed-vessels and moveth the creature to procure generation, and not to abandon this thought. And therefore persons very lecherous are by the Latinists termed *salaces*, that is to say, men who have much saltness in their seed.

Next to this, nature did another thing worthy of great consideration, namely, that to the right side of the reins, and to the right cod, she gave much heat and dryness, and to the left side of the reins, and to the left cod, much cold and moisture. Wherethrough, the seed which laboureth in the right cod issueth out hot and dry, and that of the left cod, cold and moist. What nature pretended by this variety of temperature as well in the reins as in the cods and seed-vessels is very manifest, we knowing by histories very true that at the beginning of the world, and many years after, a woman brought forth two children at a birth, whereof the one was born male, the other female, the end whereof tended that for every man there should be a wife, that mankind might take the speedier increase. She provided then that the right side of the reins should yield matter hot and dry to the right cod, and that the same with his heat and dryness should make the seed hot and dry for generation of the male. And the contrary she ordained for the forming of a woman, that the left side of the reins should send forth seed cold and moist to the left cod, and that the same with his coldness and moisture should make the seed cold and moist, whence it ensued of force that a female must be engendered. But after that the earth was replenished with people, it seemeth that this order and concert of nature was broken off, and this double child-bearing surceased, and which is worst, for one man that is begotten, six or seven women are born to the world, ordinarily. Whence we comprise that either nature is grown weary, or some error is thwarted in the minds which beareth her from working as she would. What the same is a little hereafter we will express when we may lay down the conditions which are to be observed to the end a male child (without missing) may be born. I say then that if parents will attain the end of their desire in this behalf, they are to observe six points. One of which is to eat meats hot and dry; the second, to procure that they make good digestion in the stomach; the third, to use much exercise; the fourth, not to apply themselves unto the act of generation until their seed be well ripened and seasoned; the fifth, to company with the wife four or five days before her natural course is to run; the sixth, to procure that the seed fall in the right side of the womb, which being observed (as we shall prescribe) it will grow impossible that a female should be engendered.

As touching the first condition, we must weet that albeit a good stomach do parboil and alter the meat, and spoil the same of his former quality, yet it doth never utterly deprive itself of them: for if we eat lettuce (whose quality is cold and moist) the blood engendered thereof shall be cold and moist; the whey, cold and moist; and the seed, cold and moist. And if we eat honey (whose quality is hot and dry) the blood which we breed shall be hot and dry; the whey, hot and dry; and the seed, hot and dry, for it is impossible (as Galen avoucheth) that the humours should not retain the substances and the qualities which the meat had before such time as it was eaten. Then it being true that the male sex consisteth in this, that the seed be hot and dry at the time of his forming, for certain it behoveth parents to use meats hot and dry that they may engender a male child. I grant well how in this kind of begetting there befalleth a great peril, for the seed being hot and dry we have often heretofore affirmed it followeth of force that there be born a man malicious, wily, cavilling, and addicted to many vices and evils, and such persons as these (unless they be straightly curbed) bring great danger to the commonwealth. Therefore it were better that they should not be gotten at all, but for all this there will not want parents who will say 'Let me have a boy and let him be a thief', and spare not for the iniquity of a man is more allowable than the well-doing of a woman. Howbeit this may find an easy remedy by using temperate meats which shall partake but meanly of hot and dry, or by way of preparation, seasoning the same with some spice. Such, saith Galen, are hens, partridges, turtles, doves, thrushes, blackbirds, and goats, which, by Hippocrates, must be eaten roasted to heat and dry the seed.

The bread with which the same is eaten should be white, of the finest meal, seasoned with salt and aniseed, for the brown is cold and moist (as we will prove hereafter) and very damageable to the wit. Let the drink be white wine, watered in such proportion as the stomach may allow thereof, and the water with which it is tempered should be very fresh and pure.

The second diligence which we speak of is to eat these meats in so moderate quantity as the stomach may overcome them, for albeit the meat be hot and dry of his proper nature, yet the same becometh cold and moist if the natural heat cannot digest it. Therefore though the parents eat honey, and drink white wine, these meats by this means will turn to cold seed and a female child be brought forth. For this occasion, the greater part of great and rich personages are afflicted by having more daughters than meaner folk, for they eat and drink that which their stomach cannot digest, and albeit their meat be hot and dry, sauced with sugar, spices, and honey, yet through their great quantity, then wax raw and cannot be digested. But the rawness which most endamageth generation is that of wine, for this liquor in being so vaporous and subtle occasioneth that the other meats together therewith pass to the seed-vessels raw, and that the seed falsely provoketh a man ere it be digested and seasoned.

Whereon Plato commendeth a law enacted in the Carthaginian commonwealth which forbade the married couple that they should not taste of any wine that day when they meant to perform the rights of the marriage bed, as well ware that this liquor always bred much hurt and damage to the child's bodily health, and might yield occasion that he should prove vicious and of ill conditions. Notwithstanding, if the same be moderately taken, so good seed is not engendered of any meat (for the end which we seek after) as of white wine, and especially to give wit and ability, which is that whereto we pretend.

The third diligence which we spoke of was to use exercise somewhat more than meanly, for this fretteth and consumeth the excessive moisture of the seed, and heateth and drieth the same. By this means a man becometh most fruitful and able for generation; and contrariwise, to give ourselves to our ease, and not to exercise the body, is one of the things which breedeth most coldness and moisture in the seed. Therefore, rich and dainty persons are less charged with children than the poor who take pains. Whence Hippocrates recounteth that the principal persons of Scythia were very effeminate, womanish, delicious, and inclined to do women's services, as to sweep, to rub, and to bake, and by this means were impotent for generation. And if they begot any male child, he proved either an eunuch or an hermaphrodite. Whereat they shaming and greatly aggrieved determined to make sacrifices to their God, and to offer him many gifts, beseeching him not to entreat them after that manner, but to yield them some remedy for the defect, seeing it lay in his power so to do. But Hippocrates laughed them to scorn saying that none effect betideth which seems not miraculous and divine if after that sort they fall into consideration thereof, for reducing whichsoever of them to his natural causes at last we come to end in God, by whose virtue all the agents of the world do work. But there are some effects which must be imputed to God immediately (as are those which come besides the order of nature) and others by the way of means, reckoning first as a mean the causes which are ordained to that end. The country which the Scythians inhabited, saith Hippocrates, is seated under the north, a region moist and cold beyond measure, wherethrough abundance of clouds it seems a miracle if you see the sun. The rich men sit ever on horseback, never use any exercise, eat and drink more than their natural heat can consume, all which things make the seed cold and moist. And for this cause they beget many females, and if any male were born, they proved of the condition which we have specified. 'Know you', said Hippocrates to them, 'that the remedy hereof consisteth not in sacrificing to God, neither in doing ought like that, but it behoveth withal that you walk on foot, eat little, and drink less, and not so wholy betake yourselves to your pleasures. And that you may the more plainly discern it, look upon the poor people of this country and your very slaves, who not only make no sacrifices to your God, neither offer him gifts (as wanting the means), but even blaspheme his blessed name and speak injuriously of him because he hath placed them in such estate. And yet (though so lewd and sacrilegious) they are very able for procreation, and the most part of their children prove males and strong, not cockneys,[197] not eunuchs, not hermaphrodites, as do those of yours. And the cause is for that they eat little and use much exercise, neither keep themselves always on horseback, like their masters. By which occasion, they make their seed hot and dry, and therethrough engender males and not females.' This point of philosophy was not understood by Pharaoh, nor by his council, seeing that he said in this manner: 'Come, let us keep them down with oppression that they may not multiply nor join with our enemy if war be raised against us.' And the remedy which he used to hinder that the people of Israel should not increase so fast, or at least that so many male children might not be born (which he most feared) was to keep them under with much toil of body, and to cause them for to eat leeks, garlic, and onions, which remedy took but a bad effect, as the Holy Scripture expressed: 'For the harder he held them oppressed, the more did they increase and multiply.' Yet he making reckoning that this was the surest way he could follow, doubled this their affliction of body, which prevailed so little, as if to quench a great fire he should throw thereinto much oil or grease. But if he or any of his counsellors had been seen in this point of natural philosophy, he should have given them barley bread, lettuce, melons, cucumbers, and citrons to eat, and have kept them well fed and well filled with drink, and not have suffered them to take any pain. For by this

---

[197] cockney] a child tenderly brought up, hence squeamish or effeminate.

means, their seed would have become cold and moist, and thereof more women than men been begotten, and in short time their life have been abridged. But feeding them with much flesh boiled with garlic, with leeks, and with onions, and tasking them to work so hard, he caused their seed to wax hot and dry, by which two qualities they were the more incited to procreation and ever bred issue male. For confirmation of this verity, Aristotle propoundeth a problem which saith: 'What is the cause that those who labour much and such as are subject to the fever hectic suffer many pollutions in their sleep?' Whereto verily he wist° not to shape an answer, for he telleth many things but none of them hit the truth. The right reason hereof is that the toil of the body and the hectic fever do heat and dry the seed, and these two qualities make the same tart and pricking, and for that in sleep all the natural powers are fortified, this betideth which the problem speaketh of. How fruitful and pricking the hot and dry seed is, Galen noteth in these words: 'The same is most fruitful, and soon inciteth the creature to copulation, and is lecherous and prone to lust.'

The fourth condition was not to accompany in the act of generation until the seed were settled, concocted and duly seasoned, for though the three former diligences have gone before, yet we cannot thereby know whether it have attained that perfection which it ought to have. Principally it behoveth for seven or eight days before to use the meats which we have prescribed, to the end the cods may have time to consume in their nourishment the seed which all that time was engendered of the other meats, and that this which we thus go describing may succeed.

The like diligence is to be used touching man's seed, that the same may be fruitful and apt for issue, as the gardeners do with the seeds which they will preserve, for they attend till they ripen, and cleanse, and wax dry. For if they pluck them from the stalk before they are deeply seasoned, and arrived to the point which is requisite, though they lie in the ground a whole year, they will not grow at all. For this reason, I have noted that in places where much carnal copulation is used, there is less store of children than where people are more inclined to continency. And common harlots never conceive because they stay not till the seed be digested and ripened.

It behoveth therefore to abide for some days that the seed may settle, concoct, and ripen, and be duly seasoned. For by this means is hot and dry, and the good substance which it had lost, the better recovered. But how shall we know the seed to be such as is requisite it should be, seeing the matter is of so great importance? This may easily be known if certain days have passed since the man companied with his wife, and by his continual incitement° and great desire of copulation, all which springeth for that the seed is grown fruitful and apt for procreation.

The fifth condition was that a man should meddle with his wife in the carnal act six or seven days before she have her natural course, for that the child straightways standeth in need of much food to nourish it. And the reason hereof is that the hot and dry of his temperature spendeth and consumeth not only the good blood of the mother, but also the excrements. Wherethrough Hippocrates said that the woman conceived of a male is well coloured and fair. Which groweth because the infant, through his much heat, consumeth all those excrements which are wont to disfigure the face, leaving the same as a washed cloth. And for that this is true, it is behooful that the infant be supplied with blood for his nourishment. And this experience manifesteth for it is a miracle that a male child should be engendered save upon the last days of the month.[198] The contrary befalleth when a woman goeth with a female, for through the much cold and moist of her sex, she eateth little and yieldeth store of excrements, wherethrough the woman conceived of a girl is ill favoured and full of spots and a thousand sluttishnesses stick unto her. And at the time of her delivery, she must tarry so many more days to purge herself than if she had brought a man child to the word. On the natural reason whereof, God grounded Himself when He commanded Moses that the woman who brought forth a male should remain in her bed a week and not enter into the temple until thirty three days were expired. And if she were delivered of a female, she should be unclean for the space of two weeks, and not enter into the temple until after sixty six days, in sort that when the birth is of a female, the time is doubled. Which so falleth out because in the nine months during which the child remained in the mother's womb, through the much cold and moist of her temperature, she doubly increased excrements, and the same of very malignant substance and quality, which a male infant would not have done.

---

[198] month] menstrual period.

Therefore, Hippocrates holdeth it a matter very perilous to stop the purgation of a woman who is delivered of a wench.

All this is spoken to the purpose that we must well advise ourselves of the last day of the month to the end the seed may find sufficient nourishment wherewith to relieve itself. For if the act of procreation be committed so soon as the purgation is finished, it will not take hold through defect of blood. Whereon it behoveth the parents be done to understand that if both seeds join not together at one self time (namely, that of the woman and of the man) Galen saith there will ensue no conception, although the seed of the man be never so apt for procreation. And hereof we shall render the reason to another purpose. This is very certain that all the diligences by us prescribed must also be performed on the woman's behoof,° otherwise, her seed (evil employed) will mar the conception. Therefore it is requisite they attend each to other so as at one self instant both their seeds may join together.

This, at the first coming, importeth very much, for the right cod and his seed-vessel, as Galen affirmeth, is first stirred up, and yieldeth his seed before the left, and if the generation take not effect at the first coming, it is a great haphazard[199] but that at the second a female shall be begotten. These two seeds are known first by the heat and coldness, then by the quantity of being much or little, and finally by the issuing forth speedily or slowly. The seed of the right cod cometh forth boiling, and so hot as it burneth the woman's belly, is not much in quantity and passeth out in haste. Contrariwise, the seed of the left taketh his way more temperate, is much in quantity, and for that the same is cold and gross, spendeth longer space in coming forth.

The last consideration was to procure that both the seeds (of the husband and the wife) fall into the right side of the womb, for in that place, saith Hippocrates, are males engendered, and females in the left. Galen allegeth the reason hereof saying that the right side of the womb is very hot, through the neighbourhood which it holdeth with the liver, with the right side of the reins, and with the right seed-vessel, which members we have affirmed and approved to be very hot. And seeing all the reason of working that the issue may become male consisteth in procuring that at the time of conception it partake much heat, it falleth out certain

---

[199] haphazard] matter of chance.

that it greatly importeth to bestow the seed in this place. Which the woman shall easily accomplish by resting on her right side when the act of generation is ended, with her head down and her heels up. But it behoveth her to keep her bed a day or two, for the womb doth not straightways embrace the seed but after some hours space.

The signs whereby a woman may know whether she be with child or no are manifest and plain to everyone's understanding, for if when she ariseth up on her feet the seed fall to the ground, it is certain, saith Galen, that she hath not conceived, albeit herein one point requireth consideration: that all the seed is not fruitful or apt for issue, for the one part thereof is very waterish, whose office serveth to make thin the principal seed to the end it may fare through the narrow passages, and this is that which nature sendeth forth, and it resteth when she hath conceived with the part apt for issue. It is known by that it is like water and of like quantity. That a woman rise up straightways on her feet so soon as the act of generation hath passed is a matter very perilous. Therefore Aristotle compelleth that she beforehand make evacuation of the excrements and of her urine to the end she may have no cause to rise.

The second token whereby we may know the same is that the next day following the woman will feel her belly empty, especially about the navel. Which groweth for that the womb, when it desireth to conceive, becometh very large and stretched out, for verily it suffereth the like swelling up and stiffness as doth a man's member, and when it fareth thuswise, the same occupieth much room. But at the point when it conceiveth, saith Hippocrates, suddenly the same draweth together and maketh as it were a purse to draw the seed unto it, and will not suffer it to go out, and by this means leaveth many empty places, the which women do declare saying that they have no tripes left in their belly, as if they were suddenly become lean. Moreover, forthwith they abhor carnal copulation and their husband's kindness, for the belly hath now got what it sought; but the most certain token, saith Hippocrates, is when their natural course faileth, and their breasts grow, and when they fall in loathing with meat.

What diligence is to be used that children may prove witty and wise.
4.

If we do not first know the cause whence it proceedeth that a man of great wit and sufficiency is begotten, it is impossible that the same may be reduced to art. For through conjoining and ordering his principles and causes, we grow to attain this end and by none other means. The astrologers hold that because the child is born under such an influence of the stars, he cometh to be discreet, witty, of good or ill manners, fortunate, and of those other conditions and properties which we see and consider every day in men. Which being admitted for true, it would follow a matter of impossibility to frame the same to any art, for it should be wholly a case of fortune and no way placed in men's election. The natural philosophers, as Hippocrates, Plato, Aristotle, and Galen, hold that a man receiveth the conditions of his soul at the time of his forming, and not of his birth, for then the stars do superficially alter the child, giving him heat, coldness, moisture, and drought, but not his substance, wherein the whole life relieth, as do the four elements (fire, air, earth, and water), who not only yield to the party composed (heat, cold, moisture, and dryness), but also the substance which may maintain and preserve the same qualities during all the course of life. Wherethrough that which most importeth in the engendering of children is to procure that the elements whereof they are compounded may partake the qualities which are requisite for the wit. For these, according to the weight and measure by which they enter into the composition, must always so endure in the mixture and not the alterations of heaven.

What these elements are, and in what sort they enter into the woman's womb to form the creature, Galen declareth and affirmeth them to be the same which compound all other natural things, but that the earth cometh lurking in the accustomed meats which we eat (as are flesh, bread, fish, and fruits), the water in the liquors which we drink. The air and fire, he saith, are mingled by order of nature and enter into the body by way of the pulse and of respiration. Of these four elements, mingled and digested by our natural heat, are made the two necessary principles of the infant's generation: to weet the seed, and the monthly course. But that whereof we must make greatest reckoning for the end which we inquire after are the accustomable meats whereon we feed. For these shut up the four elements in themselves, and from these the seed fetcheth more corpulency° and quality than from the water which we drink, or the fire and air which we breath in. Whence Galen saith that the parents who would beget wise children should read three books which he wrote of the faculty of the aliments, for there they should find with what kinds of meat they may effect the same. And he made no mention of the water, nor of the other elements, as materials, and of like moment. But herein he swerved from reason, for the water altereth the body much more than the air, and much less than the sound meats whereon we feed. And as touching that which concerneth the engendering of the seed, it carrieth as great importance as all the other elements together. The reason is, as Galen himself affirmeth, because the cods draw from the veins (for their nourishment) the wheyish part of the blood, and the greatest part of this whey which the veins receive partaketh of the water which we drink. And that the water worketh more alteration in the body than the air, Aristotle proveth where he demandeth what the cause is that by changing of waters we breed so great an alteration in our health, whereas if we breath a contrary air, we perceive it not. And to this he answereth that water yieldeth nourishment to the body, and so doth not the air. But he had little reason to answer after this manner, for the air also (by Hippocrates' opinion) giveth nourishment and substance as well as the water. Wherethrough Aristotle devised a better answer saying that no place nor country hath his peculiar air, for that which is now in Flanders, when the north wind bloweth, passeth within two or three days into Africa, and that in Africa, by the south is carried into the north; and that which this day is in Jerusalem, the east wind driveth into the West Indies. The which cannot betide in the waters, for they do not all issue out of the same soil, wherethrough every people hath his particular water conformable to the mine of the earth where it springeth and whence it runneth. And if a man be used to drink one kind of water, in tasting another he altereth more than by meat or air.

In sort, that the parents who have a will to beget very wise children must drink waters delicate, fresh, and of good temperature; otherwise, they shall commit error in their procreation. Aristotle saith that at the time of generation we must take heed of the south-west wind, for the same is gross and moisteneth the seed so as a female and not a male is begotten. But the west wind he highly commendeth, and advanceth it with names and titles very honourable. He calleth the same temperate, fatter of the earth, and saith that it cometh from the Elysian fields. But albeit it be true that it greatly importeth to breath an air very delicate and of good temperature, and to drink such waters, yet it standeth much more upon to use fine meats appliable to the temperature of the wit, for of these is engendered the blood and the seed, and of the seed, the creature. And if the meat be delicate and of good temperature, such is the blood made; and of such blood, such seed, and of such seed, such brain. Now, this member being temperate, and compounded of a substance subtile and delicate, Galen saith that the wit will be like thereunto: for our reasonable soul, though the same be incorruptible, yet goeth always united with the dispositions of the brain, which being not such as it is requisite they should be for discoursing and philosophizing, a man saith and doth 1000 things which are very unfitting.

The meats then which the parents are to feed on that they may engender children of great understanding (which is the ordinary wit for Spain) are, first, white bread made of the finest meal, and seasoned with salt. This is cold and dry, and of parts very subtile and delicate. There is another sort made, saith Galen, of reddish grain, which though it nourish much, and make men big limmed and of great bodily forces, yet for that the same is moist and of gross parts, it breedeth a loss in the understanding. I said 'seasoned with salt' because none of all the aliments which a man useth bettereth so much the understanding as doth this mineral. It is cold and of more dryness than any other thing, and if I remember well the sentence of Heraclitus, he said after this manner: 'A dry brightness, a wisest mind.' Then seeing that salt is so dry and so appropriate to the wit, the Scripture had good reason to term it by the name of prudence and sapience. Partridges and francolins[200] have a like substance, and the self temperature with bread of white meal, and kid, and muscatel wine. And if parents use these meats (as we have above specified) they shall breed children of great understanding. And if they would have a child of great memory, let them eight or nine days before they betake themselves to the act of generation eat trouts,

---

[200] francolin] medium-sized African and South Asian game birds resembling partridges.

1130 salmons, lampreys, and eels, by which meat they shall make their seed very moist and clammy.

These two qualities (as I have said before) make the memory easy to receive and very fast to preserve the figures a long time. By pigeons, goats, garlic, onions, leeks, rapes, pepper, vinegar, white wine, honey, and all other sorts of spices, the seed is made hot and dry, and of parts very subtle and delicate. The child who is engendered of such meat shall be of great imagination, but not of like understanding by means of the much heat, and he shall want memory through his abundance of dryness. These are wont 1140 to be very prejudicial to the commonwealth, for the heat inclineth them to many vices and evils, and giveth them a wit and mind to put the same in execution. Howbeit if we do keep them under, the commonwealth shall receive more service by these men's imagination than by the understanding and memory of the others. Hens, capons, veal, wheatears of Spain are all meats of moderate substance, for they are neither delicate not gross. I said wheatears of Spain for Galen, without making any distinction, saith that their flesh is of a gross and noisome substance which strayeth from reason, for put case that in Italy (where he wrote), it be the 1150 worst of all others, yet in this our country, through the goodness of the pastures, we may reckon the same among the meats of moderate substance. The children who are begotten on such food shall have a reasonable discourse, a reasonable memory, and a reasonable imagination, wherethrough they will not be very profoundly seen in the sciences, nor devise ought of new.

Of these we have said heretofore that they are pleasant conceited and apt, in whom may be imprinted all the rules and considerations of art, clear, obscure, easy, and difficult. But doctrine, argument, answering, doubting, and distinguishing are 1160 matters wherewith their brains can in no sort endure to be cloyed. Cow's flesh, manzo,° bread of red grain, cheese, olives, vinegar, and water alone will breed a gross seed and of faulty temperature, the son engendered upon these shall have strength like a bull, but withal be furious and of a beastly wit. Hence it proceedeth that amongst upland people it is a miracle to find one quick of capacity or towardly for learning, they are all born dull and rude for that they are begotten on meats of gross and evil substance. The contrary hereof befalleth in citizens whose children we find to be endowed with more wit and sufficiency.

1170 But if the parents carry in very deed a will to beget a son, prompt, wise, and of good conditions, let them six or seven days before their companying feed on goat's milk, for this aliment (by the opinion of all physicians) is the best and most delicate that any man can use, provided that they be sound and that it answer them in proportion. But Galen saith it behoveth to eat the same with honey, without which it is dangerous and easily corrupteth. The reason hereof is for that the milk hath no more but three elements in his composition: cheese, whey and butter. The cheese answereth the earth; the whey, the water; and the butter, the air. The fire, which mingleth the other elements and preserveth them being mingled, issuing out of the teats is exhaled, for that it is very subtle. But adjoining thereunto a little honey, which is hot and dry, in lieu of fire, the milk will so partake of all the four elements. Which being mingled and concocted by the operation of our natural heat make a seed very delicate and of good temperature. The son thus engendered shall at leastwise possess a great discourse and not be deprived of memory and imagination.

In that Aristotle wanted this doctrine he came short to answer a problem which himself propounded, demanding what the cause is that the young ones of brute beasts carry with them (for the most part) the properties and conditions of their sirs and dames, and the children of men and women not so. And we find this by experience to be true, for of wise parents are born foolish children; and of foolish parents, children very wise; of virtuous parents, lewd children; and of vicious parents, virtuous children; of hard favoured parents, fair children; and of fair parents, foul children; of white parents, brown children; and of brown parents, white and well coloured children. And amongst children of one self father and mother, one proveth simple and another witty; one foul, and another fair; one of good conditions, and another of bad; one virtuous, and another vicious. Whereas if a mare of a good harrage be covered with a horse of the like, the colt which is foaled resembleth them as well in shape and colour as in their properties.

To this problem, Aristotle shaped a very untowardly answer, saying that a man is carried away with many imaginations during the carnal act, and hence it proceedeth that the children prove so diverse. But brute beasts, because in time of procreation they are not so distraughted, neither possess so forcible an imagination as a man doth, make always their young ones after one self sort and like to themselves. This answer hath ever hitherto gone for current amongst the vulgar philosophers, and for confirmation hereof they allege the history of Jacob, which recounteth that he having placed certain rods at the watering places of the beasts, the lambs were yeaned party coloured. But little avails it them to handfast holy matters, for this history recounteth a miraculous action which God performed therein to hide some sacrament. And the answer made by Aristotle savoureth of great simplicity. And who so will not yield me credit, let him at this day cause some shepherds to try this experiment and they shall find it to be no natural matter. It is also reported in these our parts that a lady was delivered of a son more brown than was due because a black visage which was pictured fell into her imagination. Which I hold for a jest, and if perhaps it be true that she brought such a one to the world, I say that the father who begat him had the like colour to that figure. And because it may be the better known, how fromshapen[201] this philosophy is which Aristotle bringeth in, together with those that follow him, it is requisite we hold it for a thing certain that the work of generation appertaineth to the vegetative soul and not to the sensitive or reasonable, for a horse engendereth without the reasonal, and a plant without the sensitive. And if we do but mark a tree loaded with fruit, we shall find on the same a greater variety than in the children of any man. One apple will be green, another red; one little, another great; one round, another ill-shaped; one sound, another rotten; one sweet, and another bitter. And if we compare the fruit of this year with that of the last, the one will be very different and contrary to the other, which cannot be attributed to the variety of the imagination, seeing the plants do want this power.

The error of Aristotle is very manifest in his own doctrine, for he saith that the seed of the man, and not of the woman, is that which maketh the generation, and in the carnal act, the man doth nought else but scatter his seed without form or figure, as the husbandman soweth his corn in the earth. And as the grain of corn doth not by and by take root, nor formeth a stalk and leaves until some days been expired, so, saith Galen, the creature is not formed all so soon as the man's seed falleth into the woman's womb, but affirmeth that thirty or forty days are requisite ere the same can be accomplished. And if this be so, what availeth it that

---

[201] fromshapen] deformed.

the father go imagining of diverse things in the carnal act, when as the forming beginneth not until some days after? Especially when the forming is not made by the soul of the father or the mother, but by a third thing which is found in the seed itself. And the same being only vegetative, and no more, is not capable of the imagination, but followeth only the motions of the temperature and doth nothing else. After my mind, to say that men's children are born of so diverse figures through the variable imaginations of the parents is none other than to avouch that of grains some grow big, and some little, because the husbandman (when he sowed them) was distraught into sundry imaginations. Upon this so unsound opinion of Aristotle, some curious heads argue that the children of the adulterous wife resemble her husband though they be none of his. And the reason which leadeth them is manifest: for during the carnal act the adulterers settle their imagination upon the husband with fear least he come and take them napping. And for the same consideration, they conclude that the husband's children resemble the adulterer though they be not his, because the adulterous wife during the copulation with her husband always busieth herself in contemplation of the figure of her lover. And those who say that the other woman brought forth a black son because she held her imagination fixed on the picture of a black man must also grant this which by these quaint brains is inferred: for the whole carrieth one self reason, and is in my conceit a stark leasing[202] and very mockery though it be grounded on the opinion of Aristotle. Hippocrates answered this problem better when he said that the Scythians are all alike conditioned and shaped in visage, and rendereth the reason of this resemblance to be for that they all fed of one self meat, and drank of one self water, went apparelled after one self manner, and kept one self order in all things. For the same cause, the brute beasts engender young ones after their particular resemblance, because they always use the same food and have therethrough a uniform seed. But contrariwise, man, because he eateth diverse meats, every day maketh a different seed as well in substance as in temperature. The which the natural philosophers do approve in answering to a problem that saith: 'What is the cause that the excrements of brute beasts have not so unpleasant a verdure as

---

[202] leasing] falsehood.

those of mankind?' And they affirm that brute beasts use always the self meats and much exercise therewithal, but a man eateth so much meat, and of so diverse substance, as he cannot come away with them, and so they grow to corrupt. Man's seed, and that of beasts, hold one self reason and consideration for that they are both of them excrements of a third concoction.

As touching the variety of meats which man useth, it cannot be denied but must be granted that of every aliment there is made a different and particular seed. Where it falleth out apparent that the day on which a man eateth beef or bloodings, he maketh a gross seed and of bad temperature. And therefore, the son begotten thereof, shall be disfigured, foolish, black, and ill conditioned. And if he eat the carcass of a capon or of a hen, his seed shall be white, delicate, and of good temperature. Wherethrough the son so engendered shall be fair, wise, and very gentle conditioned. From hence I collect that there is no child born who partaketh not of the qualities and temperature of that meat which his parents fed upon a day before he was begotten. And if any would know of what meat he was formed, let him but consider with what meat his stomach hath most familiarity (and without all doubt) that it was. Moreover, the natural philosophers demand what the cause is that the children of the wisest men do ordinarily prove blockish and void of capacity. To which problem they answer very fondly, saying that wise men are very honest and shamefast, and therefore in companying with their wives do abstain from some diligences necessary for effecting that the child prove of that perfection which is requisite. And they confirm this by example of such parents as are foolish and ignorant, who, because they employ all their force and diligence at the time of generation, their children do all prove wise and witty. But this answer tokeneth they are slenderly seen in natural philosophy. True it is that for rendering an answer convenient it behoveth first to presuppose and prove certain points. One of which purporteth that the reasonable faculty is contrary to the wrathful and the concupiscible, in sort, that if a man be very wise, he cannot be very courageous, of much bodily forces, a great seeder, nor very able for procreation, for the natural dispositions which are requisite to the end the reasonable soul may perform his operations carry a contrariety to those which are necessary for the wrathful and the concupiscible.

Aristotle saith (and it is true) that hardiness and natural

courage consist in heat, and prudence and sapience in cold and dry. Whence we see by plain experience that the valientest persons are void of reason, spare of speech, impatient to be jested withal, and very soon ashamed, for remedy whereof they straightways set hand on their sword, as not weeting what other answer to make. But men endowed with wit have many reasons and quick answers and quippes with which they entertain the time that they may not come to blows. Of such a manner of wit Salust[203] noteth that Cicero was, telling him that he had much tongue and feet very light, wherein he had reason, for so great a wisdom, in matters of arms, could not end but in cowardice. And hence took a certain nipping proverb his original which saith: 'He is as valiant as Cicero, and as wise as Hector.' Namely, when we will note a man to be a buzzard and a cow-baby.° No less doth the natural faculty gainsay the understanding, for if a man possess great bodily forces, he cannot enjoy a good wit, and the reason is for that the force of the arms and the legs springeth from having a brain hard and earthly, and though it be true that by reason of the cold and dry of the earth he might partake a good understanding, yet in that it hath his composition of a gross substance, it ruinateth and endamageth the same. For through his coldness the courage and hardiness are quenched, wherethrough we have seen some men of great forces to be very cowards.

The contrariety which the vegetative soul hath with the reasonable is most manifest of all others, for his operations (namely, to nourish and engender) are better performed with heat and moisture than with the contrary qualities, which experience clearly manifesteth considering how powerful the same is in the age of childhood and how weak and remiss in old age. Again, in boys' estate the reasonable soul cannot use his operations, whereas in old age, which is utterly void of heat and moisture, it performeth them with great effect. In sort, that by how much the more a man is enabled for procreation, and for digestion of food, so much he leeseth of his reasonable faculty. To this alludeth that which Plato affirmeth, that there is no humour in a man which so much disturbeth the reasonable faculty as abundance of seed, only, saith he, the same yieldeth help to the art of versifying.

---

[203] Salust] Gaius Sallustius Crispus, Sallust (86–c. 35 BC) was a Roman historian and partisan of Julius Caesar. His opposition and antipathy to Cicero was no secret for his contemporaries.

Which we behold to be confirmed by daily experience: for when a man beginneth to entreat of amorous matters, suddenly he becometh a poet. And if before he were greasy and loutish, forthwith he takes it at heart to have a wrinkle in his pump, or a mote on his cape. And the reason is because these works appertain to the imagination, which increaseth and lifteth itself up from this point through the much heat, occasioned in him by this amorous passion. And that love is a hot alteration showeth apparently through the courage and hardiness which it planteth in the lover, from whom the same also reaveth all desire of meat, and will not suffer him to sleep. If the commonwealth bare an eye to these tokens, she would banish from public studies lusty scholars and great fighters, enamoured persons, poets, and those who are very neat and curious in their apparel, for they are not furnished with wit or ability for any sort of study. Out of this rule, Aristotle excepteth the melancholic by adustion, whose seed (though fruitful) reaveth not the capacity.

Finally, all the faculties which govern man, if they be very powerful, set the reasonable soul in a garboil. Hence it proceeds that if a man be very wise, he proveth a coward, of small strength of body, a spare seeder, and not very able for procreation. And this is occasioned by the qualities which make him wise, namely, coldness and dryness. And these self weaken the other powers, as appeareth in old men, who (besides their counsel and wisdom) are good for nothing else. This doctrine thus presupposed, Galen holdeth opinion that, to the end the engendering of whatsoever creature may take his perfect effect, two seeds are necessary: one which must be the agent and former, and another which must serve for nourishment, for a matter so delicate as generation cannot straightways overcome a meat so gross as is the blood until the effect be greater. And that the seed is the right aliment of the seed members, Hippocrates, Plato, and Galen do all accord, for by their opinion, if the blood be not converted into seed, it is impossible that the sinews, the veins, and the arteries can be maintained. Wherethrough Galen affirmed the difference between the veins and the cods to be that the cods do speedily make much seed, and the veins a little and in long space of time.

In sort, that nature provided for the same an aliment so like, which with light alteration and without making any excrements might maintain the other seed. And this could not be effected if the nourishment thereof had been made of the blood. The self provision, saith Galen, was made by nature in the engendering of mankind as in the forming of a chick and such other birds as come of eggs. In which we see there are two substances: one of the white, and another of the yolk, of one of which the chick is made, and by the other maintained all the time whiles the forming endureth. For the same reason are two seeds necessary in the generation of the man, one of which the creature may be made, and the other by which it may be maintained whilst the forming endureth. But Hippocrates mentioneth one thing worthy of great consideration, namely, that it is not resolved by nature which of the two seeds shall be the agent and former, and which shall serve for aliment. For many times, the seed of the woman is of greater efficacy than that of the man, and when this betideth, she maketh the generation and that of the husband serveth for aliment. Otherwhiles, that of the husband is more mighty, and that of the wife doth nought else than nourish.

This doctrine was not considered by Aristotle, who could not understand whereto the woman's seed served, and therefore uttered a thousand follies, and that the same was but a little water without virtue or force for generation. Which being granted, it would follow impossible that a woman should ever covet the conversation of man, or consent thereunto, but would shun the carnal act as being herself so honest, and the work so unclean and filthy. Wherethrough, in short space mankind would decay, and the world rest deprived of the fairest creature that ever nature formed. To this purpose Aristotle demandeth what the cause is that fleshly copulation should be an action of the greatest pleasure that nature ever ordained for the solace of living things. To which problem he answereth that nature having so desirously procured the perpetuity of mankind did therefore place so great a delight in this work to the end that they being moved by such interest might gladly apply themselves to the act of generation. And if these incitements° were wanting, no woman or man would condescend to the bands of marriage, inasmuch as the woman should reap none other benefit than to bear a burden in her belly the space of nine months, with so great travail and sorrows, and at the time of her childbirth, to undergo the hazard of forgoing her life. So would it be necessary that the commonwealth should through fear enforce women to marry to the end mankind might not come to nothing. But because nature doth her things with pleasing, she gave to a woman all the instruments necessary for making a seed inciting and apt for issue, whereby she might desire a man and take pleasure in his conversation. But if it were of that quality which Aristotle expresseth, she would rather fly and abhor him than ever love him. This self Galen proveth alleging an example of the brute beasts, wherethrough he saith that if a sow be spayed, she never desireth the boar, nor will consent that he approach unto her.

The like we do evidently see in a woman whose temperature partaketh more of coldness than is requisite, for if we tell her that she must be married, there is no word which soundeth worse in her ear. And the like befalleth to a cold man, for he wanteth the fruitful seed. Moreover, if a woman's seed were of that manner which Aristotle mentioneth, it could be no proper aliment, for to attain the last qualities of actual nutriment, a total seed is necessary whereby it may be nourished. Wherethrough, if the same come not to be concocted and semblable, it cannot perform this point, for woman's seed wanteth the instruments and places (as are the stomach, the liver, and the cods) where it may be concocted. Therefore nature provided that in the engendering of a creature two seeds should concur, which being mingled, the mightier should make the forming and the other serve for nourishment. And this is seen evidently so to be, for if a blackamoor beget a white woman with child, and a white man a negro woman, of both these unions will be born a creature partaking of either quality. Out of this doctrine I gather that to be true which many authentical histories affirm, that a dog carnally companying with a woman made her to conceive, and the like did a bear with another woman, whom he found alone in the fields. And likewise, an ape had two young ones by another. We read also of one who, walking for recreation alongst a riverside, a fish came out of the water and begat her with child. The matter herein of most difficulty for the vulgar to conceive is how it may be that these women should bring forth perfect men and partakers of the use of reason seeing the parents who engendered them were brute beasts. To this I answer that the seed of every of these women was the agent and former of the creature, as the greater in force, whence it figured the same with his accidents of man's shape. The seed of the brute beast (as not equal in strength) served for aliment and for nothing else. And that the seed of these unreasonable beasts might yield nourishment to man's seed is a matter easy to be conceived. For if any of these women had eaten a piece of bear's flesh, or of a dog, boiled or roasted, she should have received nourishment thereout, though not so good as if she had eaten mutton or partridges. The like befalleth to man's seed, that his true nourishment (in the forming of the creature) is another man's seed, but if this be wanting, the seed of some brute beast may supply the room. But a thing which these histories specify is that children born of such copulations give token in their manners and conditions that their engendering was not natural.

Out of the things already rehearsed (though we have somewhat lingered by the way therein) we may now gather the answer to that principal problem, viz., that wise men's children are well-near always formed of their mother's seed, for that of the father's (for the reasons already alleged) is not fruitful for generation and in engendering serveth only for aliment. And the man who is shaped of the woman's seed cannot be witty, nor partake ability through the much cold and moist of that sex. Whence it becometh manifest that when the child proveth discreet and prompt, the same yieldeth an infallible token that he was formed of his father's seed. And if he shew blockish and untoward, we infer that he was formed of the seed of his mother. And hereto did the wise man allude when he said: 'The wise son rejoiceth the father, but a foolish child is a grief to his mother.' It may also come to pass upon some occasion that the seed of a man may be the agent and form giver, and that of the woman serve for nourishment, but the son so begotten will prove of slender capacity, for put case that cold and dry be two qualities whereof the understanding hath need, yet it behoveth that they hold a certain quantity and measure, which once exceeded they do rather hurt than good. Even as we see men very aged, that by occasion of overmuch cold and dry, we find them become children anew and utter many follies. Let us then presuppose that to some old man there yet remain ten years of life with convenient cold and dry to discourse, in such sort as these being expired, he shall then grow a babe again.

If of such a one's seed a son be engendered, he shall till ten years' age make show of great sufficiency (for that till then, he enjoyeth the convenient cold and drought of his father), but at eleven years old, he will suddenly quail away, for that he hath outpassed the point which to these two qualities was behooful. Which we see confirmed by daily experience in children begotten in old age, who in their childhood are very advised, and

1530 afterwards in man-state prove very dullards and short of life. And this groweth because they were made of a seed cold and dry which had already outrun the one half of his race. And if the father be wise in the works of the imagination, and by means of his much heat and dryness take to wife a woman cold and moist in the third degree, the son born of such an accouplement shall be most untoward if he be formed of his father's seed, for that he made abode in a belly so cold and moist and was maintained by a blood so distemperate. The contrary betideth when the father is untoward whose seed hath ordinarily heat and excessive moisture.

1540 The son so engendered shall be dull till fifteen years of age, for that he drew part of his father's superfluous moisture. But the course of that age once spent it giveth firmness inasmuch as the foolish man's seed is more temperate and less moist. It aideth likewise the wit to continue nine months' space in a belly of so little coldness and moisture as is that of a woman cold and moist in the first degree, where it endured hunger and want. All this ordinarily befalleth for the reasons by us specified. But there is found a certain sort of men whose genitors are endowed with such force and vigour as they utterly spoil the aliments of their good

1550 qualities and convert them into their evil and gross substance. Therefore all the children whom they beget (though they have eaten delicate meats) shall prove rude and dullards. Others, contrariwise, using gross meats and of evil temperature, are so mighty in overcoming them that though they eat beef or pork, yet they make children of very delicate wit. Whence it proveth certain that there are lineages of foolish men and races of wise men, and others who of ordinary are born blunt and void of judgement.

Some doubts are encountered by those who seek to pierce into the bottom of this matter, whose answer (in the doctrine
1560 forepassed) is very easy. The first is: whence it springeth that bastard children accustomably resemble their fathers, and of a hundred lawful, ninety bear the figure and conditions of the mother? The second: why bastard children prove ordinarily deliver,[204] courageous, and very advised? The third: what the cause is that if a common strumpet conceive she never loseth her burden though she take venomous drenches to destroy the same,

---

[204] deliver] agile, active.

or be let much blood, whereas if a married woman be with child by her husband upon every light occasion the same miscarrieth?

To the first, Plato answereth saying that no man is nought of his own proper and agreeable will unless he be first incited by the viciousness of his temperature. And he gives us an example in lecherous men, who, for that they are stored with plentiful and fruitful seed, suffer great illusions and many cumbers,[205] and therefore (molested by that passion) to drive the same from them do marry wives. Of such Galen saith that they have the instruments of generation very hot and dry, and for this cause breed seed very pricking and apt for procreation. A man then who goeth seeking a woman not his own is replenished with this fruitful, digested, and well seasoned seed, whence it followeth of force that he make the generation, for where both are equal, the man's seed carrieth the greatest efficacy, and if the son be shaped of the seed of such a father, it ensueth of necessity that he resemble him. The contrary betideth in lawful children, who, for that married men have their wives ever couched by their sides, never take regard to ripen the seed, or to make it apt for procreation, but rather (upon every light enticement) yield the same from them using great violence and stirring, whereas women, abiding quiet during the carnal act, their seed-vessels yield not their seed, save when it is well concoct and seasoned. Therefore, married women do always make the engendering and their husband's seed serveth for aliment. But sometimes it comes to pass that both the seeds are matched in equal perfection and combat in such sort as both the one and the other take effect in the forming, and so is a child shaped who resembleth neither father nor mother. Another time it seemeth that they agree upon the matter, and part the likeness between them: the seed of the father maketh the nostrils and the eyes, and that of the mother, the mouth and the forehead. And which carrieth most marvel it hath so fallen out that the son hath taken one ear of his father, and another of his mother, and so the like in his eyes. But if the father's seed do altogether prevail, the child retaineth his nature and his conditions, and when the seed of the mother swayeth most, the like reason taketh effect. Therefore, the father who coveteth that his child may be made of his own seed ought to withdraw himself for some days from his

---

[205] cumbers] distress.

wife, and stay till all his seed be concocted and ripened, and then it will fall out certain that the forming shall proceed from him and the wife's seed shall serve for nourishment.

The second doubt (by means of that we have said already) beareth little difficulty, for bastard children are ordinarily made of seed hot and dry, and from this temperature (as we have oftentimes proved heretofore) spring courage, bravery, and a good imagination, whereto this wisdom of the world appertaineth. And because the seed is digested and well seasoned, nature effecteth what she likes best, and portrayeth those children as with a pencil.

To the third doubt may be answered that the conceiving of lewd women is most commonly wrought by the man's seed, and because the same is dry and very apt for issue, it fasteneth itself in the woman with very strong roots, but the child breeding of married women, being wrought by their own seed, occasioneth that the creature easily unlooseth[206] because the same was moist and watery, or as Hippocrates saith, full of mustiness.

What diligences are to be used for preserving the children's wit after they are formed.
5.

The matter whereof man is compounded proveth a thing so alterable and so subject to corruption that at the instant when he beginneth to be shaped, he likewise beginneth to be untwined and to alter, and therein can find no remedy. For it was said, 'so soon as we are born, we fail to be.' Wherethrough nature provided that in man's body there should be four natural faculties: attractive, retentive, concoctive, and expulsive. The which concocting and altering the aliments which we eat return to repair the substance that was lost, each succeeding in his place. By this we understand that it little availeth to have engendered a child of delicate seed if we make no reckoning of the meats which afterwards we feed upon. For the creation being finished, there remaineth not for the creature any part of the substance whereof it was first composed. True it is that the first seed, if the same be well concocted and

---

[206] unloose] to become loose or unfastened.

seasoned, possesseth such force that digesting and altering the meats it maketh them (though they be bad and gross) to turn to his good temperature and substance, but we may so far-forth use contrary meats as the creature shall lose those good qualities which it received from the seed whereof it was made. Therefore Plato said that one of the things which most brought man's wit and his manners to ruin was his evil bringing up in diet. For which cause he counselled that we should give unto children meats and drinks delicate and of good temperature, to the end that when they grow big, they may know how to abandon the evil and to embrace the good. The reason hereof is very clear, for if at the beginning the brain was made of delicate seed, and that this member goeth every day impairing and consuming, and must be repaired with the meats which we eat, it is certain, if these being gross and of evil temperature that using them many days together, the brain will become of the same nature. Therefore it sufficeth not that the child be born of good seed, but also it behoveth that the meat which he eateth after he is formed and born be endowed with the same qualities. What these be, it carrieth no great difficulty to manifest if you presuppose that the Greeks were the most discreet men of the world, and that, inquiring after aliments and food to make their children witty and wise, they found the best and most appropriate. For if the subtle and delicate wit consist in causing that the brain be compounded of parts subtle and of good temperature, that meat which above all others partaketh these two qualities shall be the same which it behoveth us to use for obtaining our end.

Galen and all the Greek physicians say that goat's milk boiled with honey is the best meat which any man can eat, for besides that it hath a moderate substance, therein the heat exceedeth not the cold, nor the moist the dry. Therefore we said (some few leaves past) that the parents whose will earnestly leadeth them to have a child wise, prompt, and of good conditions, must eat much goat's milk boiled with honey seven or eight days before the copulation. Albeit this aliment is so good (as Galen speaketh of) yet it falleth out a matter of importance for the wit that the meat consist of moderate substance and of subtle parts. For how much the finer the matter becometh in the nourishment of the brain, so much the more is the wit sharpened. For which cause, the Greeks drew out of the milk cheese and whey (which are the two gross aliments of his composition) and left the butter, which in nature resembleth the air. This they gave in food to their children mingled with honey with intention to make them witty and wise. And that this is the truth is plainly seen by that which Homer recounteth. Besides this meat, children did eat cracknels of white bread of very delicate water with honey and a little salt. But instead of vinegar (for that the same is very noisome and damageable to the understanding) they shall add thereunto butter of goat's milk, whose temperature and substance is appropriate for the wit. But in this regiment° grows an inconvenience very great, namely, that children using so delicate meats shall not possess sufficient strength to resist the injuries of the air, neither can defend themselves from other occasions which are wont to breed maladies. So by making them become wise, they will fall out to be unhealthful and live a small time.

This difficulty demandeth in what sort children may be brought up witty and wise, and yet the matter so handled as it may no way gainsay their healthfulness, which shall easily be effected if the parents dare to put in practice some rules and precepts which I will prescribe. And because dainty people are deceived in bringing up their children, and they treat still of this matter, I will first assign them the cause why their children, though they have schoolmasters and tutors, and themselves take such pains at their book, yet they come away so meanly with the sciences, as also in what sort they may remedy this without that they abridge their life or hazard their health. Eight things, saith Hippocrates, make man's flesh moist and fat. The first to be merry, and to live at heart's ease; the second, to sleep much; the third, to lay in a soft bed; the fourth, to fare well; the fifth, to be well apparelled and furnished; the sixth, to ride always on horseback; the seventh, to have our will; the eight, to be occupied in plays and pastimes, and in things which yield contentment and pleasure. All which is a verity so manifest as if Hippocrates had not affirmed it none durst deny the same. Only we may doubt whether delicious people do always observe this manner of life, but if it be true that they do so, we may well conclude that their seed is very moist, and that the children which they beget will of necessity overabound in superfluous moisture, which it behoveth first to be consumed, for this quality sendeth to ruin the operations of the reasonable soul, and moreover the physicians say that it maketh them to live a short space and unhealthful. By this it should seem that a good wit and a sound bodily health require one self quality, namely drought, wherethrough the precepts and rules which we are to lay down for making children wise will serve likewise to yield them much health and long life. It behoveth them so soon as a child is born of delicious parents (inasmuch as their constitution consisteth of more cold and moist than is convenient for childhood), to wash him with salt hot water, which (by the opinion of all physicians) sucketh up and drieth the flesh, and giveth soundness to the sinews, and maketh the child strong and manly, and (by consuming the overmuch moisture of his brain) enableth him with wit, and freeth him from many deadly infirmities. Contrariwise, the bath being of water fresh and hot, in that the same moisteneth the flesh, saith Hippocrates, it breedeth five annoyances. Namely, effeminating of the flesh, weakness of sinews, dullness of spirits, fluxes of blood, and baseness of stomach. But if the child issue out of his mother's belly with excessive dryness, it is requisite to wash the same with hot fresh water. Therefore Hippocrates said children are to be washed a long time with hot water to the end they may receive the less annoyance by the cramp, and that they may grow and be well coloured. But for certain this must be understood of those who come forth dry out of their mother's belly, in whom it behoveth to amend their evil temperature by applying unto them contrary qualities.

The Almains,[207] saith Galen, have a custom to wash their children in a river so soon as they are born, them seeming that as the iron which cometh burning hot out of the forge is made the stronger if it be dipped in cold water, so when the hot child is taken out of the mother's womb, it yieldeth him of greater force and vigour if he be washed in fresh water. This thing is condemned by Galen for a beastly practice, and that with great reason: for put case that by this way the skin is hardened and closed, and not easy to be altered by the injuries of the air, yet will it rest offended by the excrements which are engendered in the body, for that the same is not of force nor open so as they may be exhaled and pass forth. But the best and safest remedy is to wash the children who have superfluous moisture with hot salt water, for their excessive moisture consuming, they are the nearer to health, and the way through the skin, being stopped in them, they

---

[207] Almains] Germans.

cannot receive annoyance by any occasion. Neither are the inward excrements therefore so shut up that there are not ways left open for them where they may come out. And nature is so forcible that if they have taken from her a common way, she will seek out another to serve her turn. And when all others fail, she can skill to make new ways wherethrough to send out what doth her damage. Wherefore of two extremes it is more available for health to have a skin hard and somewhat close than thin and open.

The second thing requisite to be performed when the child shall be born is that we make him acquainted with the winds and with change of air, and not keep him still locked up in a chamber, for else it will become weak, womanish, peevish, of feeble strength, and within three or four days, give up the ghost. Nothing, saith Hippocrates, so much weakeneth the flesh as to abide still in warm places and to keep ourselves from heat and cold. Neither is there a better remedy for healthful living than to accustom our body to all winds: hot, cold, moist, and dry. Wherethrough Aristotle inquireth what the cause is that such as live in the galleys are more healthy and better coloured than those who inhabit a plashy soil. And this difficulty groweth greater considering the hard life which they lead, sleeping in their clothes in the open air, against the sun, in the cold, and the water, and faring° withal so coarsely. The like may be demanded as touching shepherds, who of all other men enjoy the soundest health, and it springeth because they have made a league with all the several qualities of the air, and their nature dismaieth at nothing. Contrariwise, we plainly see that if a man give himself to live deliciously and to beware that the sun, the cold, the evening, nor the wind offend him, within three days he shall be dispatched with a post letter to another world. Therefore, it may well be said 'he that loveth his life in this world shall leese it', for there is no man that can preserve himself from the alteration of the air, therefore it is better to accustom himself to everything to the end a man may live careless and not in suspense. The error of the vulgar consisteth in thinking that the babe is born so tender and delicate as he cannot endure to issue forth of the mother's womb (where it was so warm) into a region of the air so cold without receiving much damage. And verily they are deceived, for those of Almain (a region so cold) used to dip their children so hot in the river and though this were a beastly act, yet the same did them no hurt, nor death's harm.

The third point convenient to be accomplished is to seek out a young nurse of temperature hot and dry or (after our doctrine) cold and moist in the first degree, endured to hardness and want, to lie on the bare ground, to eat little, and to go poorly clad in wet, drought and heat. Such a one will yield a firm milk as acquainted with the alterations of the air, and the child being brought up by her for some good space will grow to possess a great firmness. And if she be discreet and advised, the same will also be of much avail for his wit, for the milk of such a one is very clean, hot, and dry, with which two qualities the much cold and moist will be corrected, which the infant brought from his mother's womb. How greatly it importeth for the strength of the creature that it suck a milk well exercised is apparently proved in horses, who being foaled by mares toiled in ploughing and harrowing prove great coursers and will abide much hardness. And if the dames run up and down idly in the pastures after the first career, they are not able to stand on their feet. The order then which should be held with the nurse is to take her into house some four or five months before the childbirth, and to give her the same meats to eat whereon the mother feedeth, that she may have time to consume the blood and bad humours which she had gathered by harmful meats that she used tofore, and to the end the child (so soon as it is born) may suck the like milk unto that which relieved it in the mother's belly, or made at least of the same meats.

The fourth is not to accustom the child to sleep in a soft bed, nor to keep him overwarm apparelled, or give him too much meat. For these three things, saith Hippocrates, scarcen and dry up the flesh, and their contraries fatten and enlarge the same. And in so doing, the child shall grow of great wit and of long life by reason of this dryness, and by the contraries he will prove fair, fat, full of blood, and blockish, which habit, Hippocrates called wrestler-like, and holdeth it for very perilous. With this self receipt and order of life was the wisest man brought up that ever the world had,[208] to weet, our saviour Christ, in that he was man, saving (for that he was born out of Nazareth) perhaps his mother had no salt water at hand wherewith she might wash him; but this was a custom of the Jews, and of all Asia besides, brought in by some skilful physicians for the good of infants. Wherethrough the

---

[208] 44. *Index*: 'With this self receipt and order of life [...] as do the other children of men' [censored].

prophet saith: 'And when thou wert born at thy birthday thy navel string was not cut off, neither wert thou for thy health's sake washed in water, nor seasoned with salt, nor wrapped in swaddling clothes.'

But as touching the other things, so soon as he was born he began to hold friendship with the cold and the other alterations of the air. His first bed was the earth, his apparel coarse, as if he would observe Hippocrates' receipt. A few days after they went with him into Egypt, a place very hot where he remained all the time that Herod lived. His mother partaking the like humours, it is certain that she must yield him a milk well exercised and acquainted with the alterations of the air. The meat which they gave him was the same which the Greeks devised to endow their children with wit and wisdom. This (I have said heretofore) was the butterish part of the milk eaten with honey. Wherefore Esay[209] saith: 'He shall eat butter and honey that he may know to eschew evil and choose the good.' By which words is seen how the prophet gave us to understand that albeit he was very God, yet he ought also to be a perfect man, and to attain natural wisdom he must apply the semblable diligences as do the other sons of men. Howbeit this seemeth difficult to be conceived, and may be also held a folly to think that because Christ our redeemer did eat butter and honey being a child, he should therefore know how to eschew evil and make choice of good, when he was elder, God being (as He is) of infinite wisdom, and having given him (as he was man), all the science infused which he could receive after his natural capacity; therefore it is certain that he knew full as much in his mother's womb as when he was thirty three years old, without eating either butter or honey, or borrowing the help of any other natural remedies requisite for human wisdom.

But for all this it is of great importance that the prophet assigned him that self meat which the Trojans and Greeks accustomably gave their children to make them witty and wise, and that he said: 'To the end he may know to shun evil and choose the good', for understanding that by means of these aliments Christ our saviour got (as he was man) more acquisite knowledge than he should have possessed if he had used other contrary meats. It behoveth us to expound this particle to the end that we

---

[209] Esay] Isaiah.

may know what he meant when he spoke in those terms. We must therefore presuppose that in Christ our redeemer were two natures, as the very truth is, and the faith so teacheth us: one divine, as he was God, and another humane, compounded of a reasonable soul and of an elemental[210] body so disposed and instrumentalized as the other children of men.

As concerning his first nature, it behoveth not to entreat of the wisdom of our saviour Christ, for it was infinite without increase or diminishment, and without dependence upon ought else, save only in that he was God, and so he was as wise in his mother's womb as when he was thirty three years of age, and so from everlasting. But in that which appertaineth to his second nature, we are to weet that the soul of Christ, even from the instant when God created it, was blessed and glorious even as now it is, and seeing it enjoyed God and his wisdom, it is certain that in him was none ignorance, but he had so much science infused as his natural capacity would bear. But withal, it is alike certain that as the glory did not communicate itself unto all the parts of the body, in respect of the redemption of mankind, no more did the wisdom infused communicate itself. For the brain was not disposed, nor instrumentalized, with the qualities and substance which are necessary to the end the soul may with such an instrument discourse and philosophize. For if you call to mind that which in the beginning of this work we delivered, the graces *gratis* given which God bestoweth upon men do ordinarily require that the instrument with which they are to be exercised, and the subject whereinto it is to be received, do partake the natural qualities requisite for every such gift. And the reason is because that the reasonable soul is an act of the body and worketh not without the service of his bodily instruments.

The brain of our redeemer Christ whilst he was a babe, and lately born, had much moisture, for in that age it was behooful so to be, and a matter natural, and therefore in that it was of such quality, his reasonable soul naturally could not discourse nor philosophize with such an instrument. Wherethrough, the science infused passed not to the bodily memory, nor to the imagination, nor the understanding, because these three are instrumental powers (as tofore we have proved) and enjoyed not that perfection

---

[210] elemental] material as opposed to spiritual.

which they were to have. But whilst the brain went drying by means of time and age, the reasonable soul went also manifesting every day more and more the infused wisdom which it had, and communicated the same to the bodily powers.

Now, besides this supernatural knowledge, he had also another which is gathered of things that they heard whilst they were children: of that which they saw, of that which they smelled, of that which they tasted, and of that which they touched. And this for certain our saviour Christ attained as other men do. And even as for discerning things perfectly, he stood in need of good eyes, and for hearing of sounds, good ears, so also he stood in need of a good brain to judge the good and the evil. Whence it is manifest that by eating those delicate meats, his head was daily better instrumentalized and attained more wisdom. In sort, that if God had taken from him his science infused, thrice in the course of his life (by seeing that which he had purchased) we shall find that at ten years he knew more than at five; at twenty, more than at ten; and at thirty three, more than at twenty. And that this doctrine is true and Catholic, the letter of the Evangelic° Text proveth saying: 'And Jesus increased in wisdom and age and grace with God and with men.' Of many Catholic senses which the Holy Scripture may receive, I hold that ever better which taketh the letter than that which reaveth the terms and words of their natural signification.

What the qualities are which the brain ought to have, and what the substance, we have already reported by the opinion of Heraclitus that dryness maketh the wisest soul. And by Galen's mind we proved that when the brain is compounded of a substance very delicate, it maketh the wit to be subtle. Christ our redeemer went purchasing more dryness by his age, for from the day that we are born, until that of our death, we daily grow to a more dryness and leesing of flesh, and a greater knowledge. The subtle and delicate parts of his brain went correcting themselves whilst he fed upon meats which the prophet speaketh of. For if every moment he had need of nourishment and restoring the substance which wasted away, and this must be performed with meats, and in none other sort, it is certain that if he had always fed on cow's beef, or pork, in few days he should have bred himself a brain gross and of evil temperature with which his reasonable soul could not have shunned evil or chosen good save by miracle and employing his divinity. But God leading him by natural means caused him to use those so delicate meats by which, the brain being maintained, the same might be made an instrument so well supplied as (even without using the divine or infused knowledge) he might naturally have eschewed evil and chosen good, as do the other children of men.

FINIS

THE EXAMINATION OF MEN'S WITS

A table of all the chapters contained in this book.

1. It is proved by example that if a child have not the disposition and ability which is requisite for that science whereunto he will addict himself, it is a superfluous labour to be instructed therein by good schoolmasters, to have store of books, and continually to study it. fol. I.[211]

2. That nature is that which makes a man of ability to learn. 13

3. What part of the body ought to be well tempered that a young man may have ability. 23

4. It is proved that the soul vegetative, sensitive, and reasonable have knowledge without that anything be taught them, if so be that they possess that convenient temperature which is requisite for their operation. 33

5. It is proved that from the three qualities (hot, moist, and dry) proceed all the differences of men's wits. 51

6. Certain doubts and arguments are propounded against the doctrine of the last chapter, and their answer. 69

7. It is showeth that though the reasonable soul have need of the temperature of the four first qualities, as well for his abiding in the body as also to discourse and syllogize, yet for all this, it followeth not that the same is corruptible and mortal. 88

8. How there may be assigned to every difference of wit his science, which shall be correspondent to him in particular, and that which is repugnant and contrary be abandoned. 102

9. How it may be proved that the eloquence and finesse of speech cannot find place in men of great understanding. 120

10. How it is proved that the theoric of divinity appertaineth to the understanding, and preaching (which is his practice) to the imagination. 126

11. That the theoric of the laws appertaineth to the memory, and pleading and judging (which are their practice) to the understanding, and the governing of a commonwealth to the imagination. 150

12. How it may be proved that of theorical physic, part appertaineth to the memory, and part to the understanding, and the practic to the imagination. 173

13. By what means it may be showed to what difference of

---

[211] These references are to the original foliation, as they appear in the 1594 edition.

ability the art of warfare appertaineth, and by what signs the man may be known who is endowed with this manner of wit. 200

14. How we may know to what difference of ability the office of a king appertaineth, and what signs he ought to have who enjoyeth this manner of wit. 238

15. In what manner parents may beget wise children, and of a wit fit for learning. 263

1. By what signs we may know in what degree of hot and dry every man resteth. 278

2. What women ought to marry with what man that they may have children. 282

3. What diligence ought to be used that children male and not female may be born. 286

4. What diligence is to be used that children may prove witty and wise. 300

5. What diligences are to be used for preserving the children's wit after they are formed. 322.

FINIS

TEXTUAL NOTES

The second proem to the Reader.

[P3. 60] 'But there are, sayth St Paul, divisions of graces'] 'sayth St Paul' is a translation by Carew of 'dice S. Paolo' (1586, *9) in the Italian 'Proemio'. This sentence does not appear in the Spanish (1575, fol. 6ᵛ) but is an addition by Camilli, who in fact includes in the body of the text the note in the margin in the Spanish text which effectively indicates the source of the quotation: 'Pau[lus] 1 ad cori[nthios]. cap. XII' (1575, fol. 6ʳ). It is particularly verses 4–11 that Huarte quotes.

CHAP. I.

[I. 34] 'in his book of Destiny'] this is a translation by Carew of the Italian 'i libri del Fato' (literally, 'the books of Destiny') (1586, A1ᵛ). This phrase does not appear in the Spanish within the body of the text, but as a note in the margin 'Lib[ro]. de fato' (1575, fol.10ʳ). Hence, it is Camilli that inserts it in the body of the text.

[I. 205–206] 'and the native of one city going to study in another'] this matches the Italian '& andando i naturali d'una Città a studiare in un'altra' (1586, A5ʳ). However, the Spanish reads 'los naturales de la ciudad de Salamāca estudiar en la villa de Alcala de Henares: y los de Alcala en Salamāca' ('the natives of the city of Salamanca study in the town of Alcalá de Henares, and those of Alcalá in Salamanca') (1575, fol.16ʳ–16ᵛ). The universities of Alcalá and Salamanca were two of the oldest universities in Spain and were well known in western Europe in the sixteenth century, so both Camilli and Carew were surely aware of their existence. The University of Salamanca is the oldest existing university in Spain, as King Alfonso IX of León founded the *Studium generale* (university studies) in 1218, and, together with Bologna, Oxford and Paris (La Sorbonne), is now one of the oldest universities in Europe still active. As for the University of Alcalá, in 1293 King Sancho IV of Castile licenced the creation of the *Studium generale* at Alcalá, although it would not be until 1499 that Cardinal Cisneros, a former student of that *Studium generale*, founded what would later become the University of Alcalá, the *Complutensis Universitas*. No doubt, the reason why Camilli removes their names from his translation and substitutes them for the general term 'città' is to adapt his text to his target readership, i.e. an Italian readership with Italian universities in mind when reading this extract.

[I. 221–222] 'in the famous places of study' matches the Italian 'ne i più celebri studii' (1586, A5^v). However, the Spanish reads 'a Salamanca' ('to Salamanca') (1575, fol. 17^r). The reason behind this modification coincides with the previous note's.

CHAP. II.

[II. 114–115] 'God doth no longer those unwanted things of the New Testament'] In Spanish there is a reference to the Old Testament too ('testamento nuevo y viejo') (1575, fol. 25^v), and in Italian, only the Old Testament is mentioned: 'testamento vecchio' (1586, B2^r).

CHAP. III.

[III. 64] 'Notwithstanding, I avouch that if his having a great head'] Camilli translates the passage as 'se bene affermò anchora che' (1586, B6^v), which means that in the Italian translation the pronoun is 'he' and not 'I', so the sentence continues exploring Galen's views and not Huarte's. In the 1575 edition of the Spanish text it reads 'aunque tambien afirmo' (1575, fol. 34^v); because of lack of accents in the 1575 edition, 'afirmo' can either be the form of the first person singular of the present indicative of the verb 'afirmar' (to state) or the third person singular of such verb in the past simple tense with its corresponding accent missing ('afirmó', 'he stated'). All modern editions in Spanish agree that in this case it is the third person singular ('afirmó'), and that, therefore, in the sentence Huarte continues to explore Galen's views.

[III. 98] 'withal the greatness of his body'] Carew fails to continue translating the rest of the sentence, which in Italian runs in the following manner: '& fragli huomini, quegli (dice egli) sono più prudenti, i quali hanno minor testa' (1586, B7^v). This corresponds to the Spanish 'y entre los hombres, aquellos (dize) son mas prudentes que tienen menor cabeça' ('and amongst men, he says, those who are more sensible have a smaller head') (1575, fol. 36^r).

[III. 111–112] 'and the fourth in the part behind the brain'] Camilli does not translate into Italian (1586, B7ᵛ) the continuation of this sentence in Spanish, 'como paresce en esta figura' ('as this image/drawing shows') (1575, fol. 36ᵛ), which sentence is therefore also missing in the English version. However, in no Spanish edition does a 'figura' (an illustration of any sort) appear; probably Huarte had planned to include one and finally did not.

CHAP. V.

[V. 445] 'Wits full of invention are by the Tuscans called goatish'] In Spanish it reads: 'A los ingenios inventivos, llam en lengua toscana, caprichosos' ('To inventive wits they call in the Tuscan tongue capricious') (1575, fol. 77ᵛ); in Italian, the term used is 'capricciosi' (1586, E5ᵛ). Carew's use of the term 'goatish' is due to the fact that he takes the word *capriccio* (which in the sixteenth-century referred to a new and strange idea in a work of art) as derived from *capra* ('goat'). José Mondéjar, 'El pensamiento lingüístico del doctor Juan Huarte de San Juan', *Revista de Filología Española*, 64.1–2 (1984), 71–128, pp. 86–87, discusses the etymology of the term *capricho*, which most likely does not derive from 'goat'.

[V. 457–458] 'These have the property of a beast who never forsakes the beaten path'] In Spanish, as in Italian, the 'beast' is a sheep: 'oveja' (1575, fol. 78ʳ), 'pecora' (1586, E6ʳ). The OED explains that 'beast' can refer to 'a domesticated animal owned and used as part of farm "stock" or cattle [French *bestiaux*, *bétail*]; at first including sheep, goats, etc.'. So the English translation is not, after all, that distant from the Spanish and the Italian.

CHAP. VI.

[VI. 276] 'Blue'] translates the Italian 'turchina' (1586, F3ᵛ), which nowhere appears in the Spanish text, where it only reads: 'papel blanco y liso' ('white and plain paper') (1575, fol. 89ʳ). See the gloss for 'blue paper'.

[VI. 314] Again, 'blue' is Carew's translation of Camilli's 'turchino' (1586, F4ᵛ). See the gloss for 'blue paper'.

[VI. 431] 'as if he should say'] The pronoun 'he' refers to Hippocrates, as the Spanish makes explicit (1575, fol. 92^r). In the Italian translation, 'Hippocrates' is also missing (1586, F5^r).

[VI. 459–460] 'clear like the agate'] this translates the Italian 'pietra agata' ('agate stone') (1586, F8^r), while in the Spanish text the adjective is 'espléndida', and the stone, the 'azabache' ('splendid like jet') (1575, fol. 95^v). Jet is the stone of melancholy par excellence. In Spain the tradition of linking the stone to melancholy dates back to the arabic tradition. Felice Gambin, *Azabache*, 2008, pp. 169, 170, 233.

CHAP. VII.

[VII. 83–85] 'But God as desirous of war, and of wisdom, having chosen a place which should produce men like unto Himself, would that the same should be first inhabited'] In the Spanish, this sentence is in Latin: 'Deus vero quasi belli ac sapientiae studiosus, loc qui viros ipsi simillimos producturus esset elect in primis incolend praebuit' (1575, fol. 101^v). Camilli interpreted 'belli' as the genitive of Latin 'bellum' (war), and hence translated 'belli' as 'della guerra': 'Dio, come studioso della guerra, & della sapienza' (1586, G2^v). Consequently, so did Carew. Although Camilli's translation is grammatically possible, given the context it is more likely that 'belli' is the neuter genitive of Latin 'bellus' (beautiful, pleasant, agreeable), and hence alludes to what is good, useful, or beautiful, and not to war.

CHAP. VIII.

[VIII. 92–93] 'by which all nations might have commerce together'] this is a translation of the Italian 'tutte le nationi potessero haver cömercio insieme' (1586, H1^v-H2^r). In Spanish, however, we read the following: 'con que todas las naciones se pudiessen comunicar' ('with which all nations could communicate') (1575, fol. 115^r).

[VIII. 232] 'Boccace'] Huarte actually refers to 'Boscán' (1575, fol. 120^v), which Camilli accurately translates as 'Boscano' (1586, H4^v). Indeed, Carew would have been familiar with Boccaccio but probably not with Juan Boscán (the translator into Spanish of Castiglione's *Il Cortegiano*), as Boscán's works had not been translated into English in the sixteenth century. John Eliot, in his *Ortho-epia Gallica: Eliots Fruits for the French* (London, 1593), confirms in the following extract that Boscán's writings were not available to the English readership of his time:

> I have read over and over again almost all his works: but who are the best Spanish Poets?
> They are Boscan, Grenade, Garcilasso and Mont-major.
> I wonder that men get them not translated into English.
> They would have no grace.
> Why so? we find them almost all translated into Latin, Italian and French.
> I believe it well, yet have they more grace in their Castilian, which is the purest Spanish dialect, in which the learned write and speak ordinarily. (Eliot 1593, G4^{r-v})

Only a few of Boscán's poems were in effect rendered into English and published in collective miscellaneous collections and anthologies: part of the poem 'Ottava Rima' ('Loves Embassy') was published in Thomas Stanley's *Poems and Translations* (London, 1651), and Sonnet 21, 'Que Strella Fue? por donde yo Cay?' ('What Cruell Starre into this World me Brought'), printed in William Drummond's *Poems* (1616). A.F. Allison, *English Translations from the Spanish and Portuguese to the Year 1700* (London: Dawsons, 1974), p. 36.

[VIII. 466] '*tra qui tantos*'] It appears like this in the Spanish too (1575, fol. 129^v). '*Tra qui tantos*' is how 'trae aquí tantos' ('bring here this many', to be completed by a noun possibly referring to objects symbolizing money) sounds when uttered fast. In Spanish '*tra qui tantos*' is followed by 'a esta mesa' ('to this table'), which Camilli does not translate (1586, I1^v), and consequently Carew neither. The name *Traquitantos* would appear in other texts published after the *Examen*, such as the third proemial discourse of the *Declaración de los siete psalmos penitenciales* (1599), or a poem preceding Francisco de Quevedo's *Sueños* (1627) entitled 'Del capitán don Joseph de Bracamonte, dialogístico soneto entre Tomumbeyo Traquitantos, alguacil de la reina Pantasilea, y Dragalvino, corchete', and in *El hidalgo de La Mancha*, the mid-seventeenth century stage version of *Don Quixote* written by Juan Bautista Diamante, Juan Crisóstomo Vélez de Guevara, and Juan de Matos Fragoso. In *El hidalgo de La Mancha*, 'Traquitantos' (line 659) is the name of a giant (a windmill) against which Don Quixote fights.

CHAP. IX.

[IX. 41–43] 'Which matter, if it had passed through the hands of any other man of good imagination and memory'] There is an allusion to Erasmus missing first in Camilli's Italian translation (1586, I2^v), and then in Carew's rendering. In Spanish the sentence reads as follows: 'L aqual materia; si tomara entre manos Erasmo, o qual quier otro höbre de buena ymaginativa y memoria como el' ('Which matter, if in the hands of Erasmus or any other man of good imagination and memory as him') (1575, fol. 132^r). Felice Gambin affirms that in the Spanish Erasmus is caricatured as a man of good imagination and memory and therefore, implicitly, of little understanding. Gambin comments that in 1520–1530s Italy there was a fierce *reductio Erasmi ad Lutherum* fostered by Italian divines, and that in the last years of the sixteenth century the name of Erasmus became almost unspeakable; in fact, neither Camilli nor Gratii mention Erasmus in their renderings possibly out of fear of the Inquisition — which may also explain the subsequent replacement of 'friars' by the initial 'F' in Chapter XIII. Felice Gambin, 'Sobre la recepción y la difusión del *Examen de ingenios para las ciencias* de Huarte de San Juan en Italia', pp. 422–24. For more on Erasmus in Italy, see S. Seidel Menchi, *Erasmo in Italia 1520–1580* (Torino: Bollati Boringhieri, 1990).

[IX. 64] '*geragnin*'] In Italian, 'gevagnin' (1586, I3^r); in Spanish, 'gevañin' (1575, fol. 133^r). Dr Josefina Rodríguez Arribas, researcher at The Warburg Institute, University of London, has kindly provided me with an illuminating etymology of the Hebrew-Spanish term 'gevañin'. Joshua 9:22 ('And Joshua called for them, and he spake unto them, saying, Wherefore have ye beguiled us, saying, We are very far from you; when ye dwell among us?') refers to the inhabitants of Gibeon as liars, as they had pretended to be what they were not before Joshua and Israel. Gibeon (גבעון), pronounced Giv'on, comes from the root ג/ב/ע, which means tall, elevated, and hence, גבעה hill, mountain. From Gibeon the adjective that refers to the inhabitants of this city is derived: Samuel II 21:1, singular גבעוני, plural גבענים (pronouned giv'oni / giv'onim). The word has nothing to do with the roots that in Hebrew mean 'liar'. However, the strong presence of Biblical culture in every day life explains that some Biblical terms were transferred to everyday language to refer to situations or people with a Biblical precedent; for instance, the use of the term 'Pharisee' in Christian cultures to denote a hypocrite. It is possible, then, that giv'oni was used to refer to a liar, one that behaves like the inhabitants of Gibeon, who deceived Israel. The word 'gevañin' could have been the final result of a process of adaptation to the Spanish pronounciation of a Hebrew term 'giv'onim' or 'giv'onin' (the final 'm' of the plural is frequently written as 'n' due to the influence of Aramaic). Biblical culture was very strong among Jews and converts in sixteenth-century Spain, and even if Huarte were not a convert himself, he could have heard the word used among converts, or maybe the word itself was not infrequent in his circle. José Mondéjar, 'El pensamiento lingüístico', 1984, p. 83, footnote 30, puts forward a different hypothesis about the origins of the term 'gevañin', for he connects the word with the plural form 'gannabim' (thieves).

CHAP. X.

[X. 414–416] 'who would be doctors in the law and yet understand not the things which they speak, nor which they avouch'] In the Spanish this paragraph does not end here but goes on:

'La vaniloquencia y parlería, delos theologos Alemanes, Ingleses, Flamencos, Franceses, y de los demas, que abitan el septentrion, echo aperder el auditorio christiano: con tanta pericia de lenguas, con tanto ornamento y gracia en el predicar: por no tener entendimiento, para alcançar la verdad. Y que estos sean faltos de entendimiento, ya lo dexamos provado atras, de opinion de Arist. aliende de otras muchas razones y experiencias, q truximos para ello. Pero si el auditorio ingles, y aleman, estuviera advertido, en lo que S. Pablo escrivio a los romanos (estando tābien ellos apretados de otros falsos predicadores) por ventura, no se engañaran tan presto. *Rogo autem vos fratres, ut observetis eos, qui dissensiones & offendicula praeter doctrinā quā vos didiscistis faciunt & declinate ab illis huiusmodi enim Christo domino nostro non serviūt sed suo vētri; & per dulces sermones et benedictiones seducunt corda inoscentium.* Como si dixera: hermanos mios, por amor de Dios os ruego q tengays cuēta particular con essos que os enseñā otra doctrina, fuera de la q aveys aprendido: y apartaos dellos: por que no sirven a nuestro señor Jesuchristo sino a sus vicios, y sensualidad: y sō tambien hablados y eloquentes, que con la dulçura de sus palabras y razones, engañan a los q poco sabē' (1575, fol. 153r–153v).

In English (my translation): 'The verbosity and garrulity of German, English, Flemish and French divines, and of the rest of them that live in the Septentrion, ruined the Christian audience with such skill of languages, with so much ornament and grace in preaching, for having no understanding to reach truth. And that these lack understanding we have already proved by opinion of Aristotle as well as of many other reasons and experiences brought for the case. But if the English and German audience were warned in what St Paul wrote to the Romans (being them also under the pressure of other false preachers) perchance they would not be so easily deceived. *Rogo autem vos fratres, ut observetis eos, qui dissensiones & offendicula praeter doctrinā quā vos didiscistis faciunt & declinate ab illis huiusmodi enim Christo domino nostro non serviūt sed suo vētri; & perdulces sermones et benedictiones seducunt corda inoscentium* [I beseech you, my brethren, to observe those who are in disagreement and put up obstacles because of the teachings that you have given, and move away from them, for in that manner they do not serve Christ, our Lord, but their own belly, and, through sweet conversation and praise, seduce the hearts of the innocent]. As if he said: 'My brethren, for the love of God I beseech you beware those that teach you a different doctrine from the one that you have learned, and walk away from them, for they do not serve our lord Jesus Christ but their own vices and sensuality, and they are so well-spoken and eloquent that with the sweetness of their words and reasons, they fool those who know little.'

Camilli translated this fragment into Italian (1586, K5^v–K6^r), which means that its absence from Carew's translation is a deliberate decision on the part of Carew to conceal Huarte's direct attack against those countries (including England, explicitly mentioned) which sided with the Reformation. For Camilli, who translated for an Italian readership, this fragment was in no way a concern, unlike for Carew.

[X. 455–456] 'Such served not our Lord Jesus Christ but their belly'] In the Spanish this paragraph does not end here but goes on as follows:

'Y assi trabajan de interpretar la escritura divina, de manera que venga bien con su inclinacion natural: dando a entender a los que poco saben que los sacerdotes se pueden casar: y q no es menester que aya cuaresma, ni ayunos ni conviene manifestar al confessor, los delictos q contra Dios cometemos. Y usando d esta maña (con escriptura mal trayda) hazen parecer virtudes, a sus malas obras y vicios; y que las gētes los tengā por sanctos' (1575, fol. 155^r–155^v).

In English (my translation): 'And thus they work in interpreting the divine Scripture, in agreement with their natural inclination, making believe those who know little that priests can marry, and that Lent is not needed, nor fastings, and that letting the confessor know our offences to God is not advisable. And with these wiles (with wrongly brought Scripture) they disguise as virtues their wrong deeds and vices, and make people think of them saints.'

Camilli translated this fragment to Italian (1586, K6^v–K7^r), which means that its absence from Carew's translation is, for the same reasons as in the previous case, intentional too.

CHAP. XI.

[XI. 296–297] 'And contrariwise, others who have studied'] In Spanish this is followed by 'mal en Salamanca' ('badly in Salamanca') (1575, fol. 174^r). In Italian, the reference to Salamanca is missing (1586, L8^v), as well as in English. The reason for this absence may be the willingness on the part of Camilli to make his translation as free from specifically Spanish geographical references as possible to enable Italian readers to relate more easily to the contents of the book.

[XI. 627] In Spanish, 'at all adventures' reads instead 'en Salamanca' ('in Salamanca') (1575, fol. 186^r). In Italian, the reference to Salamanca is missing (1586, M7^r), hence its absence in Carew's text. The reason behind this omission matches the previous case's.

CHAP. XII.

[XII. 232] 'some thousands of years'] In Spanish and Italian instead of 'some' it reads 'three': 'tre' (1586, N4^r), 'tres' (1575, fol. 191^v).

[XII. 289] 'how great enchanters the Egyptians are'] In Italian it reads 'Egittii' (1586, N5^r); in Spanish, 'gitanos' ('gipsies') (1575, fol. 198^r). At the time it was believed that the gipsies came from

Egypt, hence the origin of the Spanish term 'gitano', derived from 'egiptano' ('egipcio' being 'Egyptian' in Spanish). The etymology of 'gipsy' similarly reflects this belief, as the OED explains: 'The early form *gipcyan* is aphetic for Egyptian (…). [the gipsy people] first appeared in England about the beginning of the 16th c. and was then believed to have come from Egypt.'

[XII. 308] 'Emperor Charles the fifth'] In Italian it reads 'l'Imperator Carlo Quinto' (1586, N5^v), while in Spanish 'Emperador nuestro señor' ('the Emperor our lord') (1575, fol. 198^v). Indeed, by 'Emperador nuestro señor' Huarte meant Charles V, who had died in July 1568. The fact that Huarte nonetheless refers to him in this manner may either indicate that this fragment was written prior to Charles V's death, or that the word 'emperor' could only mean 'Charles V' to the contemporary readers of Huarte. In any case, it seems that Camilli with the expression 'l'Imperator Carlo Quinto' updates the reference. Carew follows Camilli in his translation of the formula.

[XII. 314] 'the Emperor'] In Italian, 'l'Imperatore' (1586, N5^v), in Spanish the 'emperor our lord' formula is again used (1575, fol. 199^r). The expression anaphorically refers to Charles V. Carew follows Camilli in his translation.

[XII. 314–318] 'But for all this, the Emperor gave commandment that such a physician should be sought out if any there were, though to find him they should be driven to send out of his dominions, and when none could be met withal, he sent a physician newly made a Christian'] The predominance of Jewish physicians in the late Middle Ages in peninsular Christian kingdoms suffered a setback in 1492, when the Catholic Monarchs ordered the expulsion of the Jews from the Kingdoms of Castile and Aragon through the Edict of Expulsion (also known as the Alhambra Decree). This led to sanitary trouble in various cities. Some Jewish physicians, when faced with exile, decided to convert to Christianity and hence became converts or 'new Christians'. These were often victims of the Inquisition (always suspicious of the sincerity of their conversion to Christianity), of the animosity of the common people and of other fellow physicians (i.e. their Christian competitors), at a time when anti-semitic literature, such as Alonso de Espina's popular *Fortalitium fidei contra judeos* (written in 1458, first published in

1478), was not uncommon. In 1501 the Catholic Monarchs made it compulsory to demonstrate purity of blood (*limpieza de sangre*) to practise as a physician or a surgeon, and failure to do so was punished with the confiscation of possessions. And yet, this was no impediment for some renowned convert physicians of the sixteenth century to work for nobles, kings and popes; Franciso López Villalobos, Andrés Laguna, López Pereira, Enrique Jorge Enríquez, and Cristóbal Pérez de Herrera were some of these 'new Christians'.[1] It has been speculated that Huarte may be in that group too, a deduction often based on the fact that we know so little about his life and that he often conceals his surname in the documents he signs — although he does not omit it in his book[2] — and on his open defence of the wit of the Jews for the practice of medicine. Additionally, being a 'new Christian' would have been a reason for Navarre not to accept the settlement in the kingdom of the families from the area of Ultrapuertos, from where Huarte was a native.[3] Notwithstanding, Huarte's alleged Jewish origin has also been called into doubt, arguing that members of his family had had jobs banned for Jews — some of his ancestors held public office — that moving to Andalusia was risky in case of being a new Christian, as in that region the Inquisition was on closer watch, and that otherwise he would not have dared to defend the Jewish character so openly and to get away with it unpunished. Certainly, the fact that the inquisitors

---

[1] Jon Arrizabalaga, 'The World of Iberian Converso Practitioners, from Lluís Alcanyís to Isaac Cardoso', in *Más allá de la Leyenda Negra: España y la Revolución Científica / Beyond the Black Legend: Spain and the Scientific Revolution*, ed. by Víctor Navarro and William Eamon (Valencia: Universidad de Valencia, 2007), pp. 307–22. See also José María Pérez Fernández, 'Andrés Laguna: Translation and the Early Modern Idea of Europe', *Translation and Literature*, 21 (2012), 299–318.

[2] The surname Huarte is of Basque origins (ur-artea), literally, between the waters. It may be connected to the place name of Uhart, now in France, by the rivers Nive de Arneguy and Nive de Esterençuby. At the time Uhart was a parish of San Juan de Pie de Puerto. Sáez, 'La Baeza del siglo XVI', 1989, p. 84.

[3] Granjel, *Juan Huarte y su 'Examen de ingenios'*, 1988, p. 15. Among those who believe that Huarte was of Jewish origin, we find Américo Castro, *La realidad histórica de España* (México: Porrúa, 1954), p. 549, and *El pensamiento de Cervantes* (Barcelona: Noguer, 1972), pp. 56–57; and Julio Caro Baroja, *Los judíos en la España Moderna y Contemporánea*, Vol. I (Madrid: Istmo, 1986), pp. 100–01.

did not censor Huarte's defence of the Jewish physician is indicative that they were not suspicious of him.[4]

[XII. 384–385] 'abode in Egypt 400 years'] Both in Spanish (1575, fol. 201^v) and in Italian (1586, N7^r), the figure given is 430 years.

[XII. 425] 'the men of Ethiopia and Egypt'] In Spanish it reads 'negros' ('black people') (1575, fol. 203^v); in Italian, 'huomini' ('men') (1586, N7^v), as in English.

[XII. 518] 'he had parted the Red Sea'] In the Spanish text the phrase continues with the expression 'en doze carreras' ('in twelve lanes' — one for each Tribe of Israel) (1575, fol. 207^r), which is missing in Italian (1586, O1^v), and therefore in English too.

[XII. 728] 'twelve male and twelve female Moors of Ethiopia'] In Italian it reads 'mori' ('Moors') (1586, O5^v); in Spanish, 'negros' and 'negras' ('black men', 'black women') (1575, fol. 214^v). This is so in the rest of the occurrences of this term in this chapter.

[XII. 733] 'Egyptians'] In Spanish the term is 'gitanos' ('gipsies') (1575, fol. 215^r); in this case, in Italian it reads 'zingani d'Egitto' ('gipsies of Egypt') (1586, O6^r).

CHAP. XIII.

[XIII. 115–116] 'This property to attain suddenly the means is *solertia* (quickness)'] In the Spanish text (1575, fol. 220^r), it only reads *solercia* (which means skill and cunning to do or go about something), and it is not followed by the clarifying noun in parentheses. In Italian, the word translating the Spanish *solercia* is *vivacità* (1586, O8^v), and comes without an explanatory note. Carew precisely decides to use the Latin term from which the word in Spanish chosen by Huarte is derived, and then adds a

---

[4] sé Luis Peset, 'Las críticas a la Universidad de Juan Huarte de San Juan', in *La universidades hispánicas: de la monarquía de los Austrias al centralismo liberal*, *Vol. I*, ed. by Luis E. Rodríguez-San Pedro Bezares (Salamanca: Universidad de Salamanca, 2000), pp. 390–91. José Luis Peset, *Genio y desorden* (Valladolid: Cuatro, 1999), p. 22. Also, see José Javier Biurrun Lizarazu, 'Huarte de San Juan: vida y obra en el contexto político y religioso de la España del siglo XVI', *El Basilisco: Revista de filosofía, ciencias humanas, teoría de la ciencia y de la cultura*, 21 (1996), 16–17.

descriptive synonym in brackets. This may well indicate that, on this occasion, Carew checked a copy of the book in Spanish.

[XIII. 231] 'thief'] In Spanish, 'frayle' ('friar', 'monk') (1575, fol. 225r). In Italian, 'da F.' (1586, P3r), which may stand for 'frate' (also 'friar', 'monk'). In any case, the meaning of the saying remains the same: neither a thief nor a friar wish to be discovered wandering about at night. Felice Gambin affirms that, despite Camilli's cautious decision to conceal the reference to friars behind the letter 'F', its meaning must have come across rather unproblematically to Italian readers. Felice Gambin, 'Sobre la recepción y la difusión del *Examen de ingenios para las ciencias* de Huarte de San Juan en Italia', p. 420.

[XIII. 494–495] 'and some both are, and seem so'] this was introduced by Camilli: '& altri sono, & paiono' (1586, P8r). In the Spanish this fourth class is presupposed but missing (1575, fol. 234r).

[XIII. 606] 'our Lord the Prince Don Carlos'] In Italian, 'Principe Don Carlo nostro Signore' (1586, Q2r). The fact that Charles V had died in July 1568 and Huarte still refers to him in this manner may indicate that this fragment was written prior to that date, and that Huarte forgot to update the expression.

[XIII. 623–624] 'That me seemeth was evil done to study in one university and take degree in another'] It was not infrequent for students at Salamanca (not only students of Medicine but also of other subjects) to transfer to smaller universities (such as those of Toledo, Sigüenza, Osuna, Gandía, Orihuela, Almagro, Irache, Estella and Oñate) to obtain their degree, as the fees were substantially inferior in the latter. In smaller universities, attendance control was not rigorous, and it was easy to validate previously studied subjects and soon afterwards obtain the degree. In addition, because from 1501 onwards the descendants of those condemned by the Inquisition could not graduate in Medicine from any university, and because to graduate it was requisite proof of *limpieza de sangre*, converts often studied at or graduated from smaller universities where certificates of *limpieza de sangre* were not thoroughly inspected or verified. Thus, for example, renowned physicians such as the Judeo-Portuguese João Rodríguez de Castelo Branco (also known as Amato Lusitano) graduated from Sigüenza after having studied at Salamanca. For more on this subject, see Luis Enrique Rodríguez-San Pedro Bezares and Juan Luis Polo Rodríguez, eds. *Historia de la Universidad de Salamanca. Volumen III. Saberes y confluencias* (Salamanca: Ediciones Universidad de Salamanca, 2006), pp. 328–29.

[XIII. 690–693] 'this word *hijodalgo* thereof, which importeth nought else but that such a one is descended of him who performed some notorious and virtuous action for which he deserved to be rewarded by the king or commonwealth together with all his posterity forever'] On nobility in Castile and the concept of hidalgo Richard L. Kagan, *Students and Society in Early Modern Spain* (London: Johns Hopkins University Press, 1974), p. 182, states the following:

> In Castile nobles of one sort or another accounted for as many as 10 percent of the kingdom's inhabitants, that is, about 600,000 persons. So large is this group that it is often difficult to determine who was noble and who was not. By strict, legal definition, anyone who was not a clergyman and who did not pay direct taxes was a nobleman, but among those who could be classified as noble through this definition, the differences were huge. In the Basque provinces privileges granted to the local population by medieval kings exempted everyone from direct taxation; consequently, the entire population was 'noble', including those who tilled the soil and practiced manual crafts technically prohibited for members of a privileged class whose major occupation was supposed to be war. Similarly, the kingdom of Navarre was awarded 'noble' status, and throughout the north of Castile there were countless 'nobles' who had little more than such status to their names.
>
> Simple noblemen such as these were known as hidalgos, a term derived from those individuals who had been traditionally able to keep a horse and who had also participated in the *Reconquista* against the Moors. By the sixteenth century, however, hidalgo referred broadly to the poor, often landless, noblemen who constituted the bulk of Castile's 'noble' class. In addition, there were six higher rankings of nobility of which only two, *títulos* and *grandes*, the equivalent of the English aristocracy, resembled a coherent, easily-definable group. In 1520 only sixty Castilian families boasted titles, although royal generosity in subsequent years quadrupled this number by

1641. Likewise, grandee families, set apart from títulos by their greater wealth, prestige, and certain ceremonial privileges, were limited to twenty in 1520 by order of Charles V.

[XIII. 845–846] '*de vengar quincentos sueldos*'] In Italian, 'de vengar quinientos sueldos' (1586, Q6^v); in Spanish, 'de dev gar quini tos sueldos' (1575, fol. 246^v).

[XIII. 856] 'the book of *Bezerro*'] In Italian, 'libro del Gioverico' [*sic*. 'giovenco'] (1586, Q7^r); in Spanish, 'libro dl Bezerro' (1575, fol. 247^r), literally, 'book of the calf', because of the calfskin with which it was covered. The 'libro de becerro' was the book in which in the Middle Ages the privileges and possessions of churches and monasteries were inscribed. It often became synonymous with 'libro de apeo', which was, to an extent, similar in contents: in early modern Spain it registered land and property, rights and privileges, renowned historical events and jurisdictions of a place, or of a political or religious entity. Professor Felice Gambin, from the University of Verona, confirmed to me that, to his knowledge, there was nothing in sixteenth-century Italy known as 'libro del Giovenco', which means that this translation into Italian of 'libro dl Bezerro' is word for word (*giovenco* indeed means *becerro*, young ox). Still, given the context, Italian readers might have inferred that the expression alluded to a book or a registry of privileges. The fact that Carew uses precisely the Spanish word *bezerro*, absent from Camilli's translation, may indicate that, even if he translated directly from Camilli's Italian rendering, he still had a copy of the original in Spanish before him to check if necessary. Carew might have opted for reproducing the expression in Spanish out of a feeling that the translation into Italian was a word for word rendering void of true meaning and lacking an actual referent which the Spanish, by contrast, did have.

[XIII. 856–857] 'which is kept at Salamanca'] In Spanish it reads 'Simancas' (1575, fol. 247^r); in Italian, as in English, 'Salamanca' (1586, Q7^r). Simancas is a town in the now province of Valladolid where in 1540 the first official archive of the Crown of Castile (the General Archive of Simancas) was established. Despite its importance within Spain, Simancas was probably unknown to many foreigners, while Salamanca, due to its university, was a household name. The likeness between the two names must have been at the roots of Camilli's confusion, or simply the reason behind a deliberate replacement of an obscure name for a well-known and similar-sounding one.

[XIII. 860–861] 'of what stock the young man was descended'] In Italian it similarly reads: 'di quale stirpe fosse discesso quel gio vanetto' (1586, Q7ʳ). In Spanish this sentence reads slightly different: 'de qué padres y parientes desciende este mancebo, o de qué casa en Israel' ('from which parents and relatives this lad descends, or from which house in Israel') (1575, fol. 247ᵛ).

[XIII. 861] '*solaro*'] In Italian, 'solaro' (1586, Q7ʳ); in Spanish, 'solar' (1575, fol. 247ᵛ).

[XIII. 947] 'horses'] In Italian, 'cavalli' ('horses') (1586, Q8ᵛ); in Spanish, 'corredores', literally, 'runners' (1575, fol. 250ᵛ). 'Corredores' does not necessarily refer to horses racing; it can also indicate men running.

[XIII. 986] 'muster'] In Italian, 'mostra' (1586, R1ᵛ); in Spanish, 'muestra' (1575, fol. 252ʳ). Both mean show, display, demonstration.

CHAP. XIV.

[XIV. 181] 'saw'] In Italian, 'detto' (1586, R7ʳ); in Spanish, 'dicho' (1575, fol. 262ᵛ). Both mean 'saying'.

[XIV. 285] '900 years'] In Italian (1586, S1ʳ), as well as in Spanish (1575, fol. 266ᵛ), the figure is 930.

CHAP. XV.

[XV. 435] 'what woman is hot and dry'] In Italian 'di freddo, & d'humido' (1586, V2ʳ); in Spanish, 'frialdad y humidad' (1575, fol. 298ʳ), 'cold and moist', in the two cases. This is Carew's mistake when translating.

[XV. 494] 'ribs'] In Italian, 'cosce' ('thighs') (1586, V3ʳ); in Spanish, 'muslos' ('thighs') (1575, fol. 300ʳ). It is either Carew's mistake, or his own decision to elevate the point of reference from the thighs to the ribs.

[XV. 1464] 'for woman's seed wanteth the instruments and places'] In Italian, 'dell'huomo' (men's) (1586, T6ʳ); in Spanish, 'del varón' (men's) (1575, fol. 336ᵛ). It is Carew's mistake.

[XV. 1876] 'It behoveth us to expound this particle'] In Spanish, the particle in question is *ut*: '*ut sciat reprobare malum et eligere bonum*' (1575, fol. 353ʳ). In the Italian (1586, Z6ʳ) and in the English translations this sentence does not appear because they render all quotations in Latin into the vernacular. In English, the Latin sentence corresponds to 'To the end he may know to shun evil and choose the good' (XV. 1872–1873).

[XV. 1947] 'the prophet speaketh of'] In the Spanish 'the prophet' is followed by 'Ysayas' ('Isaiah') (1575, fol. 355ᵛ); 'il Profeta' is what we read in the Italian translation, but, as in the English, no name is provided (1586, Z7ᵛ).

[XV. 1960] The following sentence marks the end of the text in Spanish: '*Laudetur Christus in aeternum*' ('Praised be Christ forever') (1575, fol. 356ʳ). In Italian it is missing (1586, Z8ʳ), which explains its absence in the translation into English too.

GLOSSARY

The glossary contains a selection of archaic, obsolete and less common words and/or meanings. Definitions are based on those provided by the current online edition of the *OED*.

A

alloy	intrinsic character or nature; quality
athwart	across the course (of anything), so as to thwart or oppose progress
auburn	of a yellowish- or brownish-white colour

B

baggage	worthless or vile fellow
behoof	benefit, advantage
behooful	beneficial; mandatory, needed
bibbing	given to drinking
bosom-sermon	one learnt by heart and recited
bunchiness	quality of being bunchy, protuberance

C

calk	to calculate, reckon
cate	provisions, food
chastice	chastisement
cod	scrotum; testicle
cod's-head	stupid fellow, blockhead
competency	rivalry, competition
con	to know
consonance	pleasing combination of sounds
converser	talker, conversationalist
corpulency	material quality or substance, density
countenanced	having a (specified) countenance
counterposition	opposition, contraposition
couplement	couplet, stanza
cow-baby	timorous person, coward
cull out	select
curl-pated	curly head of hair

GLOSSARY

D

deck	to clothe in rich garments; to adorn
delineation	portrait, painting
divinesse	divinity; divine quality
do-nought	an idler, a good-for-nothing
draught	a move at chess or any similar game
of due	by right, by just title
Dutch	German

E

ear-sore	disagreeable to the ear
endamageance	harm, injury
enfranchise	to release from obligatory payments, legal liabilities
evangelic	pertaining to the Gospel

F

fain	glad, rejoiced
far stepped	far advanced
fardel	to make into a bundle
faring	journeying, travelling
fulsomeness	sickliness or offensiveness of savour

G

gaberdine	dress, covering
gainstrive	to oppose; to make resistance

H

hammer	to debate
to give hands	to agree

I

impay	reimburse, recoup
impertinent	unsuitable, unfitted for
incitement	inciting or rousing to action
inform	to impart life or spirit to a soul
irascible	irascible appetite

J

jadish	of the nature of, with the features of, a jade (i.e. horse)
jadishness	proper characteristics of a jade (i.e. horse)

L

leese	lose

M

manzo	beef

N

numberful	numerous, multitudinous

P

palmister	palmist
polishedness	smoothness, refinement
polishment	action of polishing, improving
prognosticative	that predicts a future event

R

regiment	particular course of diet, medication
reins	kidneys
ribaldry	worthless, of little value or use

S

school divinity	scholastic theology
schoolable	able to be schooled; capable of learning
searchful	full of anxious attention; diligent in search
seed-vessel	canal for reproductive bodily fluids
sightful	endowed with sight
signore	signor
sot	foolish, stupid person
spice	kind, or species

GLOSSARY

stone	testicle
sweeten	to persuade by flattery

T

tallage	tax, levy
three-tongued	trilingual
topical	pertaining to a topic or general maxim
trucking	bartering, trafficking, bargaining

U

unconcerted	unplanned, unintended
unincensed	not kindled

W

weet	to know
well-relished	savoury
winding	tortuous, crooked
wist	to know

NEOLOGISMS

This list presents the words whose first entries in the *OED* refer to Carew's *The Examination*. The meaning of the words with an ° is explained in the glossary. Words, usages, or definitions that precede or do not appear in the *OED* are marked with *.

actuate
aggrievedness
aidable
atrabile
avouchable
behooful*
bibbing°
bunchiness°
capableness*
capricious*
competence*
competency°
continuedness
converser°
countenanced°
counterposition°
cow-baby°
curl-pated°*
decidable
divinesse°
do-nought°
endamageance°
faring°
francolin
hidalgo
hoseless
hunger-starved*
impay°
incitement°

instrumentalize
malacia*
manzo°
misunderstood
Montagnese*
numberful°
philosophize
philosophizing
pleasantry
polishedness°
polishment°
predominately
prognosticative°
satirically
scarcen
schoolable°
searchful°
seed-vessel*
signore°
simplicist
superabounding
theorist
three-tongued°
transparence
truckling°
unconcerted°
unincensed°
well-relished°

BIBLIOGRAPHY

Primary Texts

Ascham, Roger, *The Scholemaster or Plaine and Perfite Way of Teachyng Children, to Vnderstand, Write, and Speake, the Latin Tong* (London: by Iohn Daye, [1570])

Bellamy, Edward, *The Tryal of Wits: Discovering the Great Difference of Wits among Men, and What Sort of Learning Suits Best with Each Genius* (London: Richard Sare, 1698)

Camden, William, *Remaines of a Greater Worke, Concerning Britaine* (London: G[eorge] E[ld], 1605)

Camilli, Camillo, *Essame de gl'ingegni de gli huomini per apprender scienze* (Venice: Aldus, 1586)

Carew, Richard, *The Examination of Men's Wits* (London: Adam Islip for Richard Watkins, 1594)

—— 'The Excellencie of the English Tongue', in William Camden, *Remaines, Concerning Britaine* (London: John Legatt, 1614), F2^v–G2^v

—— 'XXVII. Richard Carew of Anthony to Sir Robert Cotton', in *Original Letters of Eminent Literary Men: Of the Sixteenth, Seventeenth, and Eighteenth Centuries*, ed. by Sir Henry Ellis (London: John Bowyer Nichols and Son, 1843), pp. 99–100

—— (son), 'The True and Ready Way to learn the Latine Tongue: Expressed in an Answer to a Quere, Whether the ordinary Way of teaching Latine by the Rules of Grammar, be the best Way for youths to learn it? By the late Learned and Judicious Gentleman Mr. Richard Carew of Anthony in Cornwall', in Samuel Hartlib, *The True and Readie Way to Learne the Latine Tongue* (London: R. and W. Laybourn, 1654), G3^r–H1^r

Eliot, John, *Ortho-epia Gallica: Eliots Fruits for the French* (London: Richard Field, 1593)

Feijoo, Benito Jerónimo, 'Carta XXVIII. Del descubrimiento de la circulación de la sangre, hecho por un albeytar español', in *Cartas eruditas, y curiosas: en que por la mayor parte se continúa el designio del teatro critico universal, impugnando, ò reduciendo à dudosas varias opiniones comunes*, IV vols (Madrid: Imprenta Real de la Gazeta, 1774), III, pp. 314–23

Florio, John, *A Worlde of Wordes, or Most Copious, and Exact Dictionarie in Italian and English* (London: Arnold Hatfield, 1598)

Guazzo, Stefano, *The Court of Good Counsell Wherein is Set Downe the True Rules, How a Man Should Choose a Good Wife from a Bad, and a Woman a Good Husband from a Bad* (London: Raph Blower, 1607)

Huarte de San Juan, Juan, *Examen de ingenios para las ciencias* (Baeza: Juan Bautista de Montoya, 1575)

Jonson, Ben, *Every Man out of his Humor* (London: [P. Short], 1600)

—— *The Workes of Benjamin Jonson. The Second Volume* (London: [John Beale, James Dawson, Bernard Alsop and Thomas Fawcet], 1640 [1641])

Lyly, John, *Euphues the Anatomie of Wit* (London: William Leake, 1606)

Marston, John, *Parasitaster, or The Fawne* (London: Thomas Purfoot, 1606)

Mulcaster, Richard, *Positions Concerning the Training up of Children*, ed. by William Barker (Toronto; London: University of Toronto Press, 1994)

Oudin, César, *A Grammar Spanish and English: Or A Briefe and Compendious Method, Teaching to Reade, Write, Speake, and Pronounce the Spanish Tongue* (London: John Haviland, 1622)

Percyvall, Richard, *Bibliotheca Hispanica: Containing a Grammar;*

with a Dictionarie in Spanish, English, and Latine (London: John Jackson, 1591)

Quevedo, Francisco de, *España defendida y los tiempos de ahora: de la calumnias de los noveleros y sediciosos*, ed. by R. Selden Rose (Madrid: Fortanet, 1916)

Spelman, Sir Henry, 'Henricus Spelmannus Richardo suo Careo viro praestanti Sal. P.D.', in *The English Works of Sir Henry Spelman Kt. Published in his Life-Time* (London: D. Browne, 1723), pp. 37–38

Thomas, Thomas, *Dictionarium linguae latinae et anglicanae* (London: Richard Boyle, 1587)

Thomas, William, *Principal Rules of the Italian Grammer with a Dictionarie for the Better Understandyng of Boccace, Petrarcha, and Dante* (London: Thomas Berthelet, 1550)

Vives, Juan Luis, *On Education: A Translation of the* De tradendis disciplinis *of Juan Luis Vives*, ed. by Foster Watson (Cambridge: Cambridge University Press, 1913)

SECONDARY TEXTS

Abellán, José Luis, 'Los médicos-filósofos: Juan Huarte de San Juan, Miguel Sabuco y Francisco Vallés', in *Historia crítica del pensamiento español: 2. La edad de oro: Siglo XVI* (Madrid: Espasa-Calpe, 1986), pp. 207–22

Allison, A. F., *English Translations from the Spanish and Portuguese to the Year 1700* (London: Dawsons, 1974)

Alonso Muñoyerro, Luis, *La Facultad de medicina en la Universidad de Alcalá de Henares* (Madrid: CSIC; Instituto Jerónimo Zurita, 1945)

Álvarez, María E., *La Universidad de Baeza y su tiempo: 1538–1824* (Jaén: Instituto de Estudios Jiennenses, 1958)

Arquiola, Elvira, 'Consecuencias de la obra de Huarte de San Juan en la Europa moderna', *Huarte de San Juan*, 1 (1989), 15–28

Arranz Lago, David F., 'Sobre la influencia del *Examen de ingenios* en Cervantes. Un tema revisitado', *Castilla: Estudios de Literatura*, 21 (1996), 19–38

Arrizabalaga, Jon, 'Filosofía natural, psicología de las profesiones y selección de estudiantes universitarios en la Castilla de Felipe II: la obra y el perfil intelectual de Juan Huarte de San Juan', *Huarte de San Juan*, 1 (1989), 29–58

—— 'The World of Iberian Converso Practitioners, from Lluís Alcanyís to Isaac Cardoso', in *Más allá de la Leyenda Negra: España y la Revolución Científica / Beyond the Black Legend: Spain and the Scientific Revolution*, ed. by Víctor Navarro and William Eamon (Valencia: Universidad de Valencia, 2007), pp. 307–322

Barker, William, 'Introduction', in Richard Mulcaster, *Positions Concerning the Training up of Children* (Toronto: University of Toronto Press, 1994), pp. xiii–lxxxvi

Baroja, Julio Caro, *Los judíos en la España Moderna y Contemporánea*, III vols (Madrid: Istmo, 1986)

Bartra, Roger, *Cultura y melancolía: las enfermedades del alma en la España del Siglo de Oro* (Barcelona: Anagrama, 2001), trans. by Christopher Follet, *Melancholy and Culture: Essays on the Diseases of the Soul in Golden Age Spain* (Cardiff: University of Wales Press, 2008)

Bayle, Pierre, *Dictionnaire historique et critique*, XVI vols (Paris: Desoer, 1820), VIII

Biurrun Lizarazu, José Javier, 'Huarte de San Juan: vida y obra en el contexto político y religioso de la España del siglo XVI', *El Basilisco: Revista de filosofía, ciencias humanas, teoría de la ciencia y de la cultura*, 21 (1996), 16–17

Brand, Charles Peter, *Torquato Tasso, A Study of the Poet and of his Contribution to English Literature* (Cambridge: Cambridge University Press, 1965)

Bullock, Walter L., 'Carew's Text of the *Gerusalemme Liberata*', *PMLA*, 45.1 (1930), 330–35

Calatayud Buades, Luis. *La pedagogía y los clásicos españoles. Simón de Abril, Huarte de San Juan, Saavedra Fajardo, Sabuco de Nantes* (Madrid: Imprenta de Juan Pueyo, 1925)

Carpintero, H., 'Huarte de San Juan: Reflexiones en torno a su influencia en la psicología española', *Investigaciones Psicológicas*, 11 (1992), 9–20

Castelli, Alberto, *La Gerusalemme liberata nella Inghilterra di Spenser* (Milan: Vita e Pensiero, 1936)

Castro, Américo, *La realidad histórica de España* (México: Porrúa, 1954)

—— *El pensamiento de Cervantes* (Barcelona: Noguer, 1972)

Charles-Daubert, F., 'Charron et l'Angleterre', *Recherches sur le XVIIe Siècle*, 5 (1982), 53–56

Chersa, Tommaso, *Degli illustri Toscani stati in diversi tempi a Ragusa* (Padua: tip. della Minerva, 1828)

Chomsky, Noam, *Cartesian Linguistics: A Chapter in the History of Rationalist Thought* (New York: Harper and Row, 1966)

—— *Language and Mind* (New York: Harcourt, Brace and World, 1968)

Clark, Henry C., *La Rochefoucauld and the Language of Unmasking in Seventeenth-Century France* (Geneva: Librairie Droz, 1994)

Cooper, J. P. D., 'Godolphin, Sir William (*C*.1518–1570)', in *Oxford Dictionary of National Biography* (Oxford University

Press, 2004), online edition, Jan 2008 <http://www.oxforddnb.com/view/article/67867> [accessed 22 June 2013]

Corpas Mauleón, Juan Ramón, 'Introducción', in Juan Huarte de San Juan, *Examen de ingenios para las ciencias* (Pamplona: Ediciones y Libros, 2003), pp. 9–40

Crane, Willian G., *Wit and Rhetoric in the Renaissance* (New York: Columbia University, 1937)

Cunin, Muriel, 'Corps humain, corps politique, corps poétique: *The Purple Island or The Isle of Man* de Phineas Fletcher (1633)', *Études Anglaises*, 63.3 (2010), 274–88

Dear, Peter, *Revolutionizing the Sciences: European Knowledge and its Ambitions, 1500–1700* (Basingstoke: Palgrave, 2001)

DeMolen, Richard L., 'Richard Mulcaster's Philosophy of Education', *Journal of Medieval and Renaissance Studies*, 2 (1972), 69–91

Dieckhöfer, Klemens, 'Juan Huarte de San Juan, precursor de la psicología diferencial', *XXVII Congreso Internacional de Historia de la Medicina*, II vols (Barcelona: Acadèmia de Ciències Médiques de Catalunya i Balears, 1981), I, 64–68

Dodge, R. E. N., 'The Text of the *Gerusalemme Liberata* in the Versions of Carew and Fairfax', *PMLA*, 44.3 (1929), 681–695

Duché-Gavet, Véronique, ed., *Juan Huarte au XXIe siècle. Actes de Colloque International Juan Huarte au XXIe siècle (2003. Saint Jean Pied de Port, Basse Navarre)* (Anglet: Atlantica, 2003)

Escobar, Fernando, *Huarte de San Juan y Cervantes en la locura de D. Quijote de la Mancha. Breve estudio clínico psico-somático* (Granada: Universidad de Granada, 1949)

Figuerido, César A., *La orientación profesional y el médico navarro Juan de Huarte* (Bilbao: Casa Dochao, 1930)

Franzbach, Martin, 'La influencia del *Examen de ingenios para las ciencias* (1575) de Juan Huarte de San Juan en Alemania', *Medicina Española*, 62 (1969), 450–56

—— *Lessings Huarte-Ubersetzung (1752): Die Rezeption und Wirkungsgeschichte des "Examen de ingenios para las ciencias" (1575) in Deutschland* (Hamburg: Cram, de Gruyter, 1965), trans. by Luis Ruiz Hernández, *La traducción de Huarte por Lessing (1752): recepción e historia de la influencia del "Examen de ingenios para las ciencias" (1575) en Alemania* (Pamplona: Diputación Foral de Navarra, Institución Príncipe de Viana, 1978)

Furdell, Elizabeth Lane, *Publishing and Medicine in Early Modern England* (Rochester: University of Rochester Press, 2002)

Gambin, Felice, 'Sobre la recepción y la difusión del Examen de ingenios para las ciencias de Huarte de San Juan en Italia', in *Filosofía y literatura en el mundo hispánico. Actas del IX Seminario de Historia de la Filosofía Española e Iberoamericana, Salamanca del 26 al 30 de septiembre de 1994*, ed. by Antonio Heredia Soriano and Roberto Albares Albares (Salamanca: Universidad de Salamanca, 1997), pp. 409–25

—— *Azabache. El debate sobre la melancolía en la España de los siglos de Oro*, trans. by Pilar Sánchez Otín (Madrid: Biblioteca Nueva, 2008); *Azabache: Il dibattito sulla malinconia nella Spagna dei Secoli d'Oro* (Pisa: ETS, 2005)

—— 'Il gesuita e il medico: le annotazioni alla traduzione italiana dell'*Examen de ingenios para las ciencias* di Juan Huarte de San Juan', in *Malattia e scrittura. Saperi medici, malattie e cure nelle letterature iberiche*, ed. by Silvia Monti (Verona: Cierre Grafica, 2012), pp. 147–83

García García, E. and A. Miguel Alonso, 'El *Examen de ingenios* de Huarte de San Juan en la *Bibliotheca selecta* de Antonio Possevino', *Revista de Historia de la Psicología*, 3–4 (2003), 387–997

—— 'El Examen de Ingenios de Huarte en Italia. La anatomia

ingeniorum de Antonio Zara', *Revista de historia de la psicología*, 25.4 (2004), 83–94

García Vega, Luis, 'El antifeminismo científico de Juan Huarte de San Juan, Patrón de la Psicología', *Revista de Psicología General y Aplicada*, 42, 1989, 533–42

—— and José Moya Santoyo, *Juan Huarte de San Juan: patrón de la psicología española* (Madrid: Ediciones Académicas, 1991)

Garrido Palazón, Manuel, 'El *Examen de ingenios para las ciencias* de Huarte de San Juan y el enciclopedismo retórico y didáctico de su tiempo', *Revista de Literatura*, 61.122 (1999), 349–73

Gondra Rezola, José María, 'Juan Huarte de San Juan y la eugenesia', in *Criminología y derecho penal al servicio de la persona: libro-homenaje al profesor Antonio Beristain*, ed. by J. L. Cuesta, I. Dendaluze and E. Echeburua (San Sebastián: Instituto Vasco de Criminología, 1989), pp. 199–210

—— *Huarte de San Juan: precursor de la moderna psicología de la inteligencia. Lección inaugural del curso academico 1993–1994* (Vitoria: Universidad del País Vasco, 1993)

González Manjarrés, Miguel Ángel, 'Introducción', Giovan Battista della Porta, *Fisiognomía*, II vols (Madrid: Asociación Española de Neuropsiquiatría, 2007–2008), I, pp. 7–19

Gracia Guillén, Diego, 'Judaismo, medicina y mentalidad inquisitorial en la España del siglo XVI', in *Inquisición española y mentalidad inquisitorial*, ed. by Ángel Alcalá (Barcelona: Ariel, 1984), pp. 328–52

Granda, Antonio de la, 'Juan Huarte de San Juan y Francisco Villarino', in *Estudios de Historia Social de España*, IV vols ed. by Carmelo Viñas y Mey (Madrid: CSIC; Instituto Balmes de Sociología, 1949), I, pp. 655–69

Granjel, Luis S., *Humanismo y Medicina* (Salamanca: Universidad de Salamanca, 1968)

—— *Los médicos humanistas españoles. Capítulos de la medicina española* (Salamanca: Universidad de Salamanca, 1971), pp. 13–29

—— *La medicina española renacentista* (Salamanca: Universidad de Salamanca, 1980)

—— *Juan Huarte y su "Examen de ingenios"* (Salamanca: Real Academia de Medicina, 1988)

Green, Otis H., 'El Licenciado Vidriera: Its Relation to the *Viaje del Parnaso* and the *Examen de Ingenios* de Huarte', *Linguistic and Literary Studies in Honor of Helmut A. Hatzfeld*, ed. by Alessandro S. Crisafulli (Washington: Catholic University Press, 1964), pp. 213–20

Halka, Chester S., 'Don Quijote in the Light of Huarte's *Examen de ingenios*: A Reexamination', *Anales Cervantinos*, 19 (1981), 3–13

Halliday, Frank Ernest, 'Introduction', in Richard Carew, *The Survey of Cornwall* (London: Adams & Dart, 1969), pp. 15–71

Higueras Quesada, María Dolores, 'Estudio sobre la evolución de la población de Baeza, 1550–1750', *Boletín del Instituto de Estudios Giennenses*, 176.1 (2000), 141–94

Huerga Teruelo, Álvaro, *Los alumbrados de Baeza* (Jaén: Instituto de Estudios Giennenses. Diputación Provincial, 1978)

Hull, P. L., 'Richard Carew's Discourse about the Duchy Suit (1594)', *Journal of the Royal Institution of Cornwall*, 4 (1962), 181–251

Hutchings, C. M., 'The Examen de ingenios and the Doctrine of Original Genius', *Hispania: A Journal Devoted to the Teaching of Spanish and Portuguese*, 19.2 (1936), 273–82

Iriarte, Mauricio de, *El ingenioso hidalgo y El examen de ingenios: (qué debe Cervantes al Dr. Huarte de San Juan)* (San Sebastián:

Separata de Revista Internacional de los Estudios Vascos, 24.4, 1933)

—— *El doctor Huarte de San Juan y su 'Examen de ingenios': contribución a la historia de la psicología diferencial* (Madrid: CSIC, 1948)

Jannaco, Carmine, *Storia letteraria d'Italia*, Il Seicento (Milan: Vallardi, 1986)

Kagan, Richard L., *Students and Society in Early Modern Spain* (London: Johns Hopkins University Press, 1974)

Kogel, Renée, *Pierre Charron* (Geneva: Librairie Droz, 1972)

Lea, Kathleen M. and T. M. Gang, eds., *Godfrey of Bulloigne: A Critical Edition of Edward Fairfax's Translation of Tasso's Gerusalemme Liberata* (Oxford: Clarendon Press, 1981)

Levin, Harry, ed., *Veins of Humor* (Cambridge: Harvard University Press, 1972)

Lewis, C. S., *Studies in Words* (Cambridge: Cambridge University Press, 1960)

López Piñero, José María, *Ciencia y técnica en la sociedad española de los siglos XVI y XVII* (Barcelona: Labor, 1979)

—— *Diccionario histórico de la ciencia moderna en España*, II vols (Barcelona: Península, 1983)

Mallart, José, 'Huarte y las modernas corrientes de ordenación profesional y social', in *Estudios de Historia Social de España*, IV vols, ed. by Carmelo Viñas y Mey (Madrid: CSIC; Instituto Balmes de Sociología, 1952), II, pp. 113–151

Maravall, José Antonio, *Estado moderno y mentalidad social: (siglos XV a XVII)* (Madrid: Revista de Occidente, 1972)

Márquez, Antonio, *Los alumbrados: orígenes y filosofía (1525–1559)* (Madrid: Taurus, 1980)

Martín-Araguz, A. and C. Bustamante-Martínez, 'Examen de ingenios, de Juan Huarte de San Juan, y los albores de la Neurobiología de la inteligencia en el Renacimiento español', *Revista de Neurología*, 38.12 (2004), 1176–85

Martín Ferreira, Ana Isabel, *El humanismo médico en la universidad de Alcalá (Siglo XVI)* (Alcalá de Henares: Universidad de Alcalá de Henares, 1995)

Mauriac, Pierre, 'Montesquieu connaissait Jean Huarte ...', *Figaro Littéraire*, 8 (1959), 8

McCabe, Richard, 'Wit, Eloquence, and Wisdom in *Euphues: The Anatomy of Wit*', *Studies in Philology*, 81.3 (1984), 299–324

McCrea, Brian, 'Madness and Community: Don Quixote, Huarte de San Juan's *Examen de ingenios* and Michel Foucault's *History of Insanity*', *Indiana Journal of Hispanic Literatures* 5 (1994), 213–24

McRae, Robert, 'The Unity of the Sciences: Bacon, Descartes, and Leibniz', *Journal of the History of Ideas*, 18.1 (1957), 27–48

Mehnert, Hennig, 'Der Begriff "ingenio" bei Juan Huarte und B. Gracián', *Romanische Forschungen*, 91.3 (1979), 270–80

Mendyk, S., 'Carew, Richard (1555–1620)', in *Oxford Dictionary of National Biography* (Oxford University Press, 2004) <http://www.oxforddnb.com/view/article/4635> [accessed 18 May 2013]

Menéndez y Pelayo, Marcelino, *La ciencia española* (Madrid: Librería General de Victoriano Suárez, 1933)

Miranda, T., 'Sánchez de las Brozas, Huarte de San Juan y la gramática generativa', *Alcántara*, 15 (1988), 7–31

Mondéjar, José, 'El pensamiento lingüístico del doctor Juan Huarte de San Juan', *Revista de Filología Española*, 64.1–2 (1984), 71–128

Moral, B. del., 'Estudio comparativo del "Ingenio" en Luis Vives y Huarte de San Juan', *Analecta Calasantiana*, 35–36 (1976), 65–143

Morini, Massimiliano, *Tudor Translation in Theory and Practice* (Aldershot: Ashgate, 2006)

Moya, J. and García Vega, L. 'Juan Huarte de San Juan: Padre de la Psicología Diferencial', *Revista de Historia de la Psicología*, II, 1990, 123–58

Müller, Cristina, *Ingenio y melancolía: una lectura de Huarte de San Juan*, trans. by Manuel Talens and María Pérez Harguindey (Madrid: Biblioteca Nueva, 2002)

Nash, Ralph, 'On the Indebtedness of Fairfax's Tasso to Carew', *Italica*, 34.1 (1957), 14–19

Nixon, S., 'Mr. Carew and Brawn the Beggar', *Notes and Queries*, 47.2 (2000), 180–81

O'Neill, María, 'Juan Luis Vives and Richard Mulcaster: A Humanist View of Language', *Estudios Ingleses de la Universidad Complutense*, 6 (1998), 161–75

Pastore, Renato, 'Camilli, Camillo', in *Dizionario Biografico degli Italiani* (Rome: Istituto Enciclopedia Italiana, 1974), pp. 210–12

Peltonen, Markku, ed., *The Cambridge Companion to Bacon* (Cambridge: Cambridge University Press, 1996)

Pender, Stephen, 'Introduction. Reading Physicians', in *Rhetoric and Medicine in Early Modern Europe*, ed. by Stephen Pender and Nancy S. Struever (Farnham & Burlington: Ashgate, 2012)

Pérez Fernández, José María, 'Andrés Laguna: Translation and the Early Modern Idea of Europe', *Translation and Literature*, 21 (2012), 299–318

Pérouse, Gabriel-André, *L'Examen des esprits du docteur Juan*

Huarte de San Juan. Sa diffusion et son influence en France aux XVIe et XVIIe siècles (Paris: Les Belles lettres, 1970)

—— 'Les Femmes et l'Examen des esprits du Dr Huarte (1575)', in *Les représentations de l'autre du Moyen Age au XVIIe siècle*, ed. by Evelyne Berriot-Salvadore (Saint-Etienne: Université de Saint-Etienne, 1995), pp. 273–83

—— 'Montaigne et le Dr Huarte: Avec un mot sur Pierre Charron', *Bulletin de la Société des Amis de Montaigne*, 8. 13–14 (1999), 11–22

Peset, José Luis and Elena Hernández Sandoica, *Estudiantes de Alcalá* (Alcalá de Henares: Ayuntamiento de Alcalá de Henares, 1983)

Peset, José Luis, *Genio y desorden* (Valladolid: Cuatro, 1999)

—— 'Las críticas a la Universidad de Juan Huarte de San Juan', in *La universidades hispánicas: de la monarquía de los Austrias al centralismo liberal*, II vols, ed. by Luis E. Rodríguez-San Pedro Bezares (Salamanca: Universidad de Salamanca, 2000), I, pp. 390–91

—— *Las melancolías de Sancho: humores y pasiones entre Huarte y Pinel* (Madrid: Asociación Española de Neuropsiquiatría, 2010)

Peset Llorca, Vicente, 'Tres figuras de la psicología médica del Renacimiento español (Vives, Huarte, Sabuco de Nantes)', in *La historia de la psicología y de la psiquiatría en España: desde los más remotos tiempos hasta la actualidad*, ed. by J. B. Ullersperger (Madrid: Alhambra, 1954), pp. 166–78

Pujante, David, 'La melancolía hispana, entre la enfermedad, el carácter nacional y la moda social', *Revista de la Asociación Española de Neuropsiquiatría*, 28.102 (2008), 401–18

Read, Malcolm K. and J. Trethewey, 'Juan Huarte and Pierre de Deimier: Two Views of Progress and Creativity', *Revue de Littérature Comparée*, 51.1 (1977), 40–54

—— *Juan Huarte de San Juan* (Boston: Twayne Publishers, 1981)

—— 'Ideologies of the Spanish Transition Revisited: Juan Huarte de San Juan, Juan Carlos Rodríguez, and Noam Chomsky', *Journal of Medieval and Early Modern Studies*, 34.2 (2004), 309–43

Riera, Juan, 'La literatura científica en el Renacimiento', in *Ciencia, medicina y sociedad en el Renacimiento castellano*, ed. by Juan Riera, *et al.* (Valladolid: Universidad de Valladolid, 1989), pp. 5–17

Rodríguez-San Pedro Bezares, Luis Enrique and Juan Luis Polo Rodríguez, eds. *Historia de la Universidad de Salamanca. Volumen III. Saberes y confluencias* (Salamanca: Ediciones Universidad de Salamanca, 2006)

Rogers, Carmen, 'Introduction', in Juan Huarte de San Juan, *Examen de Ingenios: The Examination of Men's Wits (1594). A Facsimile Reproduction* (Gainesville: Scholars' Facsimiles & Reprints, 1959)

Ruiz Gil, Mª Luisa, 'Juan Luis Vives y Juan Huarte de San Juan. Esquema comparativo de su doctrina psicológico-pedagógica', *Perspectivas pedagógicas*, 6.16 (1965), 64–84

Sáez, Ricardo, 'La Baeza del siglo XVI y su imborrable presencia en la obra de Huarte de San Juan', *Huarte de San Juan*, 1 (1989), 81–95

Salillas, Rafael, *Un gran inspirador de Cervantes: El Doctor Juan Huarte y su 'Examen de ingenios'* (Madrid. Imprenta de Eduardo Arias, 1905)

Sandler Berkowitz, David, *John Selden's Formative Years: Politics and Society in Early Seventeenth-century England* (Washington and London: Folger Shakespeare Library; Associated University Presses, 1988)

Sanz Serrulla, Francisco Javier, 'El doctor Huarte de San Juan, médico y catedrático en Sigüenza. Aspectos biográficos inéditos', *Anales Seguntinos*, 1.3 (1986), 309–13

Schafer, Jurgen, 'Huarte: A Marston Source', *Notes and Queries*, 18 (1971), 16–17

Seidel Menchi, S., *Erasmo in Italia 1520–1580* (Torino: Bollati Boringhieri, 1990)

Serés, Guillermo, 'Introducción', in Juan Huarte de San Juan, *Examen de ingenios para las ciencias* (Madrid: Cátedra, 1989), pp. 21–129

—— 'El ingenio de Huarte y el de Gracián: Fundamentos teóricos', *Insula: Revista de Letras y Ciencias Humanas*, 655–656 (2001), 51–53

—— 'Possevino entre Huarte y Gracián: el cultivo del ingenio y la imaginación creativa', in *La traduzione della letteratura italiana in Spagna (1300–1939). Traduzione etradizione del testo. Dalla filologia all'informatica*, ed. by M. de las Nieves Muñiz Muñiz (Florence: Franco Cesati, 2007), pp. 429–42

Shaw, Patricia, 'Wits, Fittes and Fancies: Spanish ingenio in Renaissance England', *Estudios Ingleses de la Universidad Complutense*, 12 (2004), 131–148

Sieber, Harry, 'On Juan Huarte de San Juan and Anselmo's "locura" in "El curioso impertinente"', *Revista Hispánica Moderna*, 36.1–2 (1970–1971), 1–8

Simon, Joan, *Education and Society in Tudor England* (Cambridge: Cambridge University Press, 1967)

Sortais, Gaston, *La philosophie moderne depuis Bacon jusqu'à Leibniz*, II vols (Paris: P. Lethielleux, 1920–1922)

Taavitsainen, Irma and Päivi Pahta, 'Vernacularisation of Medical Writing in English: A Corpus-Based Study of Scholasticism', *Early Science and Medicine*, 3.2 (1998), 157–85

—— eds, *Early Modern English Medical Texts: Corpus Description and Studies* (Amsterdam/Philadelphia: John Benjamins Publishing Company, 2010)

Tomita, Soko, *A Bibliographical Catalogue of Italian Books Printed in England, 1558–1603* (Farnham: Ashgate Publishing, 2009)

Tonelli Olivieri, Grazia, 'Galen and Francis Bacon: Faculties of the Soul and the Classification of Knowledge', in *The Shapes of Knowledge: From the Renaissance to the Enlightenment*, ed. by Donald R. Kelley and Richard Henry Popkin (Dordrech: Kluwer Academic, 1991), pp. 61–81

Torre, Esteban, *Ideas lingüísticas y literarias del doctor Huarte de San Juan* (Sevilla: Publicaciones de la Universidad de Sevilla, 1977)

—— *Sobre lengua y literatura en el pensamiento científico español de la segunda mitad del siglo XVI: Las aportaciones de G. Pereira, J. Huarte de San Juan y F. Sánchez el Escéptico* (Sevilla: Universidad de Sevilla, 1984)

Trustees of the British Museum, eds. *Short-title Catalogue of Books Printed in Italy and of Italian Books Printed in Other Countries from 1465 to 1600 Now in the British Library* (London: British Library, 1986)

Vázquez Fernández, A., 'Un tratado de psicología diferencial para una selección y orientación profesionales en la España del siglo XVI', *Cuadernos Salmantinos de Filosofía*, 2.1 (1975), 185–216

Virués Ortega, Javier, 'Juan Huarte de San Juan in Cartesian and Modern Psycholinguistics: An Encounter with Noam Chomsky', *Psicothema*, 17.3 (2005), 436–40

Vleeschauwer, Herman Jean de, *Autour de la classification psychologique des sciences: Juan Huarte de San Juan, Francis Bacon, Pierre Charron, D'Alembert* (Pretoria: University of South Africa, 1958)

Walsham, Alexandra, 'Richard Carew and English Topography', in *The Survey of Cornwall*, ed. by John Chynoweth, Nicholas Orme and Alexandra Walsham (Exeter: Devon and Cornwall Record Society, 2004), pp. 17–41

Watson, Foster, 'Introduction', in Juan Luis Vives, *On Education: A Translation of the* De tradendis disciplinis *of Juan Luis Vives* (Cambridge: Cambridge University Press, 1913), pp. xvii–clvii

Wear, Andrew, *Knowledge and Practice in English Medicine, 1550–1680* (Cambridge: Cambridge University Press, 2000)

Webster, C., *The Great Instauration: Science, Medicine and Reform 1626–1660* (London: Duckworth, 1975)

Weinrich, Harald, 'Prólogo', in Juan Huarte de San Juan, *Examen de ingenios para las ciencias*, ed. by Guillermo Serés (Madrid: Círculo de lectores, 1996), pp. 9–16

Weiss, Roberto, 'Introduction', *Jerusalem Delivered. The Edward Fairfax Translation* (Carbondale: Southern Illinois University Press, 1962)

Wood, D. N. C., 'Ralph Cudworth the Elder and Henri Estienne's *World of Wonders*', *English Language Notes*, 11 (1973), 93–100

—— 'Elizabethan English and Richard Carew', *Neophilologus*, 61.2 (1977), 304–15

—— 'Gerusalemme liberata — Englished by Richard Carew', *Cahiers Elisabethains: Etudes sur la pre-Renaissance et la Renaissance anglaises*, 13 (1978), 1–13

Yamada, Yumiko, 'Ben Jonson and Cervantes: The Influence of Huarte de San Juan on Their Comic Theory', in *Shakespeare and the Mediterranean*, ed. by Tom Clayton, Susan Brock, Vicente Forés and Jill Levenson (Newark: University of Delaware Press, 2004), pp. 425–36

Zuili, Marc, 'Nuevas aportaciones sobre el hispanista francés César Oudin (1560?–1625)', *Thélème: Revista Complutense de Estudios Franceses*, 20 (2005), 203–11

INDEX

Aaron 219
Abraham 88, 153–55, 217–18, 224
Abril, Pedro Simón 1, 54, 355
Achior 230
Acosta, Cristóbal de 26
Adam 79, 111, 116, 119, 168, 264–65, 274, 283
Addison, Joseph 28
Agostini, Agostino 35
Ajax 262
Allott, Robert 53
Aldus 36, 351
Alfonso V, King 226
Alfonso VIII, King 249
Alfonso IX, King 249, 329
Alfonso X, King 246
alumbradismo 20
alumbrados 20–21, 359–60
Álvarez, Diego 19
Álvarez de Toledo, Fernando 160
Amaleck 269
amplification 2–3
Anacharsis 260
anatomy 16, 39, 70, 120, 133, 145, 279
Andrewes, Lancelot 56
Antonio, Nicolás 28
Apollo 132, 162
arbitristas 6
Argentier, John 208
Aristotle 3, 44, 48, 54, 81, 87–8, 92, 95, 97–8, 100–04, 106–07, 110–11, 115–17, 123–28, 130, 133–40, 142–43, 152, 157, 161, 167–68, 170, 181–88, 190, 197, 199, 203, 205, 208, 210–11, 218, 221–22, 227–28, 230, 233–34, 240, 243, 250–53, 258–60, 262, 273–74, 276–77, 280, 285–86, 288, 298, 301–03, 306–09, 311–13, 321, 336
Artaxerxes 170
Arthur, King 55
Ascham, Roger 57, 59–60, 351
Ávila, Juan de 14, 21, 35
Azpilcueta, Martín de 35

Bacon, Sir Francis 4, 7, 65–6, 361–62, 365–66
Baglivi, Giorgio 30
Baldini, Vittorio 33
Baldus 77
Barckley, Sir Richard 53
Bellamy, Edward 8, 10, 17–8, 23, 58–59, 351
Berteud (or Bertaud) 29
Bible 68
Blanche of Castile 249–50
Boaistuau, Pierre 53
Boccaccio (Boccace) 8, 43, 162, 332, 353
Borromeo, Federico 36
Boscán, Juan 43, 332–33
Bouchet, Guillaume 30
brain 3, 18, 44–45, 64, 76, 78, 80, 86, 100–06, 108–10, 112–13, 115–17, 120–21, 123–29, 131, 133–35, 137–48, 153, 158, 164–65, 167, 169, 175, 179, 182–83, 188, 201, 214–16, 218, 222, 238, 253–56, 261, 263–65, 275, 282–83, 287, 304–05, 308, 310, 318, 320, 324–26, 331
Bright, Timothy 26, 53
Buddeus, J. F. 30
Burton, Robert 53

Cagnacini 36
Cajetan 159

INDEX

Camden, William 11, 13, 37–8, 54–55, 57, 351
Camilli, Camillo 8, 17, 24, 32, 35–37, 41–44, 47, 113, 329–34, 336–38, 341, 343, 351, 362
Carballo, Luis Alfonso de 27
Carew, George 11, 31
Carew, Juliana 11
Carew, Richard 8–13, 23, 31–44, 47, 49–54, 56–8, 64, 66–67, 113, 163, 184, 202, 247, 258, 271–72, 275, 278–79, 281, 329–34, 336–38, 340–41, 343–44, 350–51, 355–56, 359, 361–62, 367
Carew, Richard (son) 11–12, 351
Carew, Thomas 11
Carlos, Prince Don 15, 244, 341
Castro, Rodrigo de 7
Catholic monarchs 338–39
Cato the elder 171, 276
Cavalcalupo, Domenico 33
Cervantes, Miguel de 7, 26–27, 56, 339, 354–56, 359, 364, 367
Chappuys, Gabriel 24, 29
character 59, 98, 147, 260, 339, 346, 348
Charles V, Emperor 6, 14, 215, 338, 341, 343
Charron, Pierre 30, 64–66, 355, 360, 363, 366
choler adust 143, 188, 190, 217–18, 220, 222, 233–34
Chomsky, Noam 2, 355, 364, 366
Chrisippus 230
Christ 22, 48–49, 51, 150–51, 153, 171, 176, 181, 184–85, 189, 231, 236, 240, 247, 248, 252, 264, 269, 271–73, 278, 322–25, 335–36, 345
Cicero 42, 54, 83–85, 91, 113, 117–18, 144–45, 156, 158–59, 161, 165, 172–73, 175–82, 198, 205, 228, 230, 237–38, 310

Cisneros, Francisco Jiménez de, Cardinal 329
Cleantes 84
convert 14, 335, 338–39, 341
copia 3
Cotton, Sir Robert 38–39, 56–57, 351
Council of Trent 22, 160
Counter-Reformation 10
Cratippus 83
Cruz, Juan de la 20
Cudworth, Ralph, the elder 12, 367
Cueva, Juan de la 27
Culpeper, Nicholas 62

d'Alquié, François-Savinien 24
Dalibray, Charles Vion 24
Damagetus 144, 170
David 50–1, 94, 149–50, 152, 190, 251, 266, 269–71
Davies, Sir John 53
Deimier, Pierre de 30, 363
Democritus 93, 142, 144
Demosthenes 84, 145, 165, 258, 260
Descartes 2, 30, 65–6, 361
determinism 18
Diamante, Juan Bautista 333
Díaz Rengifo, Juan 27
Diezinger 30
Diogenes Cynicus 259
Divinity 19, 45–47, 76, 155–56, 158–60, 162–64, 167, 172, 174–75, 183–84, 266, 325, 327, 347–48
Drummond, William 333
Du Laurens, André 26
Durand 159

Einem, Johann Justus von 30
Elbora 284
Elias 220
Elijah 220, 269

369

Eliot, John 332–33, 351
Elyot, Sir Thomas 53, 60
Enríquez, Enrique Jorge 7, 339
Erasmus 3, 334
Esay 125, 171–72, 323
Espina, Alonso de 338
Estienne, Henri 12, 367
eugenesis 52, 61
Eve 79, 151, 185, 283, 286

Fairfax, Edward 34, 40–41, 356, 360, 362, 367
Feijoo, Benito Jerónimo 28–29, 352
Fletcher, Phineas 7, 53, 356
Florio, John 8–9, 352
Franceschi, Francesco de 32, 36
Fuentes, Alonso de 54

Gagliardelli, Domenico 30
Gale, Thomas 39, 53
Galen 15, 42, 44–45, 52–54, 66, 77, 87, 89–90, 98–101, 103–05, 107–10, 120–21, 123–26, 129, 132–35, 137, 139–40, 144–52, 166–67, 181, 193, 200, 208–10, 214, 218, 220–21, 230, 236, 253, 257–58, 260–63, 266–68, 271, 274, 277, 279, 282–83, 285–87, 289–90, 293, 295, 298, 300–07, 311–13, 316, 318, 320, 325, 330, 366
Galileo 65
Garcilaso de la Vega 333
Garve 30
Gentili, Scipione 34
Gil, Juan (doctor Egidio) 15
Godolphin, Sir Francis 31–32, 36–37, 54, 75, 355
Goethe 30
Goliath 243, 251
Gonzaga, Scipione 36
Gracián, Baltasar 7, 27, 31, 361, 365

Granada, Luis de 35
Gratii, Salustio 24, 41, 334
Guazzo, Stefano 62–63, 352
Guibelet, Jourdain 30
Guillemeau, Jacques 62
gipsy 226, 338

Hamann 30
Hannibal 232–33, 237
Harvey 65
Hatfield, Arnold 9, 34, 352
Heraclitus 124, 144, 203, 304, 325
Hercules 63, 84
Herder 30
Hereford, John Davies of 53
Herod 235, 323
Herrera, Hernando de 21
Heylyn, Peter 53
Hippocrates 15, 44, 87, 91, 93, 98, 100, 107–09, 116, 125, 128, 131, 141, 144, 170, 193, 222–23, 234, 258, 260–61, 267, 277, 279, 281–82, 289–91, 295–97, 299–303, 308, 311–12, 317, 319–23, 332
Holofernes 229–30
Holy Scripture 22, 47, 150, 173, 176, 192, 219–20, 229, 246, 265, 270, 273, 297, 325
Homer 119, 259–60, 262, 319
Horace 172
Huarte de San Juan, Juan 1–7, 9–10, 13–32, 34–35, 37–9, 41–47, 49–54, 56–59, 61–67, 73, 220, 246, 329–32, 334–36, 338–41, 352–67
Huarte, Luis 17, 22
humour 3–4, 55–56, 105, 125–26, 133, 135, 138, 140–41, 144, 149, 151, 180, 182, 188–89, 209, 211–12, 216–17, 261, 293, 295, 310, 322–23
Hunt, Christopher 32, 34–35

INDEX

Jesse 269–70

imagination 2–3, 6, 44–47, 50, 54–55, 66, 68, 78, 92, 102, 104–05, 118–20, 122, 128–30, 133–37, 139–44, 156–57, 160–64, 166–70, 172–75, 177–83, 185–91, 193–94, 199, 201, 205–07, 210–09, 222, 230–31, 233–38, 240–41, 248, 253–58, 263, 266, 268, 275, 284, 287, 305–08, 311, 315, 317, 324, 327, 334
imitation 3, 159, 179, 193, 243
Imperial, Juan 30
ingegni 8, 10, 24, 26, 32, 36, 351
ingenio 1–2, 4, 6, 8–9, 15, 19–22, 24–31, 37, 41, 49, 59, 65–66, 331, 334, 339, 341, 352, 354, 356–62, 364–65, 367
ingenium 8
Inquisition 7, 14, 17–21, 24, 174, 235, 334, 338–39, 341
Islip, Adam 8, 31, 35, 73, 351

Jacob 118–19, 307
Jaggard, John 34
James, St 145
Jeremiah 171–72
Jerome, St 171, 274
Jews 51, 150, 171, 215–17, 220, 229, 271, 273, 322, 335, 338–39
Jimeno, Pedro 16
Job 94–5
Jonas, Richard 62
Jonson, Ben 39, 55–57, 352, 367
Josephus 150, 215, 273
Judith 229, 284
Joshua 334
Julius Caesar 236–39, 310
Jupiter 220
Juvenal 161, 164, 183

Kant 30

Laguna, Andrés 339, 362
Lavater, J. C. 30
Lazarus 153–55
Lemnius, Levinus 53
Lennard, Samson 65
Leontia 259
Lessing, Gotthold Ephraim 7, 24–25, 357
Lewis, C. S. 8, 360
libre examen 22
limpieza de sangre 7, 339, 341
literature 6, 7, 20, 27, 41, 54, 57, 68–70, 338–39, 355, 361–62
Lobera de Ávila, Luis 54
López, Rodrigo 14
López de Corella, Alonso 54
López Pereira 339
López Pinciano, Alonso 27
López Villalobos, Francisco 339
Lownes, Matthew 34
Lucius Florus 236
Luke, St 154
Lyly, John 61, 352

Macrobius 237
Major, Aescatius (Joachim Caesar) 24
Malaspina, Marcantonio 33
Man, Thomas 35
Manilius 276
Marsili, Alessandro 33
Marston, John 63–46, 352, 365
Mary, St 113
Matos Fragoso, Juan de 333
Matthew, St 79, 150, 235
medical humanism 15–16
Medici, Cardinal Ferdinando de 35
medicine 3–4, 6–7, 15–17, 25, 31, 39–40, 46, 68–70, 146, 149, 211–13, 228, 339, 341, 357, 362, 365, 367
melancholy 26–7, 53, 55, 113, 125–26, 128, 135, 140–41,

143–45, 188–89, 222, 233, 293, 332, 354
memory 3, 6, 9, 44–47, 50, 66, 68, 87, 90, 96, 100, 102, 104–05, 112, 118–20, 122–23, 126–30, 133–35, 137, 139–42, 144, 153, 156–62, 164–67, 169–73, 175, 177–79, 181–84, 187–88, 190–95, 197–99, 205, 207, 209–11, 213–14, 236, 238, 245, 250, 253–55, 258, 263, 266, 268, 275, 278, 288, 292, 304–06, 324, 327, 334
Mena, Fernando 15
Mendelssohn 30
Mercado, Luis 26
Mercado, Pedro 54
Meres, Francis 3
Merola, Jerónimo 7
method 4, 16, 46, 59, 65–66, 89, 208, 352
Mexía, Pedro 54
midwifery 61, 70
Monardes, Nicolás 26
Montaigne 29–30, 363
Montemayor, Jorge de 162
Montesquieu 30, 361
Montoya, Juan Bautista de 5, 19, 22, 352
Moses 151, 219–21, 299
Mulcaster, Richard 57, 59–61, 352, 354, 356, 362

natural philosophy 25–6, 36, 53, 60, 66, 88, 95, 106, 115, 117–18, 121, 123, 136–37, 147, 149, 154, 175, 184, 218, 224–25, 243, 265, 271, 277, 290, 297, 309
nature 4–6, 16, 19, 42, 45, 50–51, 53, 56–57, 59, 60, 62–3, 66, 73, 76–80, 83, 85–7, 90–93, 95–104, 106–12, 116–19, 121–22, 128, 130, 132–33, 136, 150–52, 154, 156, 163–64, 166–68, 172, 175, 178, 181, 183, 186–87, 190–91, 199, 201, 205, 207–09, 219–20, 223, 225, 228, 233–34, 238, 241, 245, 248, 253, 256–59, 261, 263–71, 273–77, 279–83, 286, 288, 293–96, 301–02, 311–13, 316–18, 321, 324, 327, 346, 348
Nebrija (or Lebrija), Antonio de 15, 178, 250
new Christian 21, 338–39
New Testament 95, 330

Old Testament 171, 330
Orlando 162
Osanna 32, 36
Osias 284
Oudin, César 65, 352, 367
Ovid 35

Paul, St 48, 79, 118, 145, 170–72, 176, 184–85, 190, 221, 283, 329, 336
pedagogy 1
Pelagius (Pelayo) 245
Pendasio, Federico 36, 41
Percyvall, Richard 9, 352
Pereira, Gómez 27, 54, 366
Pérez de Herrera, Cristóbal 339
Peter, St 113
Philip II, King 5–6, 15, 17, 50, 76, 244, 275
Philip IV, King 14
philosophy 4, 7, 18, 25–6, 36, 46, 53, 60, 65–66, 69, 76–77, 81, 83–84, 88, 95, 106, 115, 117–18, 121, 123, 133, 136–37, 147, 149, 154, 162–63, 167, 175, 184, 186, 210, 218, 220, 224–25, 228, 238, 243, 259, 265–66, 271, 277, 279, 290, 297, 307, 309, 356

INDEX

Pindar 125
Plato 44, 54, 78, 84–85, 87, 98, 100–01, 103, 105, 107–08, 110–11, 116, 118, 121, 124, 130, 134, 139, 145–48, 161, 164, 168, 170, 173, 180, 194, 197–98, 215–16, 218, 230, 233–34, 248, 256, 258, 260, 262, 267, 273–74, 279, 296, 302, 310–11, 316, 318
poetry 27, 46, 116, 156, 160–62, 214
political medicine 6
Ponce de la Fuente, Constantino 15
Porta, Giovan Battista della 30–01, 358
positive divinity 19, 46
Possevino, Antonio 7, 26, 30–01, 36, 357, 365
Pretel, Alonso 19, 21
Primaudaye, Pierre de la 53
procreation 61–3, 97, 264, 282, 297–300, 303, 306, 309–11, 316
Protestant Reformation 47, 159
psychology 1–2, 53
Publius Lentulus 51, 272
Purchas, Samuel 56

Quevedo, Francisco de 7, 27, 333, 353
Quiroga, Gaspar de 21

Raleigh, Sir Walter 11, 13
Reformation 59, 68, 336
Requeséns, Don Luis de 160
Reynolds, Edward 53
rhetoric 3, 31, 113, 175–76, 178, 356, 362
Roderick, Don (Don Rodrigo) 245
Rodríguez de Castelo Branco, João (Amato Lusitano) 341
Roesslin, Eucharius 62

Rogers, Thomas 53

Sabuco, Miguel 1, 20, 353, 355, 363
Sadler, John 62, 371
Salicato, Altobello 36
Salust 310
Samuel 149, 269, 271, 334
Sánchez de Arévalo, Rodrigo 54
Sánchez de las Brozas, Francisco (El Brocense) 2, 361
Sancho IV, King 329
Saul 50–01, 149–50, 152, 243, 251, 269–71
Schopenhauer 30
school divinity 46–47, 155–56, 158–60, 162, 164, 167, 172, 174–75, 183–84, 348
science 1, 4–5, 24, 26, 40, 45–46, 57–58, 65–66, 68–69, 76–80, 83–89, 91, 99, 106, 110, 112–13, 118, 128, 130–32, 136–37, 149, 155–58, 160–02, 164–67, 170, 173, 175–77, 184, 194, 198, 206, 210, 215, 235, 244, 248, 251–53, 258–60, 266, 268, 305, 319, 323–25, 327, 356, 361, 365–67
Scotus 159, 164, 177
Scriptures 22, 46, 50, 154
Selden, John 56–57, 364
Sempronius 233
sex 3, 9, 70, 280, 283–84, 286, 292, 295, 299, 314
Sharp, Jane 62
Sidney, Sir Philip 11–2
Silhon, Jean de 30
Society of Antiquaries 12, 38
Socrates 85, 132, 162, 169, 258
Solomon 94, 150, 152, 215, 235, 257, 292
Solon 259
soul 3, 18–19, 22, 26, 30, 42, 44–45, 51–53, 66, 97–112,

115–21, 123–24, 131, 133, 136–39, 144, 146–55, 191, 208, 216–17, 220–21, 226, 234, 254–56, 258–59, 261–63, 265, 273, 277–78, 302, 304, 307–11, 319, 324–25, 327, 347, 354, 366
Spelman, Sir Henry 12, 353
Stanhope, George 65
Stanley, Thomas 333
Steele, Richard 28
Stoic(s) 84, 230, 239
Suárez of Toledo 244
Suetonius Tranquillus 237–38
Sulla 237

Takama, Henryk 25
Tasso, Torquato 31–36, 40–41, 354, 360, 362
Tassoni, Alessandro 30
temperament 3, 6, 22, 27, 45, 55, 64, 98
Terence 172
Thales Milesius 259
theatre 68
Theophrastus 258–59
Thomasen, Christian (Thomasius) 30
Thomas, St 159, 164, 177
Thomas, Thomas 8–9, 353
Thomas, William 8, 353
Tomkis, Thomas 53
translation 8–10, 12, 14–15, 17, 19–20, 23–25, 28–29, 31–34, 36–67, 39–44, 47, 49, 53–54, 57, 59, 62, 64–65, 67, 113, 163, 165, 329–34, 336–39, 343, 345, 353, 360, 362, 367

Ulysses 262
understanding 2–3, 6, 8, 18–19, 22–23, 44–48, 50, 53, 56, 89–90, 126, 134, 137, 139, 184, 334, 336

university 14–17, 19–21, 39–40, 58–59, 178, 244–45, 260, 329, 341–43
Urraca of Castile 249–50

Valdivia, Diego Pérez de 21
Vallés, Francisco 15, 20, 26, 54, 353
Valois, Francis of 215
Valverde de Amusco, Juan 26
Vargas, Francisco de 15
Vega, Cristóbal de 15, 26
Vegetius 228, 231, 242
Velasco, Águeda de 16
Velásquez, Andrés 26
Vélez de Guevara, Juan Crisóstomo 333
Verstegan, Richard 13, 37
Vesalius, Andreas 16
Villarino, Francisco 26, 358
Villavicencio, Fray Lorenzo de 18
Viotti, Erasmo 33
Virgil 172
Viriatus 236–37
Vives, Juan Luis 1, 4, 54, 57, 59–61, 353, 362–64, 367

Walkington, Thomas 53
Watkins, Richard 35, 73, 351
wit(s) 1–6, 8–10, 17–19, 22–23, 27, 31, 37–39, 44–45, 50, 53, 56–61, 63, 67–8, 73, 75–81, 83–92, 95, 96–106, 113, 115–16, 118, 120, 123–27, 129–132, 134–35, 140, 143, 145–46, 155–56, 161, 162–68, 170–01, 176, 186–88, 190, 193, 196, 198–200, 202, 204–05, 207, 210, 214, 216–19, 221–24, 226–28, 230–34, 236–37, 239–41, 248, 251, 253, 255, 257–62, 265–66, 268, 270–71, 273, 275, 279, 282–85, 287,

292, 295–96, 302, 304–05,
310–11, 315, 317–20, 322–23,
325, 327–28, 331, 339, 351–52,
356, 361, 364–65

Wolfe, John 34

Zara, Antonio 30–31, 358
Zenocrates 84